Biochar: A Precious Resource from Biological Waste

Shraddha Singh • Sudhakar Srivastava
Editors

Biochar: A Precious Resource from Biological Waste

Applications for Soil, Plant and Environmental Health

Springer

Editors
Shraddha Singh
Nuclear Agriculture and Biotechnology Division
Bhabha Atomic Research Centre
Mumbai, Maharashtra, India

Sudhakar Srivastava
Institute of Environment and Sustainable Development
Banaras Hindu University
Varanasi, Uttar Pradesh, India

ISBN 978-981-95-0424-4 ISBN 978-981-95-0425-1 (eBook)
https://doi.org/10.1007/978-981-95-0425-1

This Springer imprint is published by the registered company Springer Nature Singapore Pte Ltd.
The registered company address is: 152 Beach Road, #21-01/04 Gateway East, Singapore 189721, Singapore

Preface

The increasing world population needs to be fed with quality food today and in coming years amidst reducing land resources, diminishing soil quality, and climate change. Soil health is an important issue that needs to be addressed to sustain and enhance crop production. Soil characteristics for good agriculture production include organic carbon, microbial growth, optimum pH and redox conditions, availability of nutrients along with absence of toxic elements, etc. Apart from this, another problem that plagues agriculture is the generation of harvested biomass such as rice and wheat straw, extra waste vegetables, etc. In addition, plant or other waste of biological origin is generated on a daily basis at a huge scale and this includes kitchen waste, fruit peels, sewage waste, animal waste, etc. It is important to note that biological waste has trapped large amounts of nutrients including essential elements like phosphorus within it. This waste therefore needs to be sustainably used. In this aspect, biochar, a material that is generated through pyrolysis of biological waste of diversified origin, can solve the problems of waste utilization. It can be applied for different crops and fields and leads to sustainable agriculture. Biochar application has been found to improve overall soil health including pH, redox status, organic carbon, and essential nutrients of soil while reducing toxic metal bioavailability too. This book aims to discuss the role of biochar as a wonderful resource from biological waste for the prospects of application in agriculture sector.

This book aims to compile up-to-date knowledge related to general features of biochar, its synthesis methods, biochar generation from different wastes, and the utilization of biochar for beneficial purposes. The book contains a total of 14 chapters divided into two sections. The first section of the book comprises eight chapters, which present an overview of biochar and the variety of biomasses from which biochar can be generated. The chapter "Management of Biological Waste: Physical, Chemical, and Biological Methods" of the book discusses various physical, chemical, and biological methods of waste management. The chapters "Biochar: A Sustainable Solution for Agriculture and Environment" and "The Role of Biochar in Sustainable Agriculture" review the importance of biochar for sustainability of agriculture and environmental protection. The chapter "An Overview of Methods and Techniques for the Preparation and Characterization of Biochar" discusses about various techniques, methods, and principles of biochar generation and also details about the methods of biochar characterization. The chapters "Household Kitchen Waste: A Resource That Needs to Be Tapped," "Agricultural Waste Biomass

Management and Biochar Production," and "Biochar from Animal Waste and Its Applications" present prospects of biochar generation from different types of wastes like kitchen waste (the chapter "Household Kitchen Waste: A Resource That Needs to Be Tapped"), agricultural waste (the chapter "Agricultural Waste Biomass Management and Biochar Production"), and animal waste (the chapter "Biochar from Animal Waste and Its Applications"). These chapters demonstrate the vast potential of biochar research and variety of resources that are available in the form of waste in the world. The second section of the book delves around the application prospects of biochar for soil health and for the management of various abiotic stresses. The chapters "Drought Stress Mitigation with the Use of Biochar" and "Application of Biochar for Salinity Stress Management of Crop Plants" discuss in depth about the application of biochar for mitigating drought and salinity stress in plants. The chapter "Biochar as a Treasure Trove of Essential Nutrients: Combating Elemental Deficiency in Crops" tells us how biochar serves as a slow release fertilizer or a resource of essential elements when applied judiciously in soil. The chapter "Biochar-Mediated Restoration: Transforming Invasive Plants and Crop Residues into Ecological Assets" throws light on the value of biochar in ecological restoration, sustainable agriculture, and climate change mitigation. The chapter "Restoring the Soil Functions Through Biochar-Microbiome Interactions" discusses that how incorporating biochar to soils can modulate the soil microbiome and improve soil health.

The book aims to provide key information to the readers about the biological waste management and impacts of biochar on soil health. The discussion of potential waste biomass for biochar generation and various applications of biochar in the field make the book a comprehensive combination of chapters. The book will provide the reader a comprehensive view of up-to-date biochar research. This book would act as a guiding textbook for undergraduate and postgraduate students and as a mean to understand the latest research trends for doctoral students.

Mumbai, Maharashtra, India — Shraddha Singh
Varanasi, Uttar Pradesh, India — Sudhakar Srivastava

Contents

Editors and Contributors

About the Editors

Shraddha Singh is a Scientific Officer in Nuclear Agriculture & Biotechnology Division at Bhabha Atomic Research Centre (BARC), Mumbai. She is a gold medalist in M.Sc. Botany from University of Allahabad, Prayagraj. She has obtained Ph.D. degree in Botany from National Botanical Research Institute, Lucknow (CSIR-NBRI) and University of Lucknow in 2003. She has a vast research experience in the field of environmental toxicology and environmental biotechnology. Her major area of interest is phytoremediation of heavy metals, metalloids, and radionuclides from the water, soil, and waste and its implications in agriculture. She is also involved in the societal applications of her work and sustainable agriculture. She has published several research and review articles in international journals of high repute. She is also an Associate Professor at Homi Bhabha National Institute, Mumbai.

Sudhakar Srivastava is an Assistant Professor in the Institute of Environment and Sustainable Development (IESD), Banaras Hindu University (BHU). He obtained Ph.D. in Botany from CSIR-National Botanical Research Institute (CSIR-NBRI) and University of Lucknow in 2008. He served as Scientific Officer in Bhabha Atomic Research Centre (BARC), Mumbai, from 2009 to 2014 before joining BHU. He has more than 19 years of research experience in the field of plant-metal interactions. He has published more than 170 research and review articles in high-impact international and national journals. He has also published 25 book chapters and edited three books. He is NAAS Associate of National Academy of Agricultural Sciences. He has received Young Scientist Award from National Academy of Sciences, India (NASI), Allahabad (2011) and from Uttar Pradesh Council of Science & Technology (UPCST) (2013–2014). He is also a Fellow of International Society of Environmental Botanists (ISEB), Lucknow.

Contributors

Saif Ahmad Department of Botany, Plant Physiology Section, Faculty of Life Sciences, Aligarh Muslim University, Aligarh, India

Bulbul Ahmed Soil Science Division, Bangladesh Wheat and Maize Research Institute, Dinajpur, Bangladesh

S. Maqbool Ahmed Department of Botany, School of Sciences, Maulana Azad National Urdu University, Hyderabad, India

Pravej Alam Department of Biology, College of Science and Humanities, Prince Sattam Bin Abdulaziz University, Alkharj, Saudi Arabia

Annika Jahan Aonti Soil Science Division, Bangladesh Wheat and Maize Research Institute, Dinajpur, Bangladesh

Yamshi Arif Department of Botany, Plant Physiology Section, Faculty of Life Sciences, Aligarh Muslim University, Aligarh, India

Jayanta Kumar Biswas Department of Ecological Studies, University of Kalyani, Kalyani, Nadia, West Bengal, India

Nirmali Bordoloi Department of Environmental Sciences, Central University of Jharkhand, Ranchi, Jharkhand, India

Debadrita Das School of Environmental Studies, Jadavpur University, Kolkata, India

Archita Dey School of Environmental Studies, Jadavpur University, Kolkata, India

Shalini Dhyani CSIR-National Environmental Engineering Research Institute, Nehru Marg, Nagpur, Maharashtra, India

Mattaparthi Lakshmi Durga Division of Agricultural Engineering, ICAR-Indian Institute of Agricultural Research, New Delhi, India

Mohammad Faizan Department of Botany, School of Sciences, Maulana Azad National Urdu University, Hyderabad, India

Markkandan Ganesan Plant Cell Engineering Laboratory, Department of Life Sciences, Presidency University, Kolkata, India

Swetanjana Ghosh School of Environmental Studies, Jadavpur University, Kolkata, India

Abhishek Guldhe Amity Institute of Biotechnology, Amity University Maharashtra, Mumbai, India

Md. Israfil Haq Soil Science Division, Bangladesh Wheat and Maize Research Institute, Dinajpur, Bangladesh

Department of Agricultural Extension (DAE), Langadu, Rangamati, Bangladesh

Shamsul Hayat Department of Botany, Plant Physiology Section, Faculty of Life Sciences, Aligarh Muslim University, Aligarh, India

Akbar Hossain Soil Science Division, Bangladesh Wheat and Maize Research Institute, Dinajpur, Bangladesh

Farhad Jivani Amity Institute of Biotechnology, Amity University Maharashtra, Mumbai, India

Md. Parvez Kabir Soil Science Division, Bangladesh Wheat and Maize Research Institute, Dinajpur, Bangladesh

Rahul Kshirsagar Amity Institute of Biotechnology, Amity University Maharashtra, Mumbai, India

Anuj Tripathi Kumar CSIR-National Environmental Engineering Research Institute, Nehru Marg, Nagpur, Maharashtra, India

Dhananjay Kumar Laboratory of Algal Biotechnology, Department of Botany and Microbiology, H.N.B. Garhwal University, Srinagar, Garhwal, Uttarakhand, India

Vandana Kumari Advance Centre on Sericulture, Bihar Agricultural University, Sabour, Bhagalpur, Bihar, India

Urvashi Lama School of Environmental Studies, Jadavpur University, Kolkata, India

Department of Botany, Sovarani Memorial College, Howrah, India

Sarika Loni Amity Institute of Biotechnology, Amity University Maharashtra, Mumbai, India

Organic Recycling Systems Ltd., Mumbai, India

Anuradha Mahato Plant Cell Engineering Laboratory, Department of Life Sciences, Presidency University, Kolkata, India

Jayjit Majumdar Department of Ecological Studies, University of Kalyani, Kalyani, Nadia, West Bengal, India

Sharmistha Majumder School of Environmental Studies, Jadavpur University, Kolkata, India

Vineet K. Maurya Applied Microbiology Lab, Department of Botany & Microbiology, H.N.B. Garhwal University, Srinagar, Garhwal, Uttarakhand, India

Tatiana Minkina Academy of Biology and Biotechnology, Southern Federal University, Rostov-on-Don, Russia

Bipradeep Mondal Department of Ecological Studies, University of Kalyani, Kalyani, Nadia, West Bengal, India

Department of Environmental Sciences, Central University of Jharkhand, Ranchi, Jharkhand, India

Monojit Mondal School of Environmental Studies, Jadavpur University, Kolkata, India

Debojyoti Moulick Department of Environmental Science, University of Kalyani, Kalyani, Nadia, West Bengal, India

Debabrata Nath International Rice Research Institute (IRRI)-South Asia Regional Centre (ISARC), Varanasi, Uttar Pradesh, India

Rupali Nayal CSIR-National Environmental Engineering Research Institute, Nehru Marg, Nagpur, Maharashtra, India

Sankarshan Patwardhan Amity Institute of Biotechnology, Amity University Maharashtra, Mumbai, India

Md. Mobinur Rahman Maize Breeding Division, Bangladesh Wheat and Maize Research Institute, Dinajpur, Bangladesh

Soil Science Division, Bangladesh Wheat and Maize Research Institute, Dinajpur, Bangladesh

Priyadarshani Rajput Academy of Biology and Biotechnology, Southern Federal University, Rostov-on-Don, Russia

Tarit Roychowdhury School of Environmental Studies, Jadavpur University, Kolkata, India

Ajay Shinde Amity Institute of Biotechnology, Amity University Maharashtra, Mumbai, India

Manoj Shrivastava Division of Environmental Sciences, ICAR-Indian Agricultural Research Institute, New Delhi, India

Nisha Singh Department of Biochemistry, University of Lucknow, Lucknow, Uttar Pradesh, India

Renu Singh Division of Environmental Sciences, ICAR-Indian Agricultural Research Institute, New Delhi, India

Sudhanshu Singh International Rice Research Institute (IRRI)-South Asia Regional Centre (ISARC), Varanasi, Uttar Pradesh, India

Sudhakar Srivastava Institute of Environment and Sustainable Development, Banaras Hindu University, Varanasi, Uttar Pradesh, India

Penna Suprasanna Amity Institute of Biotechnology, Amity University Maharashtra, Mumbai, India

S. Saraniya CSIR-National Environmental Engineering Research Institute, Nehru Marg, Nagpur, Maharashtra, India

Manju Tanwar Organic Recycling Systems Ltd, Mumbai, India

Barun Kumar Thakur Department of Economics, FLAME University, Pune, Maharashtra, India

Manu Thomas CSIR-National Environmental Engineering Research Institute, Nehru Marg, Nagpur, Maharashtra, India

Part I

Biochar Generation: A Wealth from Waste

Management of Biological Waste: Physical, Chemical, and Biological Methods

Annika Jahan Aonti, Akbar Hossain, Bulbul Ahmed, Md. Parvez Kabir, Md. Israfil Haq, and Md. Mobinur Rahman

Abstract

Biological waste derives from biological activities, including waste from plants, animals, hospitals, homes, and municipal solid waste. It disrupts soil health as well as biodiversity, contributing to global warming and climate change. Biological waste management involves collecting waste from its origins, transporting it, processing it, disposing of it, or recycling it for further renewable energy. Biological waste management is crucial for creating a pollution-free environment and conserving nutrient resources for crop production. This waste management involves physical, chemical, and biological methods. Physical methods involve (i) incineration, (ii) pyrolysis, and (iii) gasification, which undergo dynamic chemical transformation of solid fuel into a lower amount of waste. Chemical methods include ion exchange, precipitation, oxidation, reduction, and neutralization processes for hazardous biomedical waste management. Biological methods involve the transformation of waste by living organisms to produce stable organic matter. Proper treatment of biological waste is essential for environmental conservation. The bioconversion of waste into renewable energy sources such as biogas, biodiesel, and bioethanol can reduce greenhouse gas emissions. The current chapter discusses the management of biological wastes through the use of physical, chemical, and biological methods.

A. J. Aonti · A. Hossain (✉) · B. Ahmed · M. P. Kabir · M. M. Rahman
Soil Science Division, Bangladesh Wheat and Maize Research Institute, Dinajpur, Bangladesh

M. I. Haq
Department of Agricultural Extension (DAE), Langadu, Rangamati, Bangladesh

S. Singh, S. Srivastava (eds.), *Biochar: A Precious Resource from Biological Waste*,
https://doi.org/10.1007/978-981-95-0425-1_1

Keywords

Living organisms · Biological waste · Management · Physicochemical methods · Soil health

1 Introduction

Biological waste management involves the proper disposal and treatment of waste from biological processes, including waste from plants, animals, hospitals, households, and municipal solid waste (MSW). According to the estimation of the World Bank, approximately 2.01 billion tons of solid municipal waste are produced annually worldwide, among which at least 33% of the waste is not managed in a way (Kaza et al., 2018). With the rapid increase in global solid municipal waste production, which is projected to reach approximately 70% by 2050, 3.4 billion metric tons of biological waste are produced. Effective biological waste management practices help ensure a sustainable environment by properly handling and treating biological waste generated from different sources (Fig. 1; Najib et al., 2023).

Waste management is becoming a concern, as its intensity is increasing daily due to several causes and challenges, such as the following:

- Increasing waste generation due to urbanization, industrialization, lifestyle changes, and consumption patterns.
- Approximately one-third of the food produced is lost as waste annually, resulting in hunger, economic loss, inflation, and inequality.
- Discharge of industrial liquid waste into fields affects soil health, biodiversity, and crop productivity.
- Poor infrastructure, financial constraints, and regulatory policies in developing countries lead to unprocessed waste, contributing to global warming.

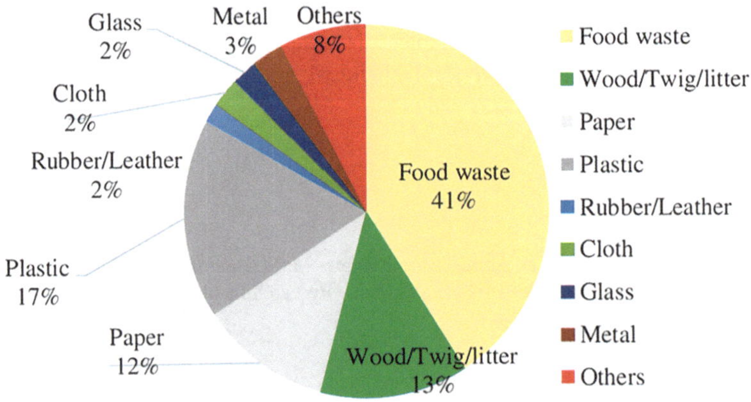

Fig. 1 Waste composition on the basis of waste type and source (Najib et al., 2023)

Waste management practices vary between developed and developing countries, urban and rural areas, and residential/industrial producers. Proper biological waste management is crucial for conserving the environment, preventing the spread of infections and reducing greenhouse gas emissions from waste. The treatment of biological waste is essential for sustainable development and environmental conservation. Proper and systematic waste management, along with improved life standards, is cost-effective and environmentally friendly. A hierarchy of biological waste management is presented in Fig. 2.

Proper education and awareness among people regarding waste generation and waste disposal are the first priorities of waste management. Educating personnel can play an important role in increasing the efficiency of the waste management system and minimizing its possible health and environmental risks. 3R techniques (reduce, reuse, and recycle) constitute another approach for waste management to produce renewable sources of energy, such as biodiesel, biogas, and biofuels, and to recover other nutrients (Demirbas, 2011). Several recycling techniques, such as physical reprocessing, biological reprocessing, and energy recovery, are being continuously developed. The bioconversion of solid waste to energy decreases the trash concentration in the environment and improves waste utilization. Waste can be converted into energy via biological and physiochemical techniques, including anaerobic digestion, esterification, and fermentation, as well as gasification, incineration, and landfilling (Meegoda et al., 2018). The concept of the amount of waste discharged $\leq$ the amount of waste generated should be introduced for a sustainable environment (Sajid et al., 2023).

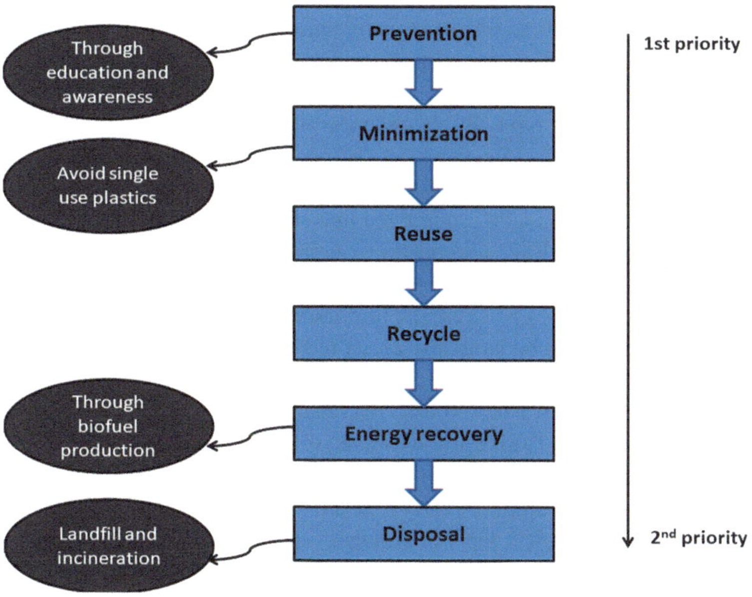

Fig. 2 The biological waste management hierarchy (Sajid et al., 2023)

2 Concept, Sources, and Classification of Biological Waste

Biowaste or biological waste results from a wide range of sources, including animal wastes, plant wastes, domestic wastes, commercial wastes, ashes, biomedical wastes, construction wastes, radioactive wastes, and industrial solid and liquid wastes. It has been steadily increasing due to the increasing human population, changes in lifestyles and consumption patterns, urbanization, food grain production, economic growth, and industrialization (Chattopadhyay & Singha, 2022). For the sustainable development of human society and environmental conservation, biological waste should be managed via a cost-effective and environmentally friendly approach.

Waste management comprises collection, transportation, pretreatment, processing, recycling or disposal, and monitoring of waste materials. The waste management concept includes several goals as follows:

(a) Reduction or lessening of the total amount of waste.
(b) Recycling and reintroduction of biological waste into suitable production cycles as secondary raw materials or energy carriers.
(c) Best possible method for reducing residual waste quantities.
(d) Flexible concept concerning fluctuations in waste quantities and the composition of domestic waste. New developments in the field of waste management must be included in the system.

The waste produced can be recycled or recovered to produce renewable sources of energy, such as biodiesel, biogas, and biofuels, and to recover other nutrients from nonrenewable sources (fossil fuel). Waste materials can be recycled via physical reprocessing (such as gasification, incineration, and landfills), biological reprocessing (such as anaerobic digestion, esterification, and fermentation), and energy recovery (Demirbas, 2011).

Minimizing waste is typically the responsibility of producers, who must prioritize it over other options. Recycling is the second most important priority, as it allows for the recovery of materials that can be repurposed for other uses. The disposal options available can be categorized into six levels, ranging from low to high: reduction, reuse, recycling, composting, incineration, and landfilling (Siddique et al., 2008). The aim of the waste hierarchy is to obtain the maximum benefit from waste management and to generate the minimum amount of waste.

Biological waste can be generally classified into several groups on the basis of sources (Fig. 3): (1) domestic waste, (2) animal waste, (3) biomedical or medical waste, (4) commercial waste, (5) ash, (6) construction waste, (7) industrial solid waste, (8) sewer waste, (9) hazardous waste, (10) agricultural residue, (11) biodegradable waste, and (12) nonbiodegradable waste [2].

Domestic wastes encompass yearly household wastes containing fabrics, plastic, paper, vegetable and fruit scraps, polythene, ceramics, etc., which are continually increasing (Kumar & Samadder, 2017). Animal or livestock waste involves livestock manure by animal category. Medical or biomedical waste is generated from the diagnosis, treatment, or immunization of animals and humans (Windfeld &

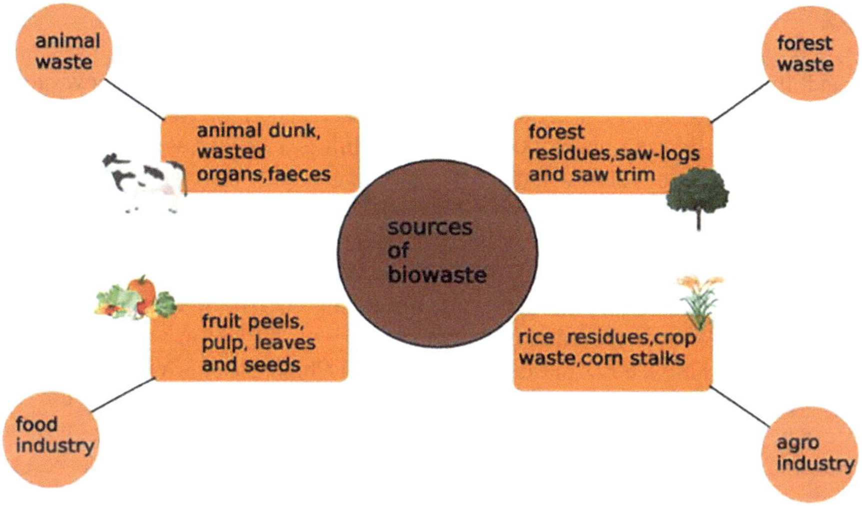

Fig. 3 Sources of biological waste (Sajid et al., 2023)

Brooks, 2015). Commercial wastes include packing materials, polythene, spoiled goods, printer papers, and vegetables, and the remaining meat is obtained from commercial places such as shops and offices. Approximately 20–50% of the waste comes from commercial sources (Salhofer, 2000). Ashes are generated as waste from the burning of wood, coal, fossil fuels, etc. Construction wastes are from construction sites, for example, cement, roofing materials, concrete, metal roads, bricks, etc. Construction waste contributes to approximately 44% of the waste (Ajayi et al., 2017). Industrial waste is generated by various small- or large-scale industries, such as the textile industry, food industry, pharmaceutical industry, and cigarette factories. Sewer waste results from sewerage, which poses various health hazards to the public. Hospitals or energy sectors possess radioactive wastes termed hazardous wastes (Demirbas & Balat, 2010). Agricultural residues include crop residue, husks, straw, wood, etc. Biodegradable (can be decomposed) wastes are from kitchen garbage, animal dung, etc. Nonbiodegradable (cannot be decomposed) wastes include plastics, glass, etc.

2.1 Disposal Methods

For proper management of biological wastes in an energy-efficient and environmentally friendly way, 3Rs (reduce, reuse, and recycle) should be followed. These eco-friendly waste disposal methods not only help in managing waste effectively but also contribute to a healthier environment and a sustainable future. Traditionally, solid and liquid waste is disposed of in landfills and incinerated.

2.2 Landfill

Landfilling refers to the process of depositing, compressing, and covering waste at designated locations. It has been a simple, adaptable, and cost-effective waste disposal solution for more than 5000 years. A landfill is a facility used to dispose of solid waste. It can be excavated into the earth and gradually filled with rubbish, or it can be piled and buried in an organized manner. Landfills or open dumpsites are the usual methods of disposing of municipal solid waste (MSW) around the world (Fig. 4). Approximately 16% of MSW is discarded by incineration, and the remaining MSW is disposed of via landfills (Demirbas et al., 2016). In the United States, 52.6% of MSW was discarded in landfills; in Brazil, 59.1%; in the Kingdom of Saudi Arabia (KSA), 85%; in Malaysia, 94.5%; in China, 79%; in Venezuela, 32%; in controlled disposal, 43%; in noncontrolled disposal or open dumps, 24%; in Mexico, 65%; in uncontrolled and open dumps, 30%; and in Thailand, 27% (Vaverková, 2019). To prevent wastewater from seeping into the ground, it is necessary to dispose of liquid waste in a manner other than landfills.

Decomposition in landfills proceeds through a series of stages, with fluctuations in specific bacterial populations and the formation and utilization of metabolic products (Sajid et al., 2023). The initial stage of decomposition involves the consumption of oxygen by aerobic bacteria within a week (Augenstein & Pacey, 1991). In the anaerobic acid stage of the second stage, cellulose, hemicellulose, proteins,

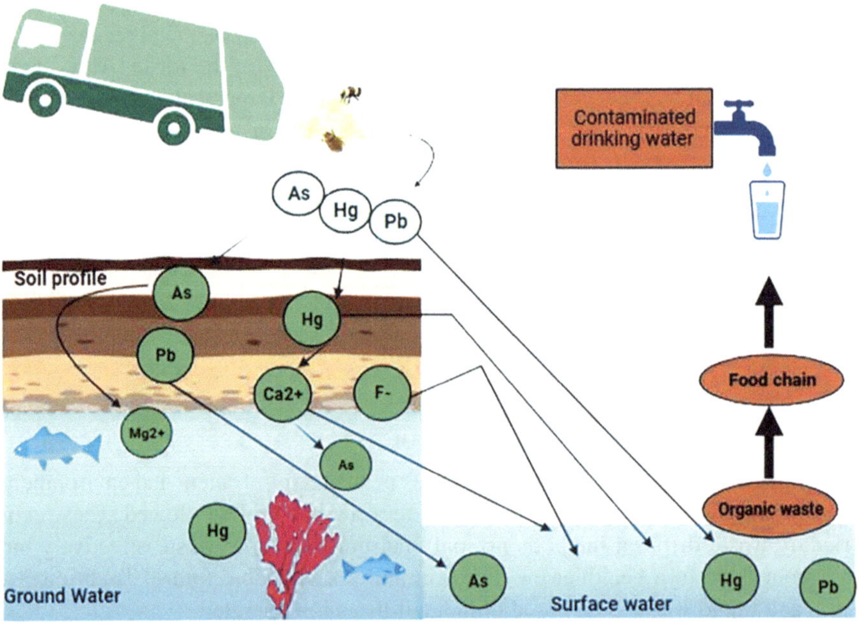

Fig. 4 Contaminants in the form of heavy metals released from municipal solid waste (MSW) landfill sites (Sajid et al., 2023)

and lipids are broken down into soluble sugars, amino acids, long-chain carboxylic acids, and glycerol. This stage is influenced by diverse populations of hydrolytic and fermentative bacteria. The final metabolic byproduct of decomposition in landfills is the generation of biogas and biofuels (Bhatia et al., 2018).

Although landfills are the most inexpensive waste management option, they are a great contributor of greenhouse gas CH_4 (1–2%) and carbon dioxide CO_2 along with heavy metals (As, Hg, Pb, and Mg) (Zhang et al., 2019). In recent years, the treatment of landfill leachate has gained significant attention because of its potential to pose a threat to both human health and the environment if it is not properly contained and managed (Uygur & Kargı, 2004). Combinations of several physical, chemical, and biological approaches are usually used to improve leachate treatment efficiency (Raghab et al., 2013). Biological treatment involves several aerobic, anaerobic, and anoxic approaches to remove biodegradable compounds from leachates. Nonbiodegradable substrates from landfill leachates are treated via physical and chemical methods.

2.3 Incineration

Incineration is a waste treatment process that involves the combustion of substances contained in waste materials. This process involves burning a combination of food and garden waste (Wernet et al., 2016). Waste containing organic materials can be managed via incineration (Cheng and Hu, 2010; Brancoli et al., 2020). It is also called thermal treatment, as it works at very high temperatures. Approximately 80–85% of the organic mass of the solid mass can be reduced by the incineration process, among which 95–96% of the volume is commendable. In the incineration process, waste materials generate ash, flue gas, and a large amount of heat. Heat can generate electric power, and flue gas (nitrogen, carbon dioxide, and sulfur dioxide) can be better utilized. Nitrogen in the form of urea increases the productivity of crops, carbon in the form of carbon dioxide can be used as a fire extinguisher, and sulfur dioxide can also be used in dental treatment. Ash is composed of Si, Fe, Ca, Al, Na, and K in the form of oxides that constitute geological materials that can be used for construction purposes. Waste incineration generates energy in the form of heat, which can be used to generate steam and electricity. Incineration is comparatively better than landfill systems in the case of hazardous waste management. A typical scheme for an incineration plant operating on real waste and with energy recovery is given in Fig. 5.

The incineration process involves several steps depending on the input materials and municipal solid waste residues.

1. Combustion.
2. Boiler/steam turbine.
3. Exhaust gas cleaning.
4. Dry or wet scrubbers.
5. Activated carbon injection.

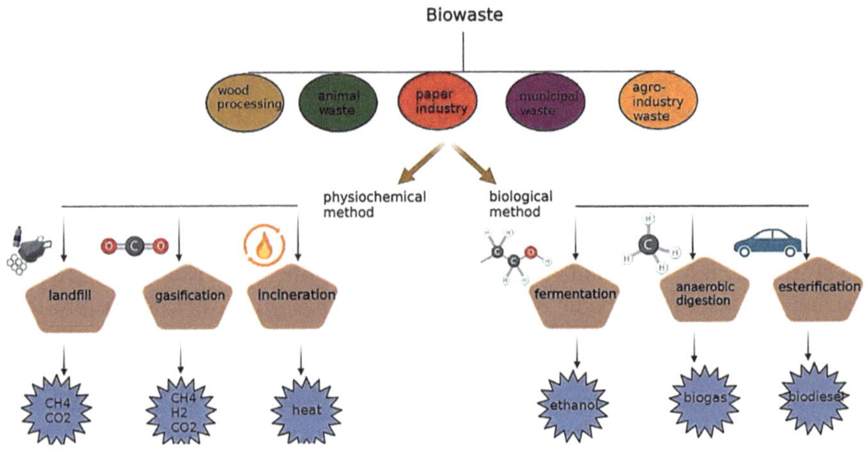

Fig. 5 Production of bioenergy from different biowaste sources (Sajid et al., 2023)

6. Bag house filter.
7. Selective noncatalytic reduction.

3 Types of Incinerators

There are three main types of combustion technologies used in commercial practice:

3.1 Rotary Kilns

Industrial and hazardous waste combustion, as well as municipal solid waste, can be incinerated by rotary kiln. A rotary kiln produces a low level of NOx and the thermal destruction of hazardous chemicals.

3.2 Moving Grate

A moving grate is designed for the complete combustion of a municipal solid waste incinerator through the introduction of turbulence.

3.3 Fluidized Bed

The application of fluidized bed combustion has recently increased in municipal solid waste incinerators, although it is still mainly used for the combustion of hazardous waste. Different types of fluidized beds (bubbling, rotating, and circulating fluidized beds) present the same design principles with higher combustion efficiency.

4 Biomass and Biowaste Resources

Biomass refers to the plant part remaining after consumption. Woody and herbaceous species, wood wastes, agricultural land, industrial residues, waste paper, municipal solid waste, biosolids, waste from food processing, animal wastes, aquatic plants, sand, algae, and so on are the main components of biowaste resources. Biomass is composed of cellulose, hemicellulose, and lignin. Biomass is a critical component of the renewable energy landscape, offering economic, environmental, and energy security benefits. Research and technological advancements are driving its potential to replace fossil fuels and contribute to climate change mitigation. However, addressing challenges related to land use, food security, and technological development is essential for the sustainable growth of biomass energy.

4.1 Agricultural Residues

Agricultural residue is a term that involves crop residues, straw and husks, olive pits, nut shells, and so on. They can be categorized into two groups: (1) Field residue: The remaining portion of the plant after the crop is harvested, such as stalks, stems, seed pods, and leaves. (2) Process residues: The usable resource of residue after processing of the crop. For example, husks, seeds, roots, bagasse, etc., are used as components of biofuel generation, whereas others are used as animal feed and for soil management. Cereal straw is the dry stalk plant part after the grain and seeds are harvested. These cereal straws are used for feed, garden mulch, and animal bedding, and the majority are used for burning into the soil. However, many countries are imposing bans on field burning practices due to environmental pollution. The nongrain and aboveground portions of corn plants are termed corn stover. Approximately 5% corn stover is used for animal bedding and feed, and the remaining portion is incorporated into the soil. It has direct burning potential and can be converted into biofuel. Rice husk is another common residue after the processing of grains from plants. They can be converted into biofuel by a gasification process. Rice husk has a high silica content and can be used in boiler combustion for slug and ash problems. Sugarcane has photosynthetic efficiency, converting 2% solar energy into biomass. In addition to the demands of herbaceous woody crops, the demands of short rotation woody crops, woody crops (poplars), and tropical grasses (elephant grass) are increasing because of their suitability (cellulose: lignin) for subsequent processing as energy crops. The techniques used to convert agricultural residues into biofuels are presented in Fig. 6.

Agricultural waste or crop residues are converted into biofuels by biochemical (enzymatic hydrolysis, anaerobic digestion, and microbial fermentation) and

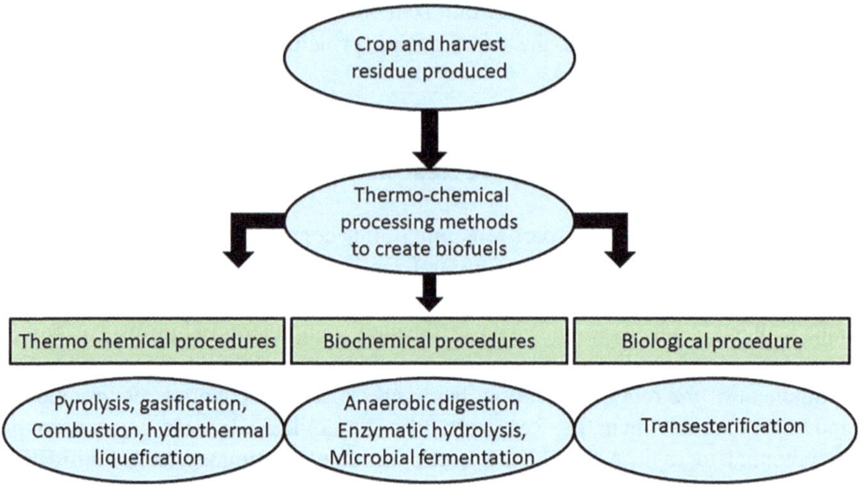

Fig. 6 Conversion techniques of agricultural residues into biofuels (Sajid et al., 2023)

thermochemical (combustion, gasification, pyrolysis, and hydrothermal liquefaction) conversion technologies. In addition, transesterification, a chemical method, is sometimes used to convert biomass (Ali et al., 2019).

Combustion is a process of burning biomasses into thermal energy in the presence of excess oxygen (Liu et al., 2017). Pyrolysis is a thermochemical decomposition process that is the opposite of the combustion process. In this process, decomposition occurred in the absence of oxygen at an increased temperature, usually at 400 °C. Rice straw, rice husk, maize straw, and wheat straw undergo this process. Pyrolysis is a stage in both the combustion and gasification processes where the thermal decomposition of organic fuel results in the production of biochar as well as tar and gas products. The aim of this process is to obtain useful forms of energy, recycle raw materials (waste polymers), and produce semiproducts constituting raw materials for further use.

The third method of converting biomass energy to useful energy is the gasification process of transforming fuel into gas containing carbon oxide and hydrogen. From medical waste to wood waste, plastic undergoes this process (Bai et al., 2019).

Anaerobic digestion (AD), a process of solid and liquid organic matter decomposition, occurs; nitrogen compounds are transformed into ammonium nitrogen; bacteria, viruses, and eggs of pathogens are destroyed; and the C/N ratio is reduced in an anaerobic environment. The process of anaerobic decomposition involves four stages: (a) hydrolysis, (b) acidogenesis, (c) acetogenesis, and (d) methanogenesis (Morales-Polo et al., 2018).

Enzymatic hydrolysis is a biochemical approach in which glucose from agricultural residues is fermented into ethanol. Under optimum pH and temperature conditions, yeast produces ethanol in a fermentation reaction. This is an eco-friendly approach for reducing air pollution (Bon & Ferrara, 2007).

Transesterification is a biological approach by which vegetable oil and animal fats can be converted into biodiesel as alternatives to petroleum diesel (Yousuf, 2012). Overall, proper utilization of agricultural residues as a source of biofuels can be a great approach to reduce the emission of greenhouse gases as well as mitigate air pollution.

4.2 Paper and Pulp Industry Waste

The paper and pulp industry operations generate vast amounts of biomass residue, including bark, leaves, needles, branches, and sludge. These nonsludges have potential for use in biofuel production processes. Most pulp mill sludge is an attractive feedstock for biofuel. When disposed of in landfills, these sludges are degraded to produce more harmful methane gas than carbon dioxide, which is the major contributor to the greenhouse effect. The alternative option is to generate biofuel production to avoid land and water contamination from the landfill system.

Black liquor, another byproduct, is highly valuable because of its dissolved lignin content. They can be combusted in the paper industry for energy production, and the remainder can be used for electricity generation. Black liquor contains several components and the formula $C_{10}H_{12.5}O_7Na_{2.4}S_{0.36}$ (Salmenoja, 1993; Backman et al., 1993), and the energy value consists of dissolved lignin and cellulosic, hemicellulosic hexose, and pentose sugar degradation products (Demirbaş, 2002).

Although the paper and pulp industry generate 300–350 million tons of waste, a smaller portion of waste is recycled, and the rest of it causes pollution, such as land pollution and water pollution blocking the water channel (Ioelovich, 2014; Gopal et al., 2019). Bioethanol, biodiesel, and petroleum products are generated from the thermochemical and biochemical conversion of waste paper and pulp. Moreover, 50–73% of the cellulose content of waste paper can be degraded into biofuels by various enzymatic processes (Wang et al., 2012).

4.3 Algae

Microalgae are a source of biofuels (biogas, biodiesel, and bioethanol) and feedstock. Pathways of biofuel production from microalgae have been stated by Sharma et al. (2025; Fig. 7). Microalgae are the fastest-growing (complete life cycle within a few days) photosynthetic organisms. Currently, these methods are receiving increasing attention because of their high biomass efficiency. Experiments on algae (containing up to 50% oil) have increasingly generated biodiesel. The oil production potential of many microalgae is much greater than that of the best-producing oil crops. The oil content of diatom algae is 40 tons of oil/ha/year, which is 7–31 times greater than that of the best performing vegetable oil plant, oil palm, and 200 times greater than that of the soybean plant (Sheehan et al., 1998; Demirbas, 2009; Amaro et al., 2012). Microalgae are an alternative source of renewable energy (biodiesel

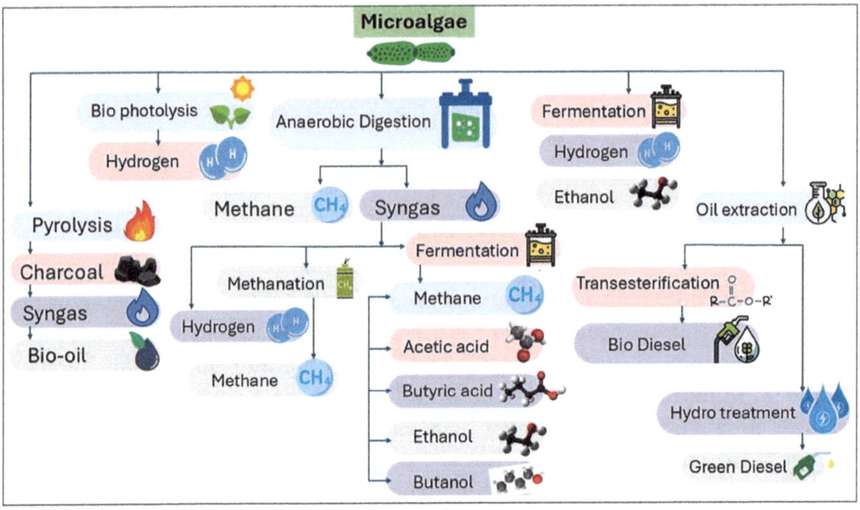

Fig. 7 Paths of biofuel production from microalgae (Source: Sharma et al., 2025)

and biofuel) because of their high oil content. Although the separation of algae from water is a costly process, no commercial activities have been undertaken yet.

5 Biogas from Biological Waste

Anaerobic digestion (AD) is a biochemical process for the decomposition of solid and liquid organic matter in an anaerobic environment with the aim of obtaining biogas (Fig. 8). Biological waste is degraded by a group of organisms in the absence of oxygen (Van Lier, 2014; Angelidaki et al., 2018; Sahota et al., 2018). Biogas is generally produced in agricultural biogas plants, where animal waste (e.g., manure), biomass from plants cultivated for this purpose (e.g., maize silage), or food waste act as substrates to generate this process (Czekała, 2018).

Biogas is a mixture of methane and carbon dioxide gas with trace amounts of other gases, such as hydrogen sulfide. Methane is a biogas that is mainly used for cooking and heating at home. Biogas can also be used as a fuel for engines, gas turbines, fuel cells, boilers, industrial heaters, other processes, or for the manufacturing of chemicals. The process of anaerobic digestion of organic matter with the participation of specialized bacterial groups involves four stages: (a) hydrolysis, (b) acidogenesis, (c) acetogenesis, and (d) methanogenesis (Morales-Polo et al., 2018).

In the first stage, microbes secrete enzymes to hydrolyze complex organic compounds (proteins, fats, and carbohydrates) into simpler structures, such as amino acids, sugars, and large-chain fatty acids. In the second step, these simple molecules are further fermented into various short-chain fatty acids (acetic, propionic, and butyric acids) along with volatile fatty acids (VFAs) and gaseous components (H_2 and CO_2) with the help of acidogenic bacteria.

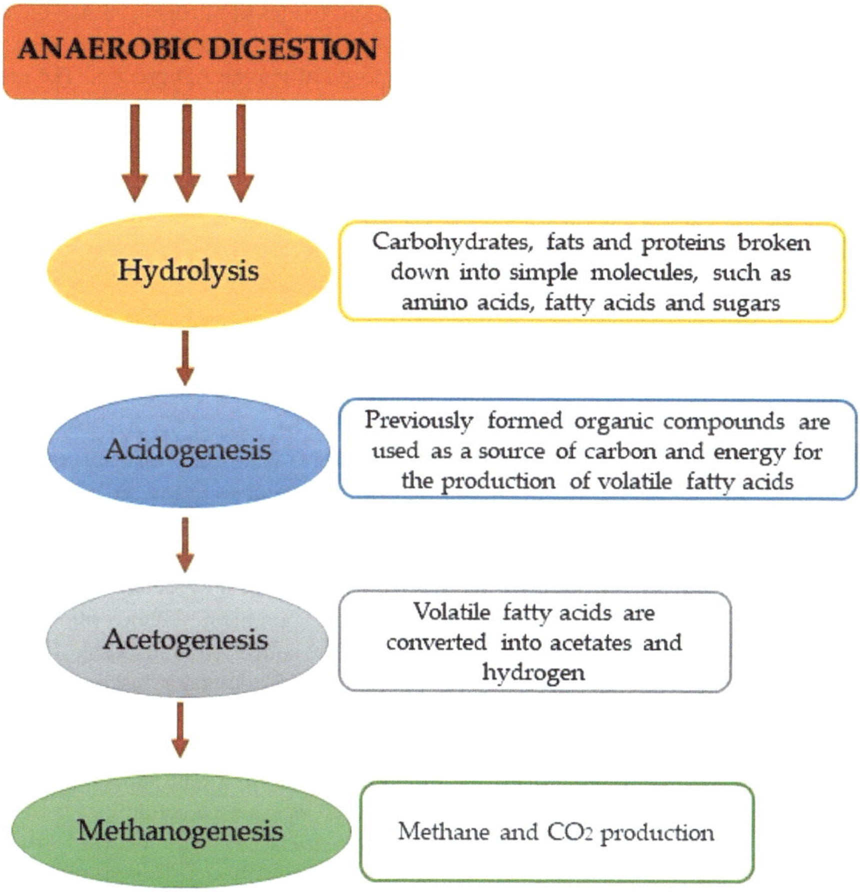

Fig. 8 Schematic diagram of the biogas production process through anaerobic digestion (Mignogna et al., 2023)

These molecules are subsequently reduced to acetic acid and hydrogen by acetogenic bacteria. The last stage is methanogenesis, in which methanogenic bacteria convert acetic acid and hydrogen into methane, CO_2, and water (Náthia-Neves et al., 2018). The methene concentration of biogas varies with the nature and composition of biological waste. For example, for pure cellulose, the biogas percentage is 50% methane and 50% carbon dioxide. In mixed waste feedstock, the methane percentage of biogas products is 40–60% by volume. Fats and oils contain approximately 70% methane (Demirbas, 2011).

In terms of environmental issues, waste is related to air and water pollution, which is harmful to human health. Furthermore, rapid urbanization and industrialization require more nonrenewable sources of energy, resulting in the depletion of energy resources. The burning of fossil fuels such as petrol, diesel, and compressed natural gas (CNG) in transportation systems is increasing daily. Burning fossil fuels releases greenhouse gases such as oxides of sulfur, oxides of nitrogen, carbon dioxide, carbon monoxide, etc. (Mahlia, 2002). All of these types of biowaste (agro-industry, animal, MSW, and food) can be disposed of via the AD process. The use of an anaerobic digestion (AD) approach not only helps solve waste management and energy problems but also generates revenue by providing side products.

6 Conclusion

Waste management faces several challenges as a result of the growing volume of biobased goods. A significant portion of solid waste is lost as a result of inappropriate waste management techniques that have the possibility of converting resources and energy. To maintain environmental, economic, and social development principles, waste and recycling management plans should be created for every building project before work begins. Many cities and towns deal with solid waste by creating landfills; some use incinerators to burn trash. Biowaste recycling and bioconversion of renewable energy sources can solve the greenhouse gas problems associated with landfill and incineration techniques. Since this information is currently unavailable, it is first necessary to ascertain how these waste products are now managed. Second, biobased products should be developed via a valorization method. One option is to separate biobased products and plastic waste after the collection or disposal of these products in separate bins. The second option for collecting biobased products together with biowaste is composting (Wojnowska-Baryła et al., 2020). Thermochemical, biochemical, and chemical conversion technologies are primarily employed to transform biomasses into biofuels such as biodiesel, bioethanol, and biogas, which are promising renewable energy sources. These technologies include combustion, pyrolysis, anaerobic digestion, and transesterification. The use of renewable energy sources through combustion is not only eco-friendly but also results in the emission of no pollutants. Additionally, producing electricity through biomass is a cost-effective solution to increase the production of electricity units while reducing the cost per unit. By properly managing and treating biological waste, we can ultimately achieve the goal of a zero-waste society (Sajid et al., 2023).

References

Ajayi, S. O., et al. (2017). Attributes of design for construction waste minimization: A case study of waste-to-energy project. *Renewable and Sustainable Energy Reviews.*, *73*, 1333–1341.

Ali, M., et al. (2019). The use of crop residues for biofuel production. In *Biomass, biopolymer-based materials, and bioenergy* (pp. 369–395). Elsevier.

Amaro, H. M., Macedo, Â. C., & Malcata, F. X. (2012). Microalgae: An alternative as sustainable source of biofuels? *Energy, 44*(1), 158–166.

Angelidaki, I., et al. (2018). Biogas upgrading and utilization: Current status and perspectives. *Biotechnology Advances, 36*(2), 452–466.

Augenstein, D., & Pacey, J. (1991). Landfill methane models. In *Proceedings from the technical sessions of SWANA'S 29th annual international solid waste exposition*. SWANA, Silver Spring, MD.

Backman, R., Frederick, W. J., & Hupa, M. (1993). Basic studies on black-liquor pyrolysis and char gasification. *Bioresource Technology, 46*, 153–158.

Bai, B., Liu, Y., Wang, Q., Zou, J., Zhang, H., Jin, H., & Li, X. (2019). Experimental investigation on gasification characteristics of plastic wastes in supercritical water. *Renewable Energy, 135*, 32–40. [CrossRef].

Bhatia, S. K., et al. (2018). Biowaste-to-bioenergy using biological methods–a mini-review. *Energy Conversion and Management., 177*, 640–660.

Bon, E. P., & Ferrara, M. A. (2007). Bioethanol production via enzymatic hydrolysis of cellulosic biomass. In *FAO seminar on the role of agricultural biotechnologies for production of bioenergy in developing countries*. ScienceDirect.

Brancoli, P., Bolton, K., & Eriksson, M. (2020). Environmental impacts of waste management and valorisation pathways for surplus bread in Sweden. *Waste Management, 117*, 136–145. https://doi.org/10.1016/j.wasman.2020.07.043

Chattopadhyay, S., & Singha, R. (2022). Biowaste: An introduction to its management. *Agriculture and Food., 4*(11), 379–381.

Cheng, H. & Hu, Y. (2010). Curbing dioxin emissions from municipal solid waste incineration in China: Re-thinking about management policies and practices. *Environmental Pollution, 158*(9), 2809–2814. https://doi.org/10.1016/j.envpol.2010.06.014

Czekała, W. (2018). Agricultural biogas plants as a chance for the development of the agri-food sector. *Journal of Ecological Engineering, 19*, 179–183. [CrossRef].

Demirbaş, A. (2002). Pyrolysis and steam gasification processes of black liquor. *Energy Conversion and Management, 43*(7), 877–884.

Demirbas, A. (2009). Production of biodiesel from algae oils. *Energy Sources, Part A: Recovery, Utilization, and Environmental Effects, 31*, 163–168.

Demirbas, A. (2011). Waste management, waste resource facilities and waste conversion processes. *Energy Conversion and Management, 52*(2), 1280–1287. https://doi.org/10.1016/j.enconman.2010.09.025

Demirbas, A., & Balat, M. J. F. E. S. (2010). Wastes energy. *Energy Conversion and Management., 2*, 1–63.

Demirbas, A., et al. (2016). Optimization of municipal solid waste (MSW) disposal in Saudi Arabia. *Energy Sources, Part A: Recovery, Utilization, and Environmental Effects., 38*(13), 1929–1937.

Gopal, P., Sivaram, N., & Barik, D. (2019). Paper industry wastes and energy generation from wastes. In *Energy from toxic organic waste for heat and power generation* (pp. 83–97). Elsevier.

Ioelovich, M. J. (2014). Waste paper as promising feedstock for production of biofuel. *Journal of Scientific Research and Reports., 3*(7), 905–917.

Kaza, S., Yao, L., Bhada-Tata, P., & Van Woerden, F. (2018). *What a Waste 2.0: A Global Snapshot of Solid Waste Management to 2050; Urban development series*. World Bank. Available online: https://openknowledge.worldbank.org/handle/10986/30317. Accessed 19 May 2021

Kumar, A., & Samadder, S. J. W. M. (2017). An empirical model for prediction of household solid waste generation rate–a case study of Dhanbad. *India. Waste Management., 68*, 3–15. https://doi.org/10.1016/j.wasman.2017.07.034

Liu, Q., Chmely, S. C., & Abdoulmoumine, N. (2017). Biomass treatment strategies for thermochemical conversion. *Energy & Fuels, 31*(4), 3525–3536. https://doi.org/10.1021/acs.energyfuels.7b00258

Mahlia, T. (2002). Emissions from electricity generation in Malaysia. *Renewable Energy, 27*(2), 293–300. https://doi.org/10.1016/S0960-1481(01)00177-X

Meegoda, J. N., Li, B., Patel, K., & Wang, L. B. (2018). A review of the processes, parameters, and optimization of anaerobic digestion. *International Journal of Environmental Research and Public Health, 15*(10), 2224. https://doi.org/10.3390/ijerph15102224

Mignogna, D., Ceci, P., Cafaro, C., Corazzi, G., & Avino, P. (2023). Production of biogas and biomethane as renewable energy sources: A review. *Applied Sciences, 13*(18), 10219. https://doi.org/10.3390/app131810219

Morales-Polo, C., Cledera-Castro, M. D. M., & Moratilla Soria, B. Y. (2018). Reviewing the anaerobic digestion of food waste: From waste generation and anaerobic process to its perspectives. *Applied Sciences, 8*(10), 1804. https://doi.org/10.3390/app8101804

Najib, N. N., Nugroho, H. Y. S. H., Isnan, W., & Saad, M. (2023). Effects of size and additives on the decomposition process and nutrient content of urban organic waste compost. In *IOP conference series: Earth and environmental science* (Vol. 1180, No. 1, p. 012043). IOP Publishing.

Náthia-Neves, G., Berni, M., Dragone, G., Mussatto, S. I., & Forster-Carneiro, T. (2018). Anaerobic digestion process: Technological aspects and recent developments. *International journal of Environmental Science and Technology, 15*, 2033–2046. https://doi.org/10.1007/s13762-018-1682-2

Raghab, S. M., Abd El Meguid, A. M., & Hegazi, H. A. (2013). Treatment of leachate from municipal solid waste landfill. *HBRC Journal, 9*(2), 187–192. https://doi.org/10.1016/j.hbrcj.2013.05.007

Sahota, S., Shah, G., Ghosh, P., Kapoor, R., Sengupta, S., Singh, P., Vijay, V., Sahay, A., Vijay, V. K., & Thakur, I. S. (2018). Review of trends in biogas upgradation technologies and future perspectives. *Bioresource Technology Reports, 1*, 79–88. https://doi.org/10.1016/j.biteb.2018.01.002

Sajid, M., Akram, A., Sajjad, S. F., Siddique, T., & Arshad, M. (2023). Biological waste management. In *Advances and challenges in hazardous waste management*. IntechOpen. https://doi.org/10.5772/intechopen.1003266

Salhofer, S. (2000). Modelling commercial/industrial waste generation: A Vienna, Austria case study. *Waste Management & Research, 18*(3), 269–282. https://doi.org/10.1177/0734242X0001800309

Salmenoja, K. (1993). Black-liquor gasification: Theoretical and experimental studies. *Bioresource Technology, 46*(1–2), 167–171. https://doi.org/10.1016/0960-8524(93)90069-N

Sharma, A. K., Jaryal, S., Sharma, S., Dhyani, A., Tewari, B. S., & Mahato, N. (2025). Biofuels from microalgae: A review on microalgae cultivation, biodiesel production techniques and storage stability. *PRO, 13*(2), 488. https://doi.org/10.3390/pr13020488

Sheehan, J., Dunahay, T., Benemann, J., & Roessler, P. (1998). *A look back at the US Department of Energy's aquatic species program—Biodiesel from algae*. National Renewable Energy Laboratory (NREL) Report: NREL/TP-580-24190.

Siddique, R., Khatib, J., & Kaur, I. (2008). Use of recycled plastic in concrete: A review. *Waste Management, 28*(10), 1835–1852. https://doi.org/10.1016/j.wasman.2007.09.011

Uygur, A., & Kargı, F. (2004). Biological nutrient removal from pre-treated landfill leachate in a sequencing batch reactor. *Journal of Environmental Management, 71*(1), 9–14. https://doi.org/10.1016/j.jenvman.2004.01.002

Van Lier, J. (2014). *Anaerobic fundamentals and COD balance*. TU Delft.

Vaverková, M. D. (2019). Landfills impact the environment. *Geosciences, 9*(10), 431. https://doi.org/10.3390/geosciences9100431

Wang, L., Templer, R., & Murphy, R. J. (2012). High-solids loading enzymatic hydrolysis of waste papers for biofuel production. *Applied Energy, 99*, 23–31. https://doi.org/10.1016/j.apenergy.2012.03.045

Wernet, G., Bauer, C., Steubing, B., Reinhard, J., Moreno-Ruiz, E. and Weidema, B. (2016). The ecoinvent database version 3 (part I): overview and methodology. *International Journal of Life cycle Assessment, 21*(9), 1218–1230. https://doi.org/10.1007/s11367-016-1087-8

Windfeld, E. S., & Brooks, M. S. L. (2015). Medical waste management–a review. *Journal of Environmental Management, 163*, 98–108. https://doi.org/10.1016/j.jenvman.2015.08.013

Wojnowska-Baryła, I., Kulikowska, D., & Bernat, K. (2020). Effect of bio-based products on waste management. *Sustainability, 12*(5), 2088. https://doi.org/10.3390/su12052088
Yousuf, A. (2012). Biodiesel from lignocellulosic biomass–prospects and challenges. *Waste Management, 32*(11), 2061–2067. https://doi.org/10.1016/j.wasman.2012.03.008
Zhang, C., Xu, T., Feng, H., & Chen, S. (2019). Greenhouse gas emissions from landfills: A review and bibliometric analysis. *Sustainability, 11*(8), 2282. https://doi.org/10.3390/su11082282

Biochar: A Sustainable Solution for Agriculture and Environment

Nisha Singh, Vineet K. Maurya, and Dhananjay Kumar

Abstract

Biochar, a carbon-rich material derived from the pyrolysis of organic biomass, is increasingly recognized as a promising candidate for its applications in the area of agriculture and environmental technology. This chapter explores the diverse benefits of biochar, focusing on its ability to improve soil health, boost plant growth, and foster sustainable agricultural practices. Biochar's porous structure and high cation exchange capacity (CEC) enhance soil fertility, water retention, and nutrient availability, leading to healthier plant growth and increased crop yields. Its unique properties also create an ideal habitat for beneficial soil microbes and fauna, such as bacteria, fungi, earthworms, and insects, which are crucial for nutrient cycling and soil aeration. Furthermore, biochar's stability and adsorptive capacity enable it to sequester carbon and immobilize pollutants, thereby mitigating climate change and reducing environmental contamination. By reducing the need for chemical fertilizers and improving soil conditions, biochar promotes more sustainable and eco-friendly farming practices. The chapter also addresses the challenges and limitations of biochar application, including economic viability, variable efficacy, and the need for standardized guidelines. Through continued research, policy support, and public awareness, biochar has the potential to revolutionize soil management and contribute significantly to global sustainability efforts.

N. Singh (✉)
Department of Biochemistry, University of Lucknow, Lucknow, Uttar Pradesh, India

V. K. Maurya
Applied Microbiology Lab, Department of Botany and Microbiology, H.N.B. Garhwal University, Srinagar, Garhwal, Uttarakhand, India

D. Kumar (✉)
Laboratory of Algal Biotechnology, Department of Botany and Microbiology, H.N.B. Garhwal University, Srinagar, Garhwal, Uttarakhand, India

S. Singh, S. Srivastava (eds.), *Biochar: A Precious Resource from Biological Waste*,
https://doi.org/10.1007/978-981-95-0425-1_2

Keywords

Biochar · Pyrolysis · Soil health · Plant growth · Sustainable agriculture · Environmental management

1 Introduction

In the quest for sustainable agricultural practices and effective climate change mitigation strategies, biochar has emerged as a promising solution. Derived from the pyrolysis of organic biomass, biochar is a carbon-rich material with unique properties that offer multiple environmental and agricultural benefits (Arif et al., 2020). This chapter provides an in-depth introduction to biochar, exploring its origins, properties, production methods, and diverse applications. By understanding the foundational aspects of biochar, we can suitably explore its potential impact on soil health, carbon sequestration, and pollution mitigation.

1.1 Historical Background of Biochar

The concept of biochar is not new; its origins can be traced back to ancient Amazonian civilizations that created terra preta, or "dark earth," by adding charred organic matter to the soil (Glaser, 2014). These fertile soils have persisted for thousands of years, demonstrating the long-term benefits of biochar. Modern interest in biochar has been rekindled by the growing need for sustainable agricultural practices and the urgent necessity to address climate change (Pillala et al., 2023).

1.2 Defining Biochar

Biochar is a stable, carbon-rich material produced through the thermal decomposition of organic biomass in an oxygen-limited environment, known as pyrolysis. The resulting product is characterized by its high carbon content, porosity, and large surface area. These properties contribute to biochar's ability to improve soil structure, enhance nutrient retention, and act as a long-term carbon sink (Ippolito et al., 2020; Lorenz et al., 2018).

1.3 Chemical Composition of Biochar

The chemical composition of biochar is influenced by the type of feedstock used and the conditions under which it is produced. Key components of biochar include:

(i) *Carbon*: Biochar is rich in carbon, typically accounting for 60–90% of its composition. The carbon in biochar is predominantly in the form of stable, aromatic structures, which contribute to its long-term persistence in the soil (Bolan et al., 2022).

(ii) *Nutrients*: Biochar contains various nutrients, including nitrogen, phosphorus, potassium, calcium, and magnesium. The availability of these nutrients depends on the feedstock and pyrolysis conditions (Saletnik et al., 2019).

(iii) *Minerals and Metals*: Depending on the source biomass, biochar can contain minerals and trace metals such as silicon, aluminium, iron, and manganese. These elements can influence soil chemistry and plant growth (Shetty & Prakash, 2020).

(iv) *Functional Groups*: Biochar surfaces are characterized by a variety of functional groups, such as carboxyl, hydroxyl, and phenolic groups. These functional groups play a crucial role in nutrient retention, cation exchange capacity, and pollutant adsorption (Fan et al., 2018).

1.4 Properties of Biochar

The unique properties of biochar are central to its wide-ranging applications. Key properties include:

(i) *Porosity and Surface Area*: Biochar, a carbon-rich material derived from biomass through pyrolysis, exhibits a complex and diverse pore structure, critical for its functionality in various applications (Jaafar et al., 2015). The pyrolysis process involves the thermal decomposition of organic materials in the absence of oxygen, resulting in the development of rudimentary pores due to the dehydration process and the release of volatile constituents from the carbon matrix. These pores, which range from nanometers to tens of micrometers in size, are categorized by the International Union of Pure and Applied Chemistry (IUPAC) into micropores (<2 nm), mesopores (2–50 nm), and macropores (>50 nm). The micropore and mesopore volumes are two major components of biochar's pore structure (Leng et al., 2021). Micropore volumes typically range from 0.012 to 0.060 cm^3/g with a median value of 0.024 cm^3/g, while mesopore volumes range from 0.007 to 0.020 cm^3/g with a median value of 0.009 cm^3/g. Statistical analysis shows that the micropore volume contributes significantly to the total pore volume, with variations ranging between 12.1% and 58.0%. These variations highlight the structural heterogeneity of biochar and its adaptability for different applications.

The surface area of biochar is a crucial parameter that determines the accessibility of surface charges, thereby affecting the cation exchange capacity. Typically, biochar's surface area ranges from 8 to 132 m^2/g, while the total pore volume varies from 0.016 to 0.083 cm^3/g. However, by selecting appropriate precursors and optimizing pyrolysis parameters, these values can be significantly enhanced. For example, the surface area and total pore volume can reach up to 490.8 m^2/g and 0.25 cm^3/g, respectively, under optimal conditions (Sun et al., 2017). Furthermore, post-treatments such as potassium hydroxide (KOH) activation can dramatically increase these values, with surface areas soaring to 3263 m^2/g and total pore volumes reaching 1.772 cm^3/g, making biochar comparable to commercial activated carbon.

(ii) *Cation Exchange Capacity (CEC)*: The cation exchange capacity (CEC) of biochar is a critical physical property that influences its ability to retain and exchange nutrients, pollutants, and other charged particles (Domingues et al., 2020). CEC measures the capacity of biochar to exchange cations (positively charged ions) with the surrounding environment, typically expressed in centimoles per kilogram (cmol/kg) or milliequivalents per 100 grams (meq/100 g). The CEC of biochar varies with the type of biomass used to produce it, with manure-derived biochars tend to have higher CEC than woody biochars. Pyrolysis temperature, surface area, and ash content also impact CEC, with higher temperatures and surface areas leading to increased CEC, and the presence of ash contributing to higher CEC values. Biochar's CEC values range from low (1–10 cmol/kg) for woody biochars to medium (10–50 cmol/kg) for grass-derived biochars, high (50–100 cmol/kg) for manure-derived biochars, and very high (>100 cmol/kg) for some non-wood-derived biochars (Hailegnaw et al., 2019). Understanding biochar's CEC property is essential for harnessing its potential for environmental remediation, soil amendment, and other applications, as it affects its ability to retain nutrients and pollutants, support microbial growth, and improve soil fertility and structure.

(iii) *Carbon Stability*: Biochar's carbon stability and content are intricately linked, forming the foundation of its properties and long-term efficacy (Ennis et al., 2012). As pyrolysis temperature increases, carbon content rises (62.2–92.4%) and ash content grows (5.7–18.7%), influencing biochar's characteristics. High carbon content indicates the presence of original organic plant residues, while increased ash content stems from concentrated inorganic constituents and combustion residues. Mineral matter forms ash, remaining in biochar after carbonization. Notably, increased carbon content leads to a more condensed carbon structure, enhancing biochar's resistance to microbial degradation (Bolan et al., 2022). However, high ash contents are associated with greater amounts of poly aromatic hydrocarbons and trace metals. By controlling feedstock materials and pyrolysis conditions, emissions of atmospheric pollutants and particulate matter can be reduced during biochar production, ensuring high-quality biochar with optimal carbon stability and content.

(iv) *pH*: Biochar is typically alkaline, which can help neutralize acidic soils and improve overall soil health. The pH of biochar is generally alkaline, ranging from 7.1 to 10.5, depending on the type of biomass used for production (Bolan et al., 2023). Biomass type significantly influences biochar pH, with wood-derived biochars tend to have lower pH values (average 2 units lower) compared to non-wood-derived biochars. Non-wood-derived biochars have higher pH values due to the presence of salts like carbonates and chlorides of potassium and calcium in the ash. The pH of biochar also correlates with its content of lignin, hemicelluloses, cellulose, and oxygen functionalities. Alkaline properties are attributed to the presence of –COO– and –O– groups and carbonate content. Oxygen-rich functional groups, such as γ-pyrone-type, chromene, diketone, or quinine groups, also contribute to biochar basicity. The formation of these groups during pyrolysis is associated with the polymerization/condensation reactions of aliphatic compounds and dehydration of the feedstock. Additionally, the pH of biochar is influenced by the presence of carboxyl groups, aliphatic O-alkylated carbons, anomeric O–C–O carbons, fused-ring aromatic structures, and aromatic C–O groups (Tomczyk et al., 2020). Understanding biochar pH is essential for its application in soil amendment and environmental remediation, as it affects nutrient availability, microbial activity, and contaminant adsorption.

(v) *Adsorptive Capacity*: The adsorption characteristics of biochar are influenced by its fabrication temperature, with higher temperatures (500–700 °C) resulting in higher sorption efficiency for antibiotics and trace/heavy metals compared to lower temperatures (250–400 °C). The higher aromaticity and reduced polarity at higher temperatures enhance the removal potential of antibiotics and trace metals. Biochar interacts with pollutants through various mechanisms, including H-bonding, electrostatic attraction, and van der Waals interactions (Ippolito et al., 2020). The functional groups in biochar generated at low temperatures facilitate sorption through H-bonding and electrostatic attraction. Biochar produced at higher temperatures retains more anionic dyes and fewer cationic dyes, and the sorption of pesticides improves with increasing production temperature. The sorption process is also pH-dependent, with the surface charge of biochar varying according to the pH, influencing electrostatic attraction and repulsion. Overall, understanding the sorption characteristics of biochar is crucial for optimizing its use in water pollution remediation, and adjusting pH conditions can enhance the removal of specific pollutants.

1.5 Production Methods

Biochar production can be carried out using various methods, and the characteristics and properties of the biochar obtained from different methods are special. The choice of production method depends on factors such as the type of biomass used, the desired biochar properties, and specific application needs.

(i) *Slow Pyrolysis*: This method involves heating biomass at relatively low temperatures (300–700 °C) for an extended period. Slow pyrolysis maximizes biochar yield and is suitable for producing biochar with high stability and nutrient retention capabilities (Tan et al., 2021a).

(ii) *Fast Pyrolysis*: This process rapidly heats biomass to higher temperatures (500–1000 °C) in a matter of seconds. This method produces less biochar but generates more bio-oil and syngas, which can be used as renewable energy sources (Tan et al., 2021a).

(iii) *Gasification*: Gasification occurs at very high temperatures (>1000 °C) and partially combusts the biomass in a controlled, oxygen-limited environment. This process produces biochar along with a high volume of syngas, making it suitable for integrated energy production systems (Braghiroli et al., 2018).
(iv) *Hydrothermal Carbonization (HTC)*: Unlike traditional pyrolysis, HTC processes biomass in a water-saturated environment at lower temperatures (180–250 °C) under high pressure. This method is particularly effective for wet biomass and produces a type of biochar known as hydrochar (Mantovani et al., 2022).

Each production method imparts distinct physical and chemical properties to the biochar, influencing its effectiveness in various applications.

1.6 Factors Influencing Biochar Properties

Several factors influence the properties and effectiveness of biochar, including feedstock type, pyrolysis temperature, and residence time.

(i) *Feedstock Type*: The type of biomass used to produce biochar significantly affects its chemical composition and physical properties. For example, woody biomass tends to produce biochar with higher carbon content and stability, while manure-based biochar is richer in nutrients (Mukome et al., 2013).
(ii) *Pyrolysis Temperature*: The temperature at which pyrolysis occurs influences the structure and properties of biochar. Higher temperatures typically result in greater carbonization, higher surface area, and increased porosity. However, they may also reduce nutrient content (Tan et al., 2021a).
(iii) *Residence Time*: The duration of the pyrolysis process, or residence time, can affect the yield and quality of biochar. Longer residence times generally lead to more complete carbonization and greater stability but may also result in lower biochar yields (Sun et al., 2017).
(iv) *Heating Rate*: The rate at which biomass is heated during pyrolysis can influence the physical structure and chemical properties of the resulting biochar. Rapid heating rates can lead to higher bio-oil production, while slower rates favor biochar formation (Ippolito et al., 2020).

2 Biological Interactions

Biochar interacts with soil biota in complex ways, influencing microbial communities, plant growth, and soil health.

2.1 Microbial Activity

Biochar plays a pivotal role in enhancing microbial activity within the soil, serving as a beneficial habitat for essential soil microbes, including bacteria and fungi. The highly porous structure of biochar creates an ideal environment for these microorganisms, offering numerous microhabitats that protect them from environmental stressors such as extreme temperatures, desiccation, and predation (Thies & Rillig, 2012). These pores also help maintain moisture, which is critical for microbial survival and activity. Moreover, biochar acts as a reservoir of nutrients and organic matter, providing a steady supply of resources that soil microbes need to thrive. The surface of biochar is rich in functional groups, which can adsorb and retain essential nutrients like nitrogen, phosphorus, and potassium. This nutrient retention benefits plants and sustains microbial populations by ensuring that microbes have continuous access to the nutrients they require for growth and metabolism (Khadem et al., 2021). The enhanced microbial activity facilitated by biochar contributes significantly to soil health. Beneficial microbes play crucial roles in nutrient cycling, organic matter decomposition, and the suppression of soilborne pathogens. By promoting a healthy and diverse microbial community, biochar helps create a more resilient and fertile soil ecosystem, ultimately leading to improved plant growth and agricultural productivity (Dai et al., 2021).

2.2 Plant Growth

Biochar significantly enhances plant growth by improving soil fertility, structure, and water retention. Its high cation exchange capacity (CEC) allows it to retain essential nutrients, such as nitrogen, phosphorus, and potassium, making them more available to plants and reducing nutrient leaching. The porous structure of biochar improves soil aeration and drainage, creating a more favorable environment for root development (Hailegnaw et al., 2019). Additionally, biochar's ability to retain moisture helps plants withstand drought conditions, ensuring more consistent growth and yields. By enhancing nutrient availability and moisture retention, biochar reduces the reliance on chemical fertilizers. This promotes more sustainable agricultural practices, as it lowers input costs for farmers and mitigates the environmental impact associated with fertilizer runoff. The improved soil structure and fertility provided by biochar lead to healthier plants and higher agricultural productivity (Schulz et al., 2013; Tanure et al., 2019).

2.3 Soil Fauna

Biochar positively impacts soil fauna, such as earthworms and insects, by enhancing soil conditions and providing a stable environment. The improved soil structure created by biochar, with its increased porosity and water-holding capacity, offers a more hospitable habitat for these organisms (Egamberdieva et al., 2021).

Earthworms, for example, benefit from the aerated and moisture-retentive soil, which supports their burrowing and feeding activities. The soil fauna plays crucial roles in nutrient cycling and soil aeration. Earthworms help decompose organic matter, releasing essential nutrients back into the soil, while their burrowing actions improve soil aeration and water infiltration. Insects contribute to breaking down organic materials and fostering microbial activity. By fostering a healthier and more active soil fauna population, biochar enhances the overall fertility and structure of the soil, promoting more robust plant growth and sustainable agricultural ecosystems (Khadem et al., 2021; Paz-Ferreiro et al., 2016).

3 Applications of Biochar

The versatility of biochar allows it to be used in various applications, spanning agriculture, environmental management, and industrial processes.

3.1 Soil Amendment

Soil is a fundamental component of agricultural productivity and environmental sustainability. The quality of the soil determines its capacity to support plant growth, retain water and nutrients, and sustain microbial life. However, many agricultural soils are degraded due to intensive farming practices, deforestation, and climate change. These factors contribute to soil erosion, nutrient depletion, and reduced water-holding capacity, ultimately compromising crop yields and ecosystem health. Biochar, a carbon-rich material produced through the pyrolysis of organic biomass, has emerged as a promising soil amendment that can address these challenges (Zahedifar, 2023). This section explores how biochar improves soil structure, increases water retention, enhances nutrient availability, and restores fertility in degraded soils, thereby promoting sustainable agricultural practices.

Soil structure refers to the arrangement of soil particles into aggregates, which influences porosity, permeability, and root penetration. Good soil structure is essential for healthy plant growth, facilitating air and water movement, root expansion, and microbial activity (Jaafar et al., 2015). Biochar contributes to improved soil structure in several ways. Biochar's porous nature enhances soil porosity, creating more space for air and water to circulate. This increased porosity improves soil aeration, which is critical for root respiration and microbial activity (Fig. 1). Additionally, biochar helps in the formation of soil aggregates by binding soil particles together, thereby improving soil stability and reducing erosion (Braghiroli et al., 2018; Ennis et al., 2012). Soil compaction is a common issue in agricultural fields, especially those subjected to heavy machinery. Compacted soils have reduced porosity and permeability, which hinder root growth and water infiltration. Incorporating biochar into the soil can alleviate compaction by increasing soil bulk density and creating a more friable soil structure. This promotes deeper root growth and enhances the soil's ability to absorb and retain water. The texture of the soil,

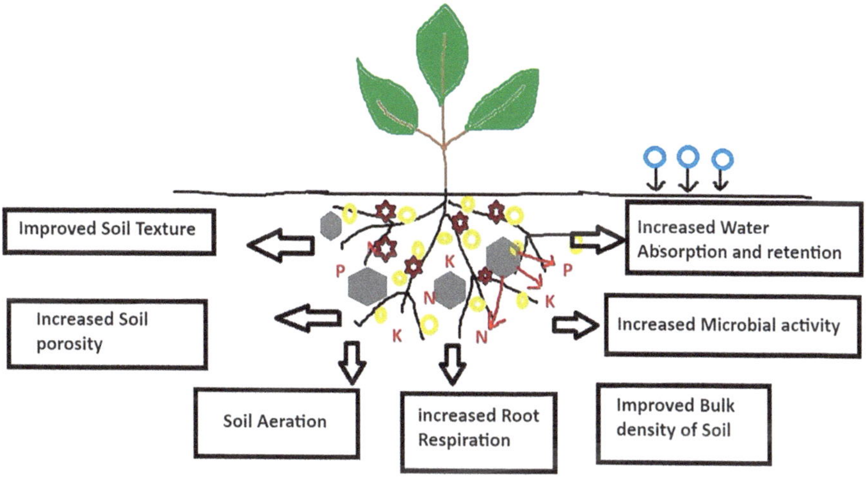

Fig. 1 Biochar enhances various soil quality parameters

determined by the proportion of sand, silt, and clay, affects its physical properties and suitability for different crops. Biochar can modify soil texture by improving the balance of these components. In sandy soils, biochar increases water and nutrient retention by adding fine particles and organic matter. In clayey soils, it enhances drainage and reduces the tendency for waterlogging (Cross et al., 2016).

Water is a vital resource for agriculture, and efficient water management is crucial for crop productivity, especially in regions prone to drought and water scarcity. Biochar significantly enhances the soil's ability to retain and manage water. Biochar's high surface area and porosity enable it to absorb and retain large quantities of water. When incorporated into the soil, biochar acts like a sponge, holding water and making it available to plants during dry periods (Tanure et al., 2019). This increased water-holding capacity is particularly beneficial in sandy soils, which typically have low water retention. By improving the soil's water retention, biochar reduces the frequency and volume of required irrigation. This not only conserves water resources but also lowers irrigation costs for farmers (Xiao et al., 2016). Efficient water use is critical for sustainable agriculture, especially in water-scarce regions. Biochar-amended soils are more resilient to drought conditions due to their enhanced ability to retain moisture. This resilience helps crops withstand periods of water stress, leading to more stable yields even in adverse climatic conditions. Biochar's role in drought mitigation is crucial for ensuring food security in the face of climate change. Besides, biochar improves soil drainage by enhancing soil structure and porosity. This is particularly important in clayey soils, which are prone to waterlogging and poor drainage. Better drainage reduces the risk of root diseases and promotes healthier plant growth (Ndede et al., 2022).

Nutrient availability is a key factor in soil fertility and crop productivity. Biochar enhances nutrient retention and availability through various mechanisms. Biochar has a high cation exchange capacity, which allows it to attract and hold onto essential nutrients such as potassium (K^+), calcium (Ca^{2+}), and magnesium (Mg^{2+}) (Hossain et al., 2020). This prevents nutrients from leaching out of the soil and makes them more readily available to plants. The increased nutrient retention reduces the need for synthetic fertilizers, promoting more sustainable nutrient management practices. Biochar is rich in stable organic carbon, which contributes to the overall organic matter content of the soil (Ding et al., 2016; Igalavithana et al., 2016). Organic matter improves soil structure, water-holding capacity, and nutrient availability. It also serves as a food source for soil microorganisms, enhancing microbial activity and nutrient cycling. Soil pH affects nutrient availability and microbial activity. Biochar is typically alkaline and can help neutralize acidic soils, improving the availability of nutrients such as phosphorus (P) and making the soil environment more conducive to plant growth. This pH regulation is particularly beneficial in regions with naturally acidic soils or where acidifying fertilizers have been extensively used. Soil microorganisms play a crucial role in nutrient cycling and organic matter decomposition. Biochar provides a habitat for beneficial microbes, increasing microbial diversity and activity (Khadem et al., 2021). Enhanced microbial activity accelerates the breakdown of organic matter and the release of nutrients, improving soil fertility and plant health.

Degraded soils, resulting from erosion, nutrient depletion, and overuse of chemical inputs, pose a significant challenge to sustainable agriculture. Biochar can restore fertility and rejuvenate degraded soils. Soil erosion is a major cause of land degradation, leading to the loss of topsoil and nutrients. Biochar helps reduce erosion by improving soil structure and stability. Its ability to enhance aggregation and reduce surface runoff protects the soil from erosion and retains valuable topsoil. Degraded soils often lack sufficient organic matter, which is essential for soil health and fertility. Biochar, rich in stable organic carbon, adds organic matter to the soil, improving its physical, chemical, and biological properties. This rebuilds soil fertility and enhances its capacity to support plant growth. Some soils are degraded due to contamination with heavy metals, pesticides, or other pollutants. Biochar's adsorptive properties enable it to bind and immobilize these contaminants, reducing their bioavailability and toxicity. This makes biochar an effective tool for soil remediation and the restoration of contaminated lands. The use of biochar as a soil amendment encourages sustainable agricultural practices by reducing reliance on chemical fertilizers and pesticides. Biochar's ability to enhance soil health and fertility promotes organic farming, conservation tillage, and other sustainable practices that protect the environment and support long-term agricultural productivity (Domingues et al., 2020; Paz-Ferreiro et al., 2016).

Numerous studies and real-world applications have demonstrated the effectiveness of biochar in improving soil health and agricultural productivity. In Vietnam, biochar produced from rice husks and straw was applied to paddy fields. Farmers observed significant improvements in soil structure, water retention, and nutrient availability. As a result, rice yields increased by 15–20%, and methane emissions from the fields were reduced by 30–40%. This case study highlights biochar's potential to enhance crop productivity and sustainability in rice cultivation (Hung et al., 2018). In Europe, biochar has been applied to various vegetable crops,

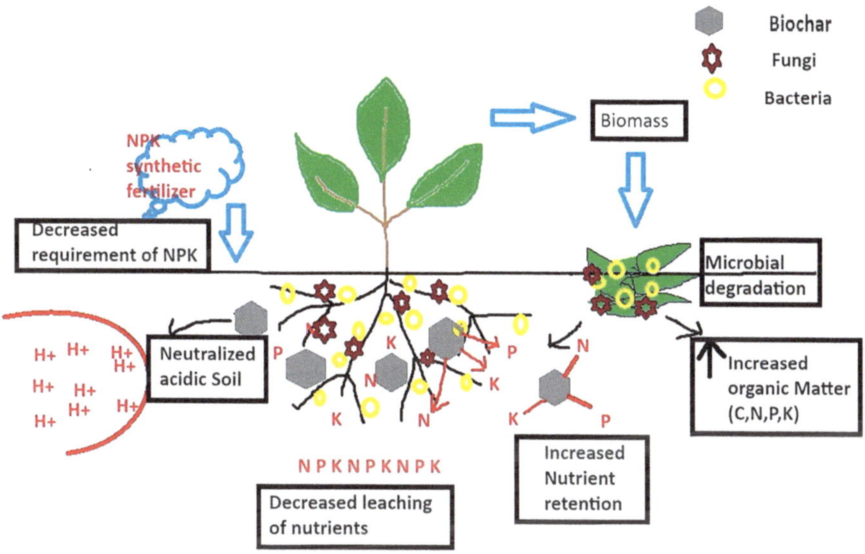

Fig. 2 Mechanism of biochar-mediated soil amendments

including tomatoes, lettuce, and carrots. Farmers reported improved soil health, increased water retention, and enhanced nutrient availability. These benefits led to higher crop yields and better-quality produce, demonstrating biochar's effectiveness in vegetable farming (Chrysargyris et al., 2020). In sub-Saharan Africa, biochar has been used to improve maize yields and soil health. Research has shown that biochar application enhances nutrient availability, boosts microbial activity, and increases water retention (Fig. 2). These improvements resulted in higher maize productivity, contributing to food security in the region (Gwenzi et al., 2015). In British Columbia, Canada, biochar produced from forestry residues was applied to degraded forest soils (Lefebvre et al., 2024). The application of biochar improved soil structure, increased organic matter content, and enhanced nutrient availability. Tree seedlings planted in biochar-amended soils showed higher survival rates and faster growth, highlighting biochar's potential in forest restoration.

3.2 Carbon Sequestration

Climate change, driven by the increased concentration of greenhouse gases (GHGs) in the atmosphere, is one of the most critical challenges facing humanity today. Carbon dioxide (CO_2) is the primary GHG contributing to global warming, primarily resulting from fossil fuel combustion, deforestation, and various industrial processes. Mitigating climate change necessitates strategies that effectively reduce atmospheric CO_2 levels. One promising approach is carbon sequestration, where carbon is captured and stored in a stable form, preventing it from reentering the atmosphere. Biochar, a carbon-rich product derived from the pyrolysis of organic biomass, has emerged as a viable method for long-term carbon sequestration, offering significant potential to mitigate climate change (Ennis et al., 2012) (Ippolito et al., 2020). Unlike other forms of organic matter that quickly break down and release CO_2 back into the atmosphere, biochar remains stable in soils for centuries to millennia. This remarkable stability makes biochar an effective tool for carbon sequestration (Khadem et al., 2021). The carbon sequestration potential of biochar lies in its ability to lock away carbon that would otherwise be released into the atmosphere through natural decomposition or combustion. When plants grow, they absorb CO_2 from the atmosphere through photosynthesis and store it in their biomass. If the biomass is left to decompose or is burned, the stored carbon is released back into the atmosphere as CO_2. However, by converting this biomass into biochar, the carbon is stabilized and sequestered in a solid form. This process not only captures atmospheric carbon but also transforms it into a material that enhances soil health and fertility (Lorenz et al., 2018). One of the key advantages of biochar as a carbon sequestration tool is its longevity. Studies have shown that biochar can remain stable in soils for hundreds to thousands of years, providing a long-term carbon sink. This stability is attributed to the chemical structure of biochar, which consists of aromatic carbon rings that are highly resistant to microbial decomposition. As a result, the carbon in biochar is effectively removed from the active carbon cycle, reducing the overall concentration of CO_2 in the atmosphere (Ennis et al., 2012).

In addition to its stability, biochar's physical properties contribute to its effectiveness as a carbon sequestration agent. Biochar has a porous structure and a high surface area, which enhance its ability to interact with soil particles and microorganisms. This interaction further stabilizes the carbon and integrates it into the soil matrix. Moreover, the porous nature of biochar improves soil aeration and water retention, which can enhance plant growth and increase the amount of carbon captured through photosynthesis. By promoting healthier and more productive plants, biochar indirectly contributes to additional carbon sequestration (Bera et al., 2024). The production and application of biochar also offer a sustainable approach to waste management. Organic waste, such as crop residues, animal manure, and forestry by-products, represents a significant source of potential CO_2 emissions if left to decompose or be burned (Yadav et al., 2023). Converting these waste materials into biochar not only sequesters carbon but also reduces methane (CH_4) and nitrous oxide (N_2O) emissions, both of which are potent GHGs. Methane emissions, in particular, are significantly reduced during the pyrolysis process compared to traditional composting or open burning of biomass.

Biochar's ability to sequester carbon and mitigate GHG emissions has garnered interest from policymakers, researchers, and environmentalists. Several studies and pilot projects have demonstrated the potential of biochar to significantly reduce the carbon footprint of agricultural and forestry operations. For example, applying biochar to agricultural soils can enhance soil fertility and crop yields while simultaneously sequestering carbon. In forestry, biochar produced from logging residues can be used to restore degraded forest soils and improve tree growth, contributing to long-term carbon storage in both the soil and biomass (Ansari et al., 2023). The economic viability of biochar as a carbon sequestration tool is also an important consideration. While the initial investment in pyrolysis equipment and biochar production can be significant, the long-term benefits of improved soil health, increased crop yields, and carbon sequestration can offset these costs. Additionally, the development of carbon markets and the potential for biochar to earn carbon credits provide financial incentives for adopting this technology. By quantifying the carbon sequestration potential of biochar and integrating it into carbon trading systems, farmers and land managers can generate additional revenue streams while contributing to climate change mitigation (Egamberdieva et al., 2021; Liu et al., 2023; Yadav et al., 2023).

3.3 Pollution Mitigation

The increasing levels of environmental pollution pose significant risks to ecosystems, human health, and agricultural productivity. Heavy metals and organic pollutants are particularly concerning due to their persistence, toxicity, and ability to bioaccumulate in the food chain. Traditional methods of pollution control often fall short of providing sustainable solutions, necessitating innovative approaches to mitigate contamination. Biochar, a carbon-rich material derived from the pyrolysis of organic biomass, has emerged as a promising tool for pollution mitigation. Its unique adsorptive properties enable biochar to capture and immobilize a wide range of pollutants, thereby reducing their bioavailability and preventing them from contaminating water sources (Majumder et al., 2023).

Biochar's effectiveness in pollution mitigation is primarily attributed to its high surface area, porosity, and functional groups on its surface. These characteristics enhance its capacity to adsorb contaminants, making it an effective medium for remediating polluted soils and water. The process of pyrolysis creates a highly porous structure in biochar, which increases its surface area and provides numerous sites for pollutant adsorption (Kumar et al., 2023). The functional groups, such as carboxyl, hydroxyl, and phenolic groups, further enhance biochar's ability to bind with various contaminants.

Heavy metals, including lead (Pb), cadmium (Cd), arsenic (As), and mercury (Hg), are common pollutants that pose serious environmental and health risks. These metals can originate from industrial activities, mining, agricultural runoff, and improper waste disposal. Once released into the environment, heavy metals can persist for long periods, contaminating soil and water and entering the food chain. Biochar's ability to adsorb heavy metals is due to its surface chemistry and high cation exchange capacity (CEC). The negatively charged surfaces of biochar attract and bind positively charged metal ions, effectively reducing their mobility and bioavailability (Nan & Wang, 2023). For instance, studies have demonstrated that biochar can significantly reduce the concentrations of lead and cadmium in contaminated soils. When biochar is applied to such soils, it binds with the heavy metals, preventing them from leaching into groundwater or being taken up by plants. This immobilization process not only protects water sources from contamination but also reduces the risk of heavy metals entering the food chain through crop uptake. Furthermore, biochar's stability ensures that these bound metals remain immobilized for extended periods, providing long-term remediation benefits (Fan et al., 2020). Organic pollutants, such as pesticides, herbicides, polycyclic aromatic hydrocarbons (PAHs), and pharmaceuticals, also pose significant environmental challenges. These contaminants can degrade water quality, harm aquatic life, and pose health risks to humans through contaminated drinking water and food. Biochar's adsorptive properties make it an effective medium for capturing and degrading these organic pollutants. The large surface area and porous structure of biochar provide ample sites for the adsorption of organic molecules, while its surface functional groups facilitate the binding and breakdown of these compounds (Qiu et al., 2021).

One of the key mechanisms by which biochar adsorbs organic pollutants is through hydrophobic interactions. Organic pollutants, particularly hydrophobic compounds, tend to adsorb onto the nonpolar surfaces of biochar. This process is enhanced by the aromatic carbon structure of biochar, which provides additional sites for the adsorption of organic molecules. Additionally, biochar can undergo surface modifications to enhance its adsorptive capacity for specific pollutants. For example, biochar can be treated with chemicals or subjected to further thermal processing to increase the number and variety of functional groups on its surface, thereby improving its ability to capture and degrade specific contaminants (Qiu et al., 2021; Tan et al., 2021b). The use of biochar in water treatment is another important application for pollution mitigation. Biochar filters can be used to remove heavy metals and organic pollutants from wastewater, stormwater, and drinking water. In wastewater treatment, biochar can be integrated into constructed wetlands, biofilters, and other treatment systems to enhance the removal of contaminants. Studies have shown that biochar-amended filtration systems can effectively reduce the concentrations of heavy metals, pesticides, and pharmaceuticals in treated water. This not only improves water quality but also reduces the environmental impact of wastewater discharge (Deng et al., 2021; Nan & Wang, 2023).

In stormwater management, biochar can be incorporated into green infrastructure, such as rain gardens and bioswales to capture and treat pollutants from urban runoff. Urban runoff often contains a mix of heavy metals, nutrients, and organic pollutants that can degrade water quality and harm aquatic ecosystems. Biochar's ability to adsorb these contaminants helps prevent them from entering rivers, lakes, and coastal waters. Moreover, biochar-amended green infrastructure enhances the infiltration and retention of stormwater, reducing the risk of flooding and improving the resilience of urban areas to extreme weather events (Domingues et al., 2020). The application of biochar for soil remediation also holds significant promise. Contaminated soils, particularly those affected by industrial activities, mining, and agricultural runoff, can be treated with biochar to immobilize pollutants and restore soil health. Biochar's ability to adsorb heavy metals and organic pollutants helps reduce their bioavailability, thereby minimizing their uptake by plants and entry into the food chain. This remediation process not only protects human health and the environment but also enhances soil fertility and productivity (Singh et al., 2020).

Applying biochar to agricultural soils contaminated with pesticides can help degrade these chemicals and reduce their toxicity. The biochar adsorbs the pesticide molecules, preventing them from leaching into groundwater or being absorbed by crops. Over time, the microbial communities in the biochar-amended soil can further break down the adsorbed pesticides, rendering them harmless. This process not only mitigates pollution but also supports sustainable agriculture by reducing reliance on chemical inputs and enhancing soil health (Egamberdieva et al., 2021). Biochar's role in pollution mitigation extends beyond its adsorptive properties. It also contributes to improving soil structure, water retention, and nutrient availability, which are essential for restoring degraded ecosystems. By enhancing soil health, biochar supports the growth of vegetation that can further stabilize contaminated sites and prevent erosion. This vegetative cover helps reduce the spread of pollutants through wind and water erosion, providing an additional layer of protection for the environment (Nan & Wang, 2023).

The economic and environmental benefits of using biochar for pollution mitigation are substantial. Biochar production can be integrated into existing agricultural and forestry practices, providing a sustainable solution for managing organic waste. By converting agricultural residues, forestry by-products, and other organic waste into biochar, we can reduce the volume of waste sent to landfills, lower GHG emissions, and produce a valuable material for environmental remediation. This approach aligns with the principles of a circular economy, where waste is minimized, and resources are efficiently utilized (Ansari et al., 2023). Moreover, biochar's application in pollution mitigation can generate financial incentives through carbon credits and environmental markets (Fig. 3). The carbon sequestration potential of biochar, combined with its pollution mitigation benefits, makes it an attractive option for offsetting carbon emissions and earning carbon credits. These financial incentives can support the adoption and scaling up of biochar technology, promoting sustainable environmental practices.

3.4 Waste Management

Effective waste management is a critical component of environmental sustainability and public health. The rapid growth of urban populations, coupled with increasing industrial and agricultural activities, has led to an escalation in the generation of waste. Traditional waste management methods, such as landfilling and incineration, pose significant environmental challenges, including greenhouse gas emissions, soil and water contamination, and the loss of valuable resources. Converting agricultural and organic waste into biochar offers a sustainable waste management solution that addresses these challenges while promoting resource recovery and environmental health (Ansari et al., 2023).

Biochar production involves the pyrolysis of organic biomass in an oxygen-limited environment, resulting in a stable, carbon-rich material. This process not only sequesters carbon but also transforms waste materials into a valuable product that can be used to enhance soil health and agricultural productivity. The versatility

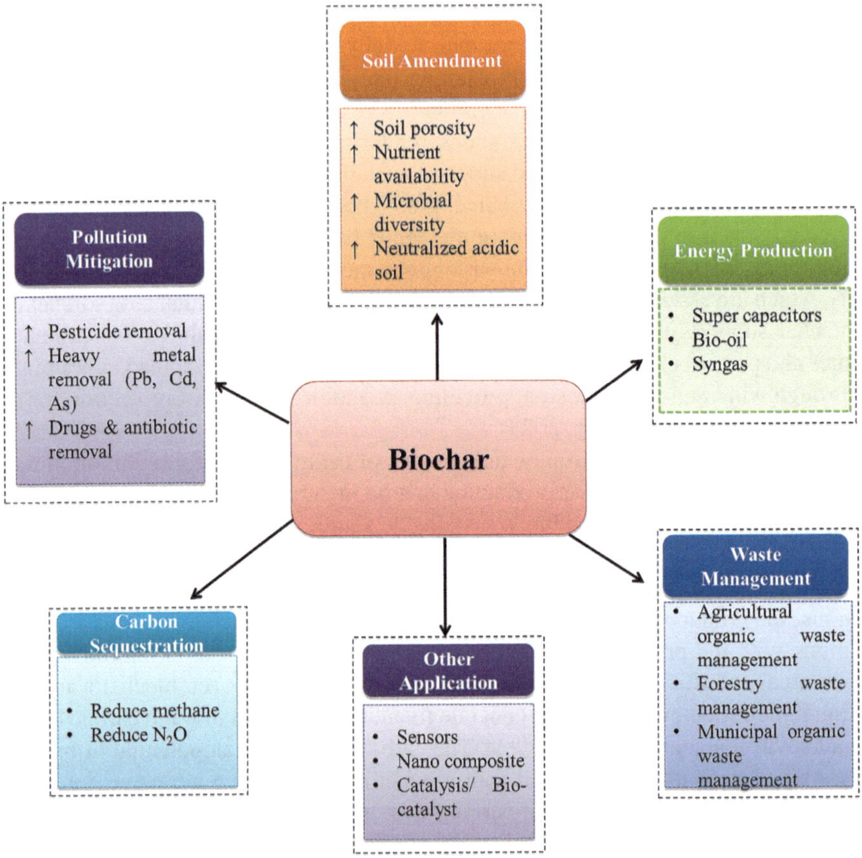

Fig. 3 Applications of biochar

and environmental benefits of biochar make it an ideal solution for managing various types of organic waste, including agricultural residues, forestry by-products, and municipal organic waste. Agricultural residues, such as crop stalks, husks, and straw, represent a significant source of organic waste. Traditionally, these residues are often burned in the fields, contributing to air pollution and greenhouse gas emissions. Alternatively, they may be left to decompose, which can lead to methane emissions, a potent greenhouse gas. Converting these residues into biochar provides a sustainable alternative that mitigates these environmental impacts. The pyrolysis process reduces the volume of the waste, stabilizes the carbon content, and produces biochar that can be returned to the soil to improve fertility and structure (Arif et al., 2020; Saletnik et al., 2019; Gurwick et al., 2013).

Forestry operations generate substantial amounts of waste, including branches, bark, and sawdust. These by-products can pose disposal challenges and contribute to environmental degradation if not managed properly. Biochar production offers an effective means of utilizing forestry waste, transforming it into a valuable resource for soil amendment and carbon sequestration. Mobile pyrolysis units can be deployed directly in forestry areas, allowing for on-site conversion of waste into biochar. This approach minimizes transportation costs and emissions, promoting a circular economy within the forestry sector. Municipal organic waste, including food scraps, yard waste, and paper, is another significant contributor to landfill volumes. Organic waste in landfills decomposes anaerobically, producing methane and leachate that can contaminate groundwater. Diverting municipal organic waste from landfills and converting it into biochar provides a sustainable waste management solution that reduces greenhouse gas emissions and landfill use. Municipalities can implement centralized biochar production facilities or integrate biochar production into existing waste management infrastructure. The resulting biochar can be used in urban agriculture, landscaping, and soil remediation projects, enhancing the sustainability and resilience of urban environments (Chen et al., 2021).

The benefits of converting organic waste into biochar extend beyond waste reduction and carbon sequestration. Biochar production promotes resource recovery by transforming waste materials into a valuable soil amendment. This process aligns with the principles of a circular economy, where waste is minimized, and resources are efficiently utilized. By returning biochar to the soil, nutrients and organic matter are recycled, improving soil health and productivity. This approach reduces the reliance on chemical fertilizers and enhances the sustainability of agricultural practices (Bolan et al., 2022). Biochar's ability to improve soil health is well documented. When applied to soil, biochar enhances its structure, increases water retention, and boosts nutrient availability. These benefits are particularly important in degraded soils where nutrient depletion and poor structure limit agricultural productivity. By incorporating biochar into such soils, farmers can restore fertility, reduce erosion, and enhance crop yields. This not only supports sustainable agriculture but also contributes to food security and rural development.

The integration of biochar production into agricultural and forestry practices also offers economic benefits. Farmers and forest managers can reduce waste disposal costs and generate additional revenue streams by producing and selling biochar. Moreover, the improved soil health and increased crop yields associated with biochar application can enhance the profitability of agricultural operations. The development of biochar markets and carbon credits further incentivizes the adoption of this technology, providing financial rewards for sustainable practices (Bera et al., 2024). In addition to its agricultural applications, biochar can be used in various environmental remediation projects. Contaminated soils, particularly those affected by heavy metals and organic pollutants, can be treated with biochar to immobilize contaminants and restore soil health. Biochar's adsorptive properties enable it to capture and bind pollutants, reducing their bioavailability and toxicity. This makes biochar an effective tool for soil remediation and the restoration of contaminated sites. By incorporating biochar into remediation strategies, we can enhance the effectiveness and sustainability of these efforts.

3.5 Energy Production

The growing demand for renewable energy sources and sustainable waste management solutions has led to increased interest in innovative technologies that address both energy production and environmental sustainability. One such technology is the integrated production of biochar with energy recovery systems. By harnessing the pyrolysis process, which converts organic biomass into biochar, bio-oil, and syngas, it is possible to create renewable energy sources while simultaneously producing valuable biochar. This approach not only maximizes resource utilization but also contributes to energy security, waste reduction, and climate change mitigation (Bhatia et al., 2021).

Pyrolysis is a thermochemical process that decomposes organic materials at elevated temperatures in the absence of oxygen. This process produces three main products: biochar, bio-oil, and syngas. Biochar is the solid, carbon-rich residue that remains after pyrolysis, while bio-oil and syngas are the liquid and gaseous by-products, respectively. Each of these products has distinct properties and potential applications, making pyrolysis a versatile technology for integrated energy production and waste management (Domingues et al., 2020). Biochar is well known for its benefits as a soil amendment, improving soil structure, water retention, and nutrient availability. However, the production of biochar through pyrolysis also generates bio-oil and syngas, both of which can be utilized as renewable energy sources. Bio-oil is a complex mixture of organic compounds that can be used as a fuel or further refined into chemicals and transportation fuels. Syngas, primarily composed of hydrogen, carbon monoxide, and methane, can be used for heat and power generation or as a feedstock for producing synthetic fuels and chemicals (Ippolito et al., 2020).

Integrating biochar production with energy recovery systems leverages the full potential of the pyrolysis process. This integrated approach not only produces valuable biochar for agricultural and environmental applications but also generates renewable energy, reducing reliance on fossil fuels and lowering greenhouse gas emissions. The energy produced from bio-oil and syngas can be used to power the pyrolysis process itself, making it more energy efficient and sustainable (Qiu et al., 2021). One of the key advantages of this integrated approach is its ability to utilize a wide range of biomass feedstocks, including agricultural residues, forestry by-products, and organic waste. These feedstocks are often abundant and underutilized, representing a significant source of renewable energy. By converting these materials into biochar, bio-oil, and syngas, it is possible to create a closed-loop system that maximizes resource recovery and minimizes waste (Mukome et al., 2013). For instance, agricultural residues such as corn stover, wheat straw, and rice husks can be used as feedstocks for pyrolysis. These residues are typically left in the fields to decompose or are burned, both of which release carbon dioxide and other greenhouse gases into the atmosphere. By diverting these residues to pyrolysis, farmers can produce biochar to improve soil health and productivity while also generating bio-oil and syngas for renewable energy. This not only enhances the sustainability of agricultural practices but also provides additional revenue streams for farmers

(Ippolito et al., 2020). Forestry operations also generate substantial amounts of waste, including branches, bark, and sawdust. These by-products can pose disposal challenges and contribute to environmental degradation if not managed properly. Mobile pyrolysis units can be deployed directly in forestry areas to convert this waste into biochar, bio-oil, and syngas on-site. The biochar can be used to restore soil health and support reforestation efforts, while the bio-oil and syngas can be used to generate heat and power for remote forestry operations. This approach reduces transportation costs and emissions, promoting a circular economy within the forestry sector (Mukome et al., 2013).

Municipal organic waste, including food scraps, yard waste, and paper, is another significant contributor to landfill volumes. Organic waste in landfills decomposes anaerobically, producing methane, a potent greenhouse gas. Diverting municipal organic waste from landfills to pyrolysis facilities can significantly reduce methane emissions and produce valuable biochar, bio-oil, and syngas. Municipalities can implement centralized pyrolysis facilities or integrate pyrolysis into existing waste management infrastructure. The renewable energy produced can be used to power municipal operations, while the biochar can be used in urban agriculture, landscaping, and soil remediation projects (Chen et al., 2021). The integration of biochar production with energy recovery systems also offers economic benefits. The bio-oil and syngas produced can be used to generate electricity, heat, or transportation fuels, providing a sustainable and cost-effective alternative to fossil fuels. This can reduce energy costs for agricultural, forestry, and municipal operations, enhancing their overall economic viability. Additionally, the development of biochar markets and carbon credits provides financial incentives for adopting this technology, further supporting its commercialization and scaling.

3.6 Other Applications

Biochar, a carbon-rich product derived from the pyrolysis of organic materials, is increasingly recognized for its potential applications in sensor nanocomposites and biocatalysts. Its unique properties, such as high surface area, porosity, and stability, make it an attractive material for these advanced technologies. This discussion explores the integration of biochar in sensor nanocomposites and biocatalysts, highlighting its advantages and applications in these fields.

3.6.1 Biochar in Sensor Nanocomposites

One of the primary advantages of using biochar in sensor nanocomposites is its enhanced sensitivity and selectivity. Biochar's high surface area and porous structure provide numerous active sites for the adsorption of target analytes, allowing for a higher degree of interaction between the sensor and the analyte. This characteristic is particularly beneficial for sensors designed to detect trace amounts of substances. Additionally, the porosity of biochar facilitates rapid diffusion of analytes into the material, improving the sensors' response time (Krajčovičová et al., 2023).

Biochar can significantly enhance the conductivity of sensor nanocomposites when combined with conductive materials such as graphene, carbon nanotubes, or metal oxides. The integration of biochar with these materials creates a synergistic effect, resulting in nanocomposites with superior electrical properties, crucial for developing electrochemical sensors where detection mechanisms often rely on changes in electrical signals (Zheng et al., 2023). Another significant advantage of biochar is its stability and biocompatibility. Biochar is chemically stable and can withstand harsh environmental conditions, making it suitable for sensors that need to operate in extreme environments or over extended periods. Furthermore, its biocompatibility makes it an excellent candidate for developing biosensors that interact with biological systems. For instance, biochar-based nanocomposites can be used in medical diagnostics to detect biomolecules or pathogens in bodily fluids (Zhuo et al., 2023). Cost-effectiveness is another notable benefit of biochar. Since it is derived from biomass, biochar is relatively inexpensive compared to other advanced materials. This cost advantage is particularly relevant for large-scale production of sensor nanocomposites, where the economic feasibility of the materials used can significantly impact the overall cost of the sensors (Ashfaq et al., 2024).

3.6.2 Biochar in Biocatalysts

Biochar also shows great promise as a support material in biocatalysis. One of its primary roles in this context is to immobilize enzymes and microorganisms. The high surface area and porosity of biochar provide an excellent platform for attaching and stabilizing these biocatalytic agents. Immobilizing enzymes or microorganisms on biochar can enhance their stability and reusability, critical factors for the efficiency and sustainability of biocatalytic processes (Pandey et al., 2020). The porous structure of biochar can facilitate better mass transfer of substrates and products during catalytic reactions, increasing the catalytic activity and overall efficiency of the biocatalysts (Cao et al., 2017). Additionally, biochar can adsorb inhibitory compounds present in the reaction medium, protecting the biocatalysts and maintaining their activity. This protective role can be particularly beneficial in industrial applications where the reaction media often contain various impurities and inhibitors (Faheem et al., 2020). Environmental applications of biochar-based biocatalysts are particularly noteworthy. Biochar's ability to adsorb and concentrate pollutants around the catalytic sites makes it an ideal material for environmental remediation. For example, biochar-supported biocatalysts can be used to degrade organic pollutants in wastewater or transform biomass into valuable chemicals and biofuels (Manikandan et al., 2023).

4 Challenges and Limitations

Despite its promising potential, the effective large-scale utilization of biochar faces several significant challenges and limitations. Addressing these issues is essential to maximize biochar's benefits and ensure its sustainability as a soil amendment and environmental management tool. One of the primary challenges is the economic viability of biochar production. The cost of producing biochar can be prohibitively high, depending on the feedstock used and the production methods employed. High-quality feedstocks and advanced pyrolysis technologies often require substantial initial investment and operational costs. For small-scale farmers and rural communities, these expenses can be a significant barrier to application. To make biochar more accessible and affordable, economic incentives such as subsidies, grants, and carbon credits are necessary. In this context, advancements in scalable production technologies, including mobile pyrolysis units and integrated biochar-bioenergy systems, can reduce costs and enhance the economic feasibility of biochar production (Braghiroli et al., 2018).

Another critical issue is the variable efficacy of biochar. The performance of biochar can differ significantly based on several factors, including soil type, climate conditions, and crop species. For example, biochar that works well in sandy soils may not produce the same benefits in clayey soils. Similarly, climatic conditions, such as temperature and precipitation, can influence biochar's effectiveness in enhancing soil health and crop yields. Crop-specific responses to biochar also vary, requiring tailored application strategies. Localized research and field trials are essential to understand these variations and develop region-specific guidelines for biochar use. By conducting detailed studies on the interactions between biochar, soil properties, and plant species, researchers can optimize biochar applications to suit specific agricultural and environmental conditions.

The potential negative effects of biochar also warrant careful consideration. While biochar generally enhances soil properties, it can sometimes alter soil pH, which may not always be beneficial. For instance, biochar is often alkaline and can raise the pH of acidic soils, improving nutrient availability. However, in already alkaline soils, further increases in pH can lead to nutrient imbalances and hinder plant growth. Moreover, biochar can introduce contaminants if the feedstock used is contaminated with heavy metals or other pollutants. Additionally, the introduction of biochar into soil ecosystems can disrupt existing microbial communities, potentially affecting soil health and plant-microbe interactions. To mitigate these risks, a thorough understanding of biochar properties and careful management practices are necessary. Selecting appropriate feedstocks, ensuring clean production processes, and monitoring soil conditions post-application can help prevent adverse outcomes (Bolan et al., 2022; Cross et al., 2016).

Regulatory and standardization issues further complicate a wider acceptance of biochar technology. Currently, there is a lack of standardized guidelines and best practices for biochar production and applications. This can lead to inconsistent results and variability in biochar quality, hindering its applicability and use. Developing comprehensive regulatory frameworks is crucial to ensure the safe and effective use of biochar. These frameworks should include standards for feedstock selection, production methods, and quality control measures. Additionally, guidelines for application rates, based on soil type and crop requirements, can help optimize biochar's benefits while minimizing potential risks. Regulatory oversight can also facilitate market development for biochar, promoting its commercialization and integration into mainstream agricultural practices (Hossain et al., 2020).

To address these challenges, a multifaceted approach involving research, policy support, and stakeholder collaboration is necessary. Continued research is vital to improve our understanding of biochar's interactions with different soil types, climates, and crops. This research should focus on long-term field studies, assessing the sustainability and cumulative effects of biochar over time. Policymakers can support biochar adoption through financial incentives, regulatory frameworks, and public awareness campaigns. By providing subsidies, tax breaks, and funding for biochar projects, governments can lower the economic barriers to entry and encourage wider use. Public awareness and education initiatives can highlight the environmental and economic benefits of biochar, fostering acceptance and enthusiasm among farmers, foresters, and waste management professionals.

5 Future Prospects of Biochar

The future of biochar is promising, with significant potential to address environmental challenges and promote sustainable practices (Vuppaladadiyam et al., 2023). To fully realize its benefits, several key areas must be prioritized: research and development, policy and economic incentives, integration with other sustainable practices, and public awareness and education. Continued research and development are essential for understanding the long-term effects of biochar on various soil types and environmental conditions. Advancements in pyrolysis technologies and the utilization of diverse feedstocks, such as agricultural residues, forestry by-products, and municipal organic waste, can enhance biochar's efficacy and sustainability. Innovations like mobile pyrolysis units will make production more efficient and cost-effective, facilitating wider adoption. Policy and economic incentives are crucial for encouraging investment in biochar production and application. Governments and policymakers can support this by providing subsidies, grants, and carbon credits, which reduce financial barriers and make biochar more accessible. Developing regulatory frameworks and best practice guidelines will ensure consistent quality and safe usage, fostering market confidence and broader acceptance. Integrating biochar with other sustainable practices can amplify its environmental benefits. Combining biochar with composting enhances nutrient cycling and soil health, while organic farming practices that incorporate biochar reduce reliance on chemical fertilizers and pesticides. Linking biochar production with renewable energy systems, such as bio-oil and syngas generation, can provide additional energy resources and further environmental benefits.

6 Conclusion

In summary, biochar shows great promise for agricultural use and environmental management. Its unique properties improve soil health, enhance plant growth, and support sustainable agricultural practices while also capturing carbon and immobilizing pollutants. By addressing the challenges and limitations of biochar application, we can unleash its full potential to transform soil management and make a significant contribution to global sustainability efforts. However, further research, policy support, and public awareness are needed to effectively achieve this benchmark.

References

Ansari, S. A., Shakeel, A., Sawarkar, R., Maddalwar, S., Khan, D., & Singh, L. (2023). Additive facilitated co-composting of lignocellulosic biomass waste, approach towards minimizing greenhouse gas emissions: An up to date review. *Environmental Research, 224*, 115529.

Arif, M., Jan, T., Riaz, M., Fahad, S., Adnan, M., Amanullah, et al. (2020). Biochar; a remedy for climate change. In *Environment, climate, plant and vegetation growth* (pp. 151–171). Springer.

Ashfaq, M., Chauhan, D., & Talreja, N. (2024). Nanobiochar: An emerging material for the environment, energy, and biomedical applications. In Biochar-Based Catalysts: Preparation, Characterization and Applications (pp. 1-16). Singapore: Springer Nature Singapore.

Bera, T., Samui, S., Dey, A., & Ankireddypalli, J. (2024). Soil carbon sequestration in the context of climate change. In *Climate change impacts on soil-plant-atmosphere continuum* (pp. 63–106). Springer.

Bhatia, S. K., Palai, A. K., Kumar, A., Bhatia, R. K., Patel, A. K., Thakur, V. K., et al. (2021). Trends in renewable energy production employing biomass-based biochar. *Bioresource Technology, 340*, 125644.

Bolan, N., Hoang, S. A., Beiyuan, J., Gupta, S., Hou, D., Karakoti, A., et al. (2022). Multifunctional applications of biochar beyond carbon storage. *International Materials Reviews, 67*(2), 150–200.

Bolan, N., Sarmah, A. K., Bordoloi, S., Bolan, S., Padhye, L. P., Van Zwieten, L., et al. (2023). Soil acidification and the liming potential of biochar. *Environmental Pollution, 317*, 120632.

Braghiroli, F. L., Bouafif, H., Hamza, N., Bouslimi, B., Neculita, C. M., & Koubaa, A. (2018). The influence of pilot-scale pyro-gasification and activation conditions on porosity development in activated biochars. *Biomass and Bioenergy, 118*, 105–114.

Cao, X., Sun, S., & Sun, R. (2017). Application of biochar-based catalysts in biomass upgrading: a review. RSC advances, 7(77), 48793–48805.

Chen, G., Zhang, R., Guo, X., Wu, W., Guo, Q., Zhang, Y., et al. (2021). Comparative evaluation on municipal sewage sludge utilization processes for sustainable management in Tibet. *Science of the Total Environment, 765*, 142676.

Chrysargyris, A., Prasad, M., Kavanagh, A., & Tzortzakis, N. (2020). Biochar type, ratio, and nutrient levels in growing media affects seedling production and plant performance. *Agronomy, 10*(9), 1421.

Cross, A., Zwart, K., Shackley, S., & Ruysschaert, G. (2016). The role of biochar in agricultural soils. In *Biochar in European soils and agriculture* (pp. 95–120). Routledge.

Dai, Z., Xiong, X., Zhu, H., Xu, H., Leng, P., Li, J., et al. (2021). Association of biochar properties with changes in soil bacterial, fungal and fauna communities and nutrient cycling processes. *Biochar, 3*, 239–254.

Deng, S., Chen, J., & Chang, J. (2021). Application of biochar as an innovative substrate in constructed wetlands/biofilters for wastewater treatment: Performance and ecological benefits. *Journal of Cleaner Production, 293*, 126156.

Ding, Y., Liu, Y., Liu, S., Li, Z., Tan, X., Huang, X., et al. (2016). Biochar to improve soil fertility. A review. *Agronomy for Sustainable Development, 36*, 1–18.

Domingues, R. R., Sánchez-Monedero, M. A., Spokas, K. A., Melo, L. C. A., Trugilho, P. F., Valenciano, M. N., et al. (2020). Enhancing cation exchange capacity of weathered soils using biochar: Feedstock, pyrolysis conditions and addition rate. *Agronomy, 10*(6), 824.

Egamberdieva, D., Jabbarov, Z., Arora, N. K., Wirth, S., & Bellingrath-Kimura, S. D. (2021). Biochar mitigates effects of pesticides on soil biological activities. *Environmental Sustainability, 4*(2), 335–342.

Ennis, C. J., Evans, A. G., Islam, M., Ralebitso-Senior, T. K., & Senior, E. (2012). Biochar: Carbon sequestration, land remediation, and impacts on soil microbiology. *Critical Reviews in Environmental Science and Technology, 42*(22), 2311–2364.

Fan, Q., Sun, J., Chu, L., Cui, L., Quan, G., Yan, J., et al. (2018). Effects of chemical oxidation on surface oxygen-containing functional groups and adsorption behavior of biochar. *Chemosphere, 207*, 33–40.

Fan, J., Cai, C., Chi, H., Reid, B. J., Coulon, F., Zhang, Y., et al. (2020). Remediation of cadmium and lead polluted soil using thiol-modified biochar. *Journal of Hazardous Materials, 388*, 122037.

Faheem, Du, J., Kim, S. H., Hassan, M. A., Irshad, S., & Bao, J. (2020). Application of biochar in advanced oxidation processes: supportive, adsorptive, and catalytic role. Environmental Science and Pollution Research, 27(30), 37286–37312.

Glaser, B. (2014). Soil biogeochemistry:From molecular to ecosystem level using terra preta and biochar as examples. In *Agroecology, ecosystems, and sustainability* (1st ed., pp. 1–40). CRC Press.

Gurwick, N. P., Moore, L. A., Kelly, C., & Elias, P. (2013). A systematic review of biochar research, with a focus on its stability in situ and its promise as a climate mitigation strategy. *PLoS One, 8*(9), e75932.

Gwenzi, W., Chaukura, N., Mukome, F. N. D., Machado, S., & Nyamasoka, B. (2015). Biochar production and applications in sub-Saharan Africa: Opportunities, constraints, risks and uncertainties. *Journal of Environmental Management, 150*, 250–261.

Hailegnaw, N. S., Mercl, F., Pračke, K., Száková, J., & Tlustoš, P. (2019). Mutual relationships of biochar and soil pH, CEC, and exchangeable base cations in a model laboratory experiment. *Journal of Soils and Sediments, 19*, 2405–2416.

Hossain, M. Z., Bahar, M. M., Sarkar, B., Donne, S. W., Ok, Y. S., Palansooriya, K. N., et al. (2020). Biochar and its importance on nutrient dynamics in soil and plant. *Biochar, 2*, 379–420.

Hung, N. T. Q., Thong, L. K., Nguyen, M. K., & Han, L. T. N. (2018). Potential of biochar production from agriculture residues at household scale: A case study in Go Cong Tay District, Tien Giang Province, Vietnam. *Environment and Natural Resources Journal, 16*(2), 68–78.

Igalavithana, A. D., Ok, Y. S., Usman, A. R. A., Al-Wabel, M. I., Oleszczuk, P., & Lee, S. S. (2016). The effects of biochar amendment on soil fertility. In *Agricultural and environmental applications of biochar: Advances and barriers* (Vol. 63, pp. 123–144). SSSA.

Ippolito, J. A., Cui, L., Kammann, C., Wrage-Mönnig, N., Estavillo, J. M., Fuertes-Mendizabal, T., et al. (2020). Feedstock choice, pyrolysis temperature and type influence biochar characteristics: A comprehensive meta-data analysis review. *Biochar, 2*, 421–438.

Jaafar, N. M., Clode, P. L., & Abbott, L. K. (2015). Soil microbial responses to biochars varying in particle size, surface and pore properties. *Pedosphere, 25*(5), 770–780.

Khadem, A., Raiesi, F., Besharati, H., & Khalaj, M. A. (2021). The effects of biochar on soil nutrients status, microbial activity and carbon sequestration potential in two calcareous soils. *Biochar, 3*(1), 105–116.

Kumar, K., Kumar, R., Kaushal, S., Thakur, N., Umar, A., Akbar, S., et al. (2023). Biomass waste-derived carbon materials for sustainable remediation of polluted environments: A comprehensive review. *Chemosphere, 345*, 140419.

Krajčovičová, T. E., Hatala, M., Gemeiner, P., Híveš, J., Mackuľak, T., Nemčeková, K., & Svitková, V. (2023). Biochar for Water Pollution Control: From Sensing to Decontamination. Chemosensors, 11(7), 394.

Lefebvre, D., Cornelis, J., Meersmans, J., Edgar, J., Hamilton, M., & Bi, X. (2024). Environmental factors controlling biochar climate change mitigation potential in British Columbia's agricultural soils. *GCB Bioenergy, 16*(1), e13109.

Leng, L., Xiong, Q., Yang, L., Li, H., Zhou, Y., Zhang, W., ... & Huang, H. (2021). An overview on engineering the surface area and porosity of biochar. Science of the total Environment, 763, 144204.

Liu, Q., He, X., Wang, K., & Li, D. (2023). Biochar drives humus formation during composting by regulating the specialized metabolic features of microbiome. *Chemical Engineering Journal, 458*, 141380.

Lorenz, K., Lal, R., Lorenz, K., & Lal, R. (2018). Biochar. In *Carbon sequestration in agricultural ecosystems* (pp. 301–355). Springer.

Manikandan, S. K., Pallavi, P., Shetty, K., Bhattacharjee, D., Giannakoudakis, D. A., Katsoyiannis, I. A., & Nair, V. (2023). Effective usage of biochar and microorganisms for the removal of heavy metal ions and pesticides. Molecules, 28(2), 719.

Majumder, S., Sharma, P., Singh, S. P., Nadda, A. K., Sahoo, P. K., Xia, C., et al. (2023). Engineered biochar for the effective sorption and remediation of emerging pollutants in the environment. *Journal of Environmental Chemical Engineering, 11*(2), 109590.

Mantovani, M., Collina, E., Lasagni, M., & Mezzanotte, V. (2022). *Hydrothermal carbonization of microalgal biomass: An eco-sustainable process to improve properties and effectiveness of microalgae in different applications*. Research Square.

Mukome, F. N. D., Zhang, X., Silva, L. C. R., Six, J., & Parikh, S. J. (2013). Use of chemical and physical characteristics to investigate trends in biochar feedstocks. *Journal of Agricultural and Food Chemistry, 61*(9), 2196–2204.

Nan, H., & Wang, C. (2023). Biochar-based materials for adsorption of heavy metals from wastewater. In *Resource recovery in drinking water treatment* (pp. 265–288). Elsevier.

Ndede, E. O., Kurebito, S., Idowu, O., Tokunari, T., & Jindo, K. (2022). The potential of biochar to enhance the water retention properties of sandy agricultural soils. *Agronomy, 12*(2), 311.

Paz-Ferreiro, J., Méndez, A., & Gascó, G. (2016). Application of biochar for soil biological improvement. In *Agricultural and environmental applications of biochar: Advances and barriers* (Vol. 63, pp. 145–173). SSSA.

Pillala, R., Sowmya, N. S. L., Gajbhiye, P., & Upadhyay, H. (2023). Impact of biochar in sustainable crop production and soil health management: A review. *The Pharma Innovation Journal, 12*, 569–577.

Pandey, D., Daverey, A., & Arunachalam, K. (2020). Biochar: Production, properties and emerging role as a support for enzyme immobilization. Journal of Cleaner Production, 255, 120267.

Qiu, M., Hu, B., Chen, Z., Yang, H., Zhuang, L., & Wang, X. (2021). Challenges of organic pollutant photocatalysis by biochar-based catalysts. *Biochar, 3*(2), 117–123.

Saletnik, B., Zaguła, G., Bajcar, M., Tarapatskyy, M., Bobula, G., & Puchalski, C. (2019). Biochar as a multifunctional component of the environment—A review. *Applied Sciences, 9*(6), 1139.

Schulz, H., Dunst, G., & Glaser, B. (2013). Positive effects of composted biochar on plant growth and soil fertility. *Agronomy for Sustainable Development, 33*, 817–827.

Shetty R, Prakash NB. Effect of different biochars on acid soil and growth parameters of rice plants under aluminium toxicity. Scientific Reports; 2020;10(1):12249.

Singh, C., Tiwari, S., & Singh, J. S. (2020). Biochar: A sustainable tool in soil pollutant bioremediation. In *Bioremediation of industrial waste for environmental safety: Volume II: Biological agents and methods for industrial waste management* (pp. 475–494). Springer.

Sun, J., He, F., Pan, Y., & Zhang, Z. (2017). Effects of pyrolysis temperature and residence time on physicochemical properties of different biochar types. *Acta Agriculturae Scandinavica, Section B—Soil & Plant Science, 67*(1), 12–22.

Tan, H., Lee, C. T., Ong, P. Y., Wong, K. Y., Bong, C. P. C., Li, C., et al. (2021a). A review on the comparison between slow pyrolysis and fast pyrolysis on the quality of lignocellulosic and lignin-based biochar. *IOP Conference Series: Materials Science and Engineering, 1051*, 012075.

Tan, X.-F., Zhu, S.-S., Wang, R.-P., Chen, Y.-D., Show, P.-L., Zhang, F.-F., et al. (2021b). Role of biochar surface characteristics in the adsorption of aromatic compounds: Pore structure and functional groups. *Chinese Chemical Letters, 32*(10), 2939–2946.

Tanure, M. M. C., da Costa, L. M., Huiz, H. A., Fernandes, R. B. A., Cecon, P. R., Junior, J. D. P., et al. (2019). Soil water retention, physiological characteristics, and growth of maize plants in response to biochar application to soil. *Soil and Tillage Research, 192*, 164–173.

Thies, J. E., & Rillig, M. C. (2012). Characteristics of biochar: Biological properties. In *Biochar for environmental management* (pp. 117–138). Routledge.

Tomczyk, A., Sokołowska, Z., & Boguta, P. (2020). Biochar physicochemical properties: pyrolysis temperature and feedstock kind effects. Reviews in Environmental Science and Bio/Technology, 19(1), 191–215.

Vuppaladadiyam, A. K., Vuppaladadiyam, S. S. V., Sahoo, A., Murugavelh, S., Anthony, E., Bhaskar, T., et al. (2023). Bio-oil and biochar from the pyrolytic conversion of biomass: A current and future perspective on the trade-off between economic, environmental, and technical indicators. *Science of the Total Environment, 857*, 159155.

Xiao, Q., Zhu, L., Shen, Y., & Li, S. (2016). Sensitivity of soil water retention and availability to biochar addition in rainfed semi-arid farmland during a three-year field experiment. *Field Crops Research, 196*, 284–293.

Yadav, S. P. S., Bhandari, S., Bhatta, D., Poudel, A., Bhattarai, S., Yadav, P., et al. (2023). Biochar application: A sustainable approach to improve soil health. *Journal of Agriculture and Food Research, 11*, 100498.

Zahedifar, M. (2023). Assessing alteration of soil quality, degradation, and resistance indices under different land uses through network and factor analysis. *CATENA, 222*, 106807.

Zheng, Y., Yu, C., & Fu, L. (2023). Biochar-based materials for electroanalytical applications: An overview. Green Analytical Chemistry, 7, 100081.

Zhuo, Q., Liang, Y., Hu, Y., Shi, M., Zhao, C., & Zhang, S. (2023). Applications of biochar in medical and related environmental fields: current status and future perspectives. Carbon Research, 2(1), 32.

The Role of Biochar in Sustainable Agriculture

Manoj Shrivastava, Renu Singh, and Sudhakar Srivastava

Abstract

Biochar, a carbonized biomass derived from sustainable sources, offers significant potential for improving soil fertility and mitigating climate change. This chapter explores biochar production from urban waste, focusing on its agricultural applications. Urban waste, particularly municipal solid waste, presents a viable feedstock due to its abundant organic and inorganic components. The chapter details the pyrolysis process, a critical method for biochar production, highlighting its efficiency and environmental benefits. Biochar's porous nature enhances soil water retention, nutrient availability, and microbial activity, thus improving crop yields and quality. Furthermore, biochar sequesters carbon, reduces greenhouse gas emissions, and decreases reliance on chemical fertilizers. Traditional and modern biochar production techniques are reviewed, emphasizing optimizing conditions to maximize yield and quality upon agricultural use. Challenges such as lack of awareness, technical barriers, and environmental concerns are discussed alongside potential solutions. The chapter underscores biochar's role in sustainable agriculture, especially in developing countries, where it can contribute to food security, reduce deforestation, and promote cleaner cooking practices. Integrating biochar systems into agricultural practices can realize significant environmental, economic, and social benefits.

M. Shrivastava (✉) · R. Singh
Division of Environmental Sciences, ICAR-Indian Agricultural Research Institute, New Delhi, India

S. Srivastava
Institute of Environment & Sustainable Development Banaras Hindu University, Varanasi, India

S. Singh, S. Srivastava (eds.), *Biochar: A Precious Resource from Biological Waste*,
https://doi.org/10.1007/978-981-95-0425-1_3

Keywords

Pyrolysis · Yield · Quality · Urban waste · Greenhouse gas emission · Agriculture

1 Introduction

The population growth, industrial developments, and urbanization have led to the generation of huge quantities of urban waste (Kumar & Agrawal, 2020) that has become a significant environmental challenge for whole of the world. Urban waste comprises of various organic and inorganic components; organic ones contain both perishable (food remains, fruit and vegetable peels, paper, textiles, etc.) and nonperishable (plastic bags, bottles) things. The inorganic fraction contains the elements of glass, electrical debris, metals, sandstone, etc. Most organic and inorganic fractions can be recycled, while decaying organic matter decomposes biologically. Urban waste comes from various sources and can be divided into municipal solid waste (MSW), sewage, livestock waste, and industrial waste. The systematic developments in the urban areas, job opportunities, and good facilities for living drive people to the cities leading to an increase in the urban population and thus increasing urban waste generation. Municipal corporations are responsible for the proper management of such waste. According to a report, approximately 90 million tons of MSW are produced annually in India, with an average of 0.337 t per capita rate. The natural component of the urban waste, which ranges from 30% to 45%, can be utilized for production of energy, compost, and biochar (Gabhane et al., 2020). MSW is one of the key components of urban waste that is available in huge quantities and can be used for biochar production and as an adsorbent for extracting dyes, minerals, impurities, toxic substances, etc.

The main thermochemical processes for the conversion of biomass into useful products like biofuels, syngas, and biochar are pyrolysis, carbonization, and gasification (Fig. 1). Other possible uses of biomass include biochemical conversion and transesterification (Fig. 1). Thermochemical decomposition of biomass is a biochar production technology that is performed at temperature ranging from 200 to 900 °C (Lehmann & Joseph, 2009). Torrefaction and gasification are not commonly used processes for biochar generation using MSW, which is principally done through pyrolysis. The conditions of the process are selected on the basis of the properties of biomass resource used and as per the requirement of final product of interest. The feedstock and pyrolysis temperature greatly influence biochar yields, composition, functional groups, etc. The biochar yield is high (35%) when pyrolysis is done for a few minutes to hours at moderately high temperatures. However, during rapid pyrolysis carried out at high temperatures (around 1000 °C), the yield of biochar is low (about 10%), while the main product is bio-oil, which accounts for approximately 70% (Mohan et al., 2014).

Biochar is carbonized biomass obtained from waste and thus sustainable sources (Lehmann & Joseph, 2015) that when applied to soil enhance its agricultural and environmental value. Biochar is a porous material and therefore, it can enhance the

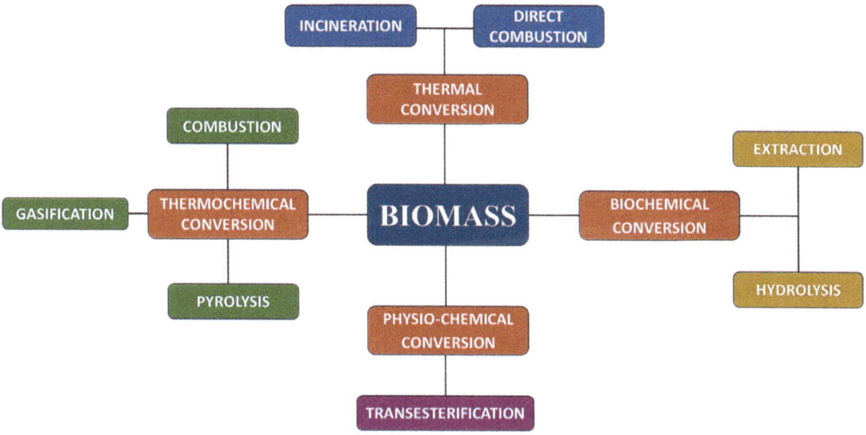

Fig. 1 Conversion route of biomass to energy

potential of soil to retain water and nutrients and improve growth of plants. The growing interest in biochar production from waste biomass is due to the fact that this approach manages waste material along with the generation of a nutrient-rich resource (Gabhane et al., 2020). Biochar production techniques are under constant development so as to improve productivity and quality of biochar and to also reduce the cost of production. The availability of feedstock of uniform composition is a key factor for production of biochar ensuring quality criteria (Gabhane et al., 2020). Although stocks are widely available, their proper classification is essential for biochar generation for specific further use. The waste materials used for biochar production include agricultural residues, straw, woody plant biomass (dead leaves and twigs), urban and industrial waste, and food and kitchen waste. The process of pyrolysis for biochar production can be done either with a simple fire in the field in a crude manner and also with sophisticated equipment in biorefinery (Fuchs et al., 2014). While the crude method of biochar generation does not ensure similar quality during each cycle of production, the modern technologies ensure quality of biochar of similar composition every time. However, the cost of crude method is very low and therefore, it would be suitable for largescale use. The need exists to design and develop low-cost technological interventions for wide-scale use of biomass pyrolysis for managing waste biomass at the site of generation. This will reduce the cost of biochar production as no transport of feedstock would be necessary and will also increase the application possibilities of biochar.

2 Importance of Biochar

Biochar is a crucial ingredient in a new carbon-negative strategy to solve many environmental, economic, and energy problems. Biochar production is associated with less carbon dioxide emissions and is considered as negative carbon technology.

Its production can be linked to carbon credits and therefore, the generation of biochar under standardized conditions can help gain economic benefits and assist in climate change alleviation (Shareef, 2017). Biochar provides several benefits as summarized in points below (Nematian et al., 2021):

- Soil health improvement
- Improvement of soil structure, water infiltration, and ventilation
- Alters pH of the soil
- Supports microbial growth
- Increase in agricultural productivity
- Carbon sequestration
- Reduction in the availability of toxic elements to plants
- Acts as slow-release fertilizer
- Reduction in the use of pesticides and fertilizers
- Decrease in the methane and nitrous oxide emissions from the soil
- Supports local energy production and distribution
- Building jobs and sustaining economic growth at local level

3 Characterization of Biochar

Various factors profoundly impact the physicochemical properties of biochar. These factors include the mode of pyrolysis, pyrolysis temperature, heating rate, and residence time. Among these, pyrolysis temperature significantly influences biochar's elemental composition, carbon structure, and functional group composition. Notably, an increase in temperature leads to a reduction in surface oxygen-containing functional groups and an elevation in the pH value. This indicates that pyrolysis temperature plays a critical role in shaping the properties of biochar.

The torrefaction process involves altering the properties of biomass, including particle size, moisture content, and surface area. Biochar with smaller particle sizes has a larger surface area, resulting in more effective removal of toxic substances and requiring less time for removal (Yaashikaa et al., 2020). It is essential to consider various characteristics when dealing with biochar for the purpose of removal of toxic compounds. These include porosity, particle size, and chemical properties such as pH, cation exchange capacity (CEC), and nutrient content. Biochar has high sorption properties due to its porous surface effectively capturing and retaining water during dehydration. Biochar with smaller pore sizes may not effectively adsorb pesticide molecules, regardless of their polarity or charges. Biochar after preparation may also be modified with chemical reactions to add more functional groups as per the requirement. Hence, the effective surface area of biochar may also differ between modified and raw biochar. The activated biochar has higher surface area with more functional groups than the raw biochar. The activation of biochar may be done by physical and chemical methods (Kim et al., 2012).

The conditions of biochar preparation affect its performance for a specific purpose. For example, in an experiment, biochar prepared at 300 °C had higher iodine adsorption capacity than that prepared at 600 °C (Liang et al., 2016). Structural and elemental analyses play a crucial role in assessing the environmental impact of biochar. Additionally, the interaction of metals with biochar is influenced by factors such as pH. For instance, biochar's function and the speciation of metal contaminant ions vary depending on the pH levels (Yaashikaa et al., 2020). These unique characteristics of biochar demonstrate its efficacy as an exceptional adsorbent for removing various soil pollutants including toxic metals (Hu et al., 2020). The techniques used to characterize biochar, such as scanning electron microscopy (SEM), primarily focus on analyzing its structure, surface functional groups, and elemental composition (Brewer et al., 2014).

The aging of biochar as it is exposed to environmental processes can adversely affect the environment. This can lead to changes in its properties, impacting the environment and ecological components. A cyclic movement of biochar can occur among soil, water, and air during its residence in the environment. While leaching from soil can take biochar into the water, runoff transports biochar from water to soil. Similarly, while wind can take with it fine biochar particles into the atmosphere, it can return to soil or water through settling and precipitation. Therefore, it is crucial to thoroughly examine the negative environmental impacts of biochar across different environmental components to mitigate potential risks. Researchers have observed a complex interplay of carbon emissions in laboratory and field experiments investigating biochar application. Generally, biochar application to soil has been found to significantly reduce the emissions of CO_2, N_2O, and CH_4, three major greenhouse gases. As a result, biochar application has garnered attention as a potential strategy for carbon sequestration and as a means to mitigate the effects of global warming (Lin et al., 2021).

Studies focusing on organic waste composting have examined the impact of biochar addition, with the addition ratio typically falling from 2% to 20%. The effectiveness of biochar in waste composting varies based on its specific type, particularly concerning the removal levels of nitrogenous compounds. Additionally, it has been suggested that compost enriched with essential metals can be effective biofertilizer. Biochar also serves as an alternative pH buffer. Furthermore, incorporating biochar into the solid-state anaerobic digestion has shown promising results in increasing volatile fatty acids and methane production from agricultural wastes (Zhou et al., 2021).

4 Biochar Production Process: Pyrolysis Technology

Pyrolysis is made from Greek words pyro (fire) and lysis (loosening). It denotes the heat-dependent decomposition of biomass into a carbon-rich solid residue (char). The process requires less oxygen or can be performed under nitrogen gas flow (Fig. 2). The proportions of major components of biomass, viz., cellulose, hemicellulose, lignin, water, and minerals (ash), vary for different types of source materials. For example, the content of water can be as high as 60% in freshly cut wood or other agricultural residue while dried material would contain as low as 12–20% water.

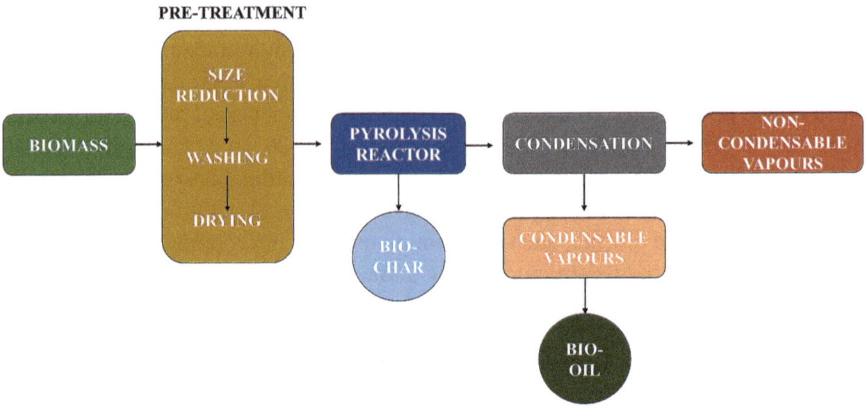

Fig. 2 Schematic diagram of the fast pyrolysis process

Due to endothermic nature of the process mentioned above, the input of heat is necessary to raise the temperature. Upon heating, water is removed firstly at temperatures of above 100 °C. As the temperature increases further and goes above 150 °C, the biomass begins to break down and softens; this is known as conditioning. At this stage, the water that is chemically bound within the biomass is released, along with gases including volatile organic compounds.

At temperature in the range of 200–280 °C, chemical bonds of molecules within biomass begin to break. This stage releases methanol, acetic acid, and other volatile organic compounds, along with CO_2 and CO. Torrefied biomass becomes brittle, which makes it easier and less energy intensive to grind for uses such as boiler fuel. It can also be stored easily as it is resistant to biological degradation and does not uptake water. The vapors coming out from the biomass pyrolysis are condensed as liquid which is known by different names such as wood vinegar, smoke water, liquid smoke, bio-oil, or pyrolysis oil. This liquid can be used as flavoring agent, fungicide, and plant growth promoter. It can also enhance composting process and the biochar efficiency.

The biomass undergoes thermal decomposition at temperatures ranging from 250 to 300 °C, releasing an explosive mixture including H_2, CO, CH_4, CO_2, other hydrocarbons, and tars. The process becomes exothermic due to the breakdown of large polymers of the biomass releasing energy. As biomass is heated, some oxygen is released, reacting with gases and char to produce energy. This energy generates heat, breaking down the biomass further. Theoretically, the process can sustain itself up to around 400 °C, resulting in a carbon-rich, oxygen-depleted residue similar to charcoal. The external heat input remains necessary even during this step as heat continues to release. The ash content of different biochar materials varies; for example, wood biochar gives about 1.5–5% ash along with volatiles (25–35%) and fixed carbon (60–70%). Even after exothermic pyrolysis, the volatile compounds are still left in the biomass. Hence, heating is continued so as to increase the fixed carbon content, surface area, and porosity by decomposing more volatiles. If the pyrolysis of wood is done at temperature of 550–600 °C, fixed carbon content reaches up to 80–85%. The properties of the biochar depend on feedstock, pyrolysis temperature and duration, the speed of temperature increase at every step, and the oxygen-free or oxygen-less conditions. In conditions of temperatures above 600 °C, if some amount of air and steam is present, it increases surface temperature of biochar to 700–800 °C and induces activation and gasification processes. Activation enhances the surface area of biochar and gasification increases the production of clean gas that can be used to generate electricity. Depending on the speed of temperature increase, pyrolysis can be slow to fast. If the initial material is broken down to fine particles and then subjected to pyrolysis, the process occurs rapidly as high temperatures easily act on small size particles. This process is known as fast pyrolysis. The char produced in fast pyrolysis may contain higher condensed volatiles, which might impact its use in soil improvement. In contrast, large biomass like wood logs would take days to weeks to completely carbonize. As a result of the pyrolysis of biomass, three products are produced: one liquid bio-oil, one solid biochar, and one gaseous syngas. The percentage of these products is based upon several factors. Bio-oil yields of 60–70 wt% from a typical biomass feedstock may be reached in fast pyrolysis (the heating rate of 1000 °C/s) with 15–25 wt% biochar yields. The remaining 10–15 wt% is syngas. Slow pyrolysis means slower heating rates and yields biochar as the major end product. Table 1 describes the types of pyrolysis processes, the desired conditions, and the product profile.

Table 1 Type of pyrolysis process

Types	Residence time	Temperatures range (°C)	Pressure	Major product	Other products
Slow pyrolysis	>1 h	300 and 600	–	Biochar (35–45%)	Bio-oil (25–35%) and syngas (20–30%)
Fast pyrolysis	1–10 s	>500	–	Bio-oil (60%)	Biochar (20%) and syngas (20%)
Flash pyrolysis	≤1 min	1000	–	Bio-oil	Biochar
Intermediate pyrolysis	5 and 17 min	500 and 650	1 bar	Bio-oil (40–60%)	Biochar (15–25%) and syngas (20–30%)
Vacuum pyrolysis	2 min	450–600	0.5 and 2 bar	Bio-oil (80%)	Biochar (10%) and syngas (10%)
Hydropyrolysis	Over 15 s	350–600	50–200 bar	Bio-oil (90%)	Biochar (5%) and syngas (5%)

5 Parameters Affecting Biochar Quality

5.1 Biochar Preparation Method

Biochar had come into use some 2500 years ago as per the archaeological evidences from Amazon Basin of South America (Glaser, 2007), where carbonized biomass named Terra Preta was used to provide organic matter and nutrient levels to soil (Ruthiraan et al., 2017). The woods were piled up in pits covered with soil and then burned slowly without air (Barrow, 2012). The liquid products of biochar were also used for various purposes, such as preserving corpses and meat, painting the house, sealing wood barrels, building an ark, and attaching arrows to the shaft of a spear. With human scientific development, improved methods were developed for biochar generation such as brick kilns or clay burners. These methods were commonly used until the end of the nineteenth century. Presently, sophisticated machines like steel ovens, and pyrolyzers are used for biochar preparation. The thermochemical modification processes of biochar generation nowadays include pyrolysis, hydrothermal carbonization, gasification, and torrefaction (Pang, 2019). Presently, biochar is being prepared from traditional to modern methods and choice of method affects the properties of biochar.

5.2 Feedstock Characteristics

The type of feedstock affects final biochar characteristics due to differences in the content of water and mass in different materials, for example, aquatic plants containing more water than terrestrial ones. The biomass can be divided into two main categories, woody and nonwoody. Woody biomass, coming mainly from trees, has low moisture and ash, and high calorific and bulk density. Nonwoody biomass comes from crop residues, animal waste, and solid waste and opposite characteristics like high moisture and ash, and low calorific value and bulk density (Amin et al., 2016).

5.3 Pretreatment

The pretreatment of biomass also influences the characteristics of final biochar product. Biomass pretreatment falls into four main categories: physical, physio-chemical/thermal, chemical, and biological. Physical pretreatment involves mechanical methods like milling and grinding to change biomass properties, with particle size reduction being a key approach. Pretreatment also influences the speed of biochar production and avoids extensive energy inputs (Armah et al., 2022).

5.4 Residence Time

The duration of pyrolysis, or residence time, significantly influences the degree of carbonization and biochar yield. Longer residence times lead to increased carbonization, resulting in a more stable and combustible biochar. It also reduces vulnerability of the biochar to microbial degradation and increases surface area. Optimizing residence time is crucial for producing high-quality biochar with desired properties (Chang et al., 2021).

6 Agricultural Applications of Biochar

6.1 Supports Microbial Growth

The inclusion of biochar in soil has a significant impact on soil microbes due to the various physical and chemical properties it possesses. The process of biochar addition to soil is known as "charosphere," which significantly influences soil quality by modifying the habitat of microorganisms (Pingree et al., 2022; Chen et al., 2023). Biochar enhances soil organic carbon (SOC) content by fostering a favorable environment for soil microbes, leading to various microbial activities. The intricate structure of biochar provides many benefits, including increased porosity, bulk density, soil tensile strength, water retention, and favorable pH, all of which contribute to creating conducive habitats for soil microbes. In addition to creating a suitable environment for microbial growth, biochar's high surface area plays a crucial role in facilitating the adhesion of soil microbes within its pores. The biomass used to make biochar contains basic cations that are transferred into the soil upon application. This action helps increase the soil's ability to absorb cations and raises the soil's pH. Additionally, biochar contains essential nutrients such as calcium, potassium, nitrogen, and phosphorus, which contribute to the soil's fertility and can support microbial communities. Moreover, using biochar as a soil amendment can increase pore space—using biomass to produce biochar results in incorporating basic cations (Tan et al., 2017).

6.2 Nutrient Augmentation

The nutrient concentration in biochar undergoes alteration and condensation into aromatic structures during pyrolysis, resulting in the provision of essential nutrients such as phosphorus (P), calcium (Ca), magnesium (Mg), and silicon (Si) to the soil (Bolan et al., 2022). Moreover, biochar also contains considerable trace elements that contribute to its overall nutrient value. During the pyrolysis process, the nitrogen content in biochar may increase due to the condensation of the biomass. At the same time, phosphorus concentration exhibits variability based on the feedstock material and pyrolytic temperatures. Furthermore, the physical properties of biochar (surface area, porosity, cation and anion exchange capacity, and alkaline pH) enhance retention of nutrients in the soil. Biochar improves soil structure and increases water retention capacity of soil also (Hossain et al., 2020).

7 Integrated Soil Fertility Management

Biochar is commonly applied alongside total rates of either inorganic or organic fertilizers or a combination of both to preserve soil fertility. Multiple meta-analyses have demonstrated that applying biochar can lead to an approximate 10% increase in crop yield. Furthermore, the most recent meta-analysis on biochar revealed that when biochar was co-applied with inorganic fertilizers within 1 year, there was a substantial 48% increase in crop yield (Bai et al., 2022). Biochar profoundly impacts enzymatic activity in soil due to its rich nutrient content, including nitrogen, phosphorus, potassium, magnesium, and trace elements crucial for supporting the soil microbial community. Researchers have observed that increasing the frequency and concentration of biochar application can significantly improve the activity of urease, an enzyme responsible for breaking down urea in the soil. This enhancement in enzymatic activity is attributed to the higher levels of organic matter and essential nutrients introduced by the biochar. Additionally, biochar has been found to increase alkaline phosphatase activity, primarily due to the improved absorption of phosphorus in the biochar produced at lower temperatures. Moreover, various studies have documented increased activity of C cycling enzymes, such as β-glucosidase, following biochar application. This heightened enzymatic activity can be linked to biochar's ability to elevate the levels of soluble phenolic compounds, microbial biomass, and overall microbial activity. Nonetheless, it is worth noting that an inverse relationship has been reported between the activity of C-cycling enzymes and the pyrolytic temperatures at which biochar is produced. As pyrolytic temperatures increase, there is a corresponding decrease in the activity of these enzymes, which is attributed to the altered C/N ratio of the biochar (Rahmanian & Khadem, 2024).

7.1 Remediation of Toxic Contaminants

Biochar can also remediate contaminated soil by absorbing toxic pollutants like heavy metals, pesticides, and hydrocarbons. Increasing biochar concentration in soil enhances the degradation and adsorption of organic pollutants, decreasing pesticide content. The porous structure and functional groups of biochar contribute to pesticide adsorption. Optimizing biochar quantity is crucial for improved pollutant adsorption. Removal mechanisms involve physisorption and chemisorption processes, as well as chemical transformation, partitioning, and biodegradation (Xiong et al., 2019). Land degradation is a major agricultural problem worldwide. To address this growing problem, researchers suggest that biochar be used in degraded soils to improve quality (Spears, 2018). Biochar application to agricultural soils would depend on the specific soil conditions and level of contaminants in the soil (Zhang et al., 2022; Bolan et al., 2023).

8 Economic Feasibility

Biochar's economic viability and market competitiveness are essential for driving broader-scale production and adoption in the agricultural sector within the circular bioeconomy concept. Entrepreneurs can gain a competitive advantage by better understanding production costs, which can help alleviate the fear of failure and drive demand. Additionally, technological innovation significantly reduces production time, improves cost competitiveness, and increases profitability. Nematian et al. (2021) recently conducted a detailed techno-economic analysis on converting orchard waste into biochar, showcasing its potential to enhance soil health. Despite being an experimental product with limited market demand, the study demonstrated the economic growth potential of biochar produced from orchard waste. The study also revealed a probable range of biochar production costs through a stochastic Monte Carlo simulation, providing valuable insights for entrepreneurs and stakeholders in the biochar industry.

The cost of making biochar is a crucial part of selling and using it. When biochar is the main product, the cost has to cover operating expenses like production, maintenance, feedstock, transport, labor, and distribution to make sure the business stays successful. In other cases, biochar might be a by-product of processes aimed at improving agricultural or land management operations and heating systems. Whether biochar is a main product or a by-product, its value depends on the price customers are willing to pay. It is assumed that the economic value of biochar stays the same across a feasible temperature range and that market prices wouldn't change based on quality (Yoder et al., 2011).

When considering the local deployment of biochar, it is important to develop specific strategies and protocols to ensure that its implementation only occurs when negative side effects are minimal. It is essential to acknowledge that the large-scale deployment of biochar technologies, as envisioned in various climate change mitigation scenarios, carries the potential for adverse impacts on food security and natural ecosystems. This risk stems from the potential expansion of dedicated biomass plantations for biochar feedstocks, which could come at the expense of croplands or forests. To address these potential challenges, cross-sectoral integrated policies should be established to safeguard against the future growth of the biochar market leading to competition with food production or triggering deforestation. Additionally, it is crucial to recognize that biochar is intended to remain in soils for extended periods, spanning decades or even centuries, with the long-term aging and effects not yet fully comprehended. The theoretical removal of biochar from soils would necessitate significant soil disturbance, further underscoring the need for careful consideration of the implications associated with its deployment (Tisserant & Cherubini, 2019).

The environmental benefits of biochar play a crucial role in mitigating potential damage to larger ecosystems, which could otherwise lead to substantial economic costs borne by businesses or cleanup expenses for the government. For instance, biochar application has been shown to help prevent problems such as algal blooms, drinking water contamination, harm to fisheries, and damage to tourism industries (EPA, 2015). It's important to note that many of these environmental benefits, which are referred to as positive externalities, are not assigned a specific price and therefore don't have a direct market value. This creates a scenario where a market failure occurs due to the neglect of these positive externalities in cost-benefit analyses for significant investment decisions. In order for the agricultural sector to consider biochar as a viable option, it may be necessary to introduce regulatory and financial incentives that facilitate the inclusion of the environmental cost savings derived from biochar. At present, there is a lack of policy incentives for biochar production and application that would allow for the realization of its proven benefits through monetization.

9 Future Prospects and Research Directions

Biochar offers significant benefits in agriculture and environmental conservation, including soil improvement and carbon sequestration, mitigating greenhouse gas emissions and environmental remediation. However, biochar's effectiveness depends on factors like type, soil properties, and management practices. Advances in nanotechnology could enhance biochar's potential by increasing surface area and thus improving reactivity and adsorption capacity. Nano-biochar may offer new opportunities for pollution control and remediation. Further research is needed to explore nano-biochar's suitability for carbon sequestration and greenhouse gas mitigation. With continued development, biochar could become a valuable tool for minimizing environmental risks and protecting human health (Nepal et al., 2023).

In order to fully understand the long-term effects of biochar on agroecosystems, several knowledge gaps need to be addressed. These include determining the optimal dose of biochar for different soil types and environmental conditions. It is necessary to identify the mechanisms through which biochar affects soil biotic properties. Further, its effects on carbon sequestration and greenhouse gas emissions should be quantitatively assessed. Additionally, it is crucial to investigate how biochar can be modified for various environmental applications and to evaluate the associated long-term environmental and human risks (Amonette et al., 2021).

10 Research Gaps and Opportunities

1. Investigate the co-location of carbon and other nutrients in biochar from different sources and study how they enhance microbial activity and functions.
2. Monitor how biochar affects the formation of stable macroaggregates in different types of soil and assess its importance in providing a habitat for soil microbial growth.

3. Study the mechanisms through which biochar works in different environmental and soil conditions, and assess its practical effects in agriculture and its potential applications. Identify any limitations and suggest ways to overcome them.
4. Measure the extent to which biochar reduces the availability of soil contaminants, allowing for improved microbial growth and function.

11 Conclusion

Biochar production involves using different types of biomass as feedstock, which are then pyrolyzed through various processes. Biochar has the potential to play a significant role in rural economic development and to be an affordable adsorbent in environmental applications. This review aims to bridge the gap between biochar strategies and production and to encourage the use of biochar in environmental applications. The biochar production and applications highlight its diverse uses, such as an adsorbent, nutrient source, and soil amendment for suppressing plant diseases. Preparation methods and feedstock type influence biochar's properties and applications. However, further research is needed to address key challenges, such as evaluating field efficiency and economic feasibility.

References

Amin, F. R., Huang, Y., He, Y., Zhang, R., Liu, G., & Chen, C. (2016). Biochar applications and modern techniques for characterization. *Clean Technologies and Environmental Policy, 18*, 1457–1473.

Amonette, J. E., Blanco-Canqui, H., Hassebrook, C., Laird, D. A., Lal, R., Lehmann, J., & Page-Dumroese, D. (2021). Integrated biochar research: A roadmap. *Journal of Soil and Water Conservation, 76*(1), 24A–29A.

Armah, E. K., Chetty, M., Adedeji, J. A., Estrice, D. E., Mutsvene, B., Singh, N., & Tshemese, Z. (2022). Biochar: Production, application and the future. In *Biochar-productive technologies, properties and applications*. IntechOpen.

Bai, S. H., Omidvar, N., Gallart, M., Kämper, W., Tahmasbian, I., Farrar, M. B., et al. (2022). Combined effects of biochar and fertilizer applications on yield: A review and meta-analysis. *Science of the Total Environment, 808*, 152073.

Barrow, C. J. (2012). Biochar: Potential for countering land degradation and for improving agriculture. *Applied Geography, 34*, 21. https://doi.org/10.1016/J.APGEOG.2011.09.008

Bolan, N., Hoang, S. A., Beiyuan, J., Gupta, S., Hou, D., Karakoti, A., et al. (2022). Multifunctional applications of biochar beyond carbon storage. *International Materials Reviews, 67*(2), 150–200.

Bolan, S., Sharma, S., Mukherjee, S., Kumar, M., Rao, C. S., Nataraj, K. C., et al. (2023). Biochar modulating soil biological health: A review. *Science of the Total Environment, 914*, 169585.

Brewer, C. E., Chuang, V. J., Masiello, C. A., Gonnermann, H., Gao, X., Dugan, B., et al. (2014). New approaches to measuring biochar density and porosity. *Biomass and Bioenergy, 66*, 176–185.

Chang, C., Liu, Z., Li, P., Wang, X., Song, J., Fang, S., & Pang, S. (2021). Study on products characteristics from catalytic fast pyrolysis of biomass based on the effects of modified biochars. *Energy, 229*, 120818.

Chen, Y., Xu, M., Yang, L., Jing, H., Mao, W., Liu, J., et al. (2023). A critical review of biochar application for the remediation of greenhouse gas emissions and nutrient loss in rice paddies: Characteristics, mechanisms, and future recommendations. *Agronomy, 13*(3), 893.

EPA. (2015). *A compilation of cost data associated with the impacts and control of nutrient pollution* (US EPA, Reports and Assessments, 3) (pp. 1–25). EPA.

Fuchs, M., Garcia-Perez, M., Small, P., & Flora, G. (2014) Campfire Lessons - breaking down the combustion process to understand biochar production, the Biochar Journal, Arbaz, Switzerland, www.biocharjournal. org/en/ct/47. Accessed: 22.07.2025

Gabhane, J. W., Bhange, V. P., Patil, P. D., Bankar, S. T., & Kumar, S. (2020). Recent trends in biochar production methods and its application as a soil health conditioner: A review. *SN Applied Sciences, 2*, 1–21.

Glaser, B. (2007). Prehistorically modified soils of central Amazonia: A model for sustainable agriculture in the twenty-first century. *Philosophical Transactions of the Royal Society of London. Series B, Biological Sciences, 362*(1478), 187–196. https://doi.org/10.1098/rstb.2006.1978

Hossain, M. Z., Bahar, M. M., Sarkar, B., Donne, S. W., Ok, Y. S., Palansooriya, K. N., et al. (2020). Biochar and its importance on nutrient dynamics in soil and plant. *Biochar, 2*, 379–420.

Kim, K. H., Kim, J., Cho, T., & Choi, J. W. (2012). Influence of pyrolysis temperature on physicochemical properties of biochar obtained from the fast pyrolysis of pitch pine (Pinus rigida). *Bioresource Technology, 118*, 158–162.

Kumar, A., & Agrawal, A. (2020). Recent trends in solid waste management status, challenges, and potential for the future Indian cities – A review, Current Research in Environmental Sustainability, 2, 100011, https://doi.org/10.1016/j.crsust.2020.100011.

Liang, C., Gascó, G., Fu, S., Méndez, A., & Paz-Ferreiro, J. (2016). Biochar from pruning residues as a soil amendment: Effects of pyrolysis temperature and particle size. *Soil and Tillage Research, 164*, 3–10.

Lin, H., Sun, W., Yu, Y., Ding, Y., Yang, Y., Zhang, Z., & Ma, J. (2021). Simultaneous reductions in antibiotics and heavy metal pollution during manure composting. *Science of the Total Environment, 788*, 147830.

Lehmann, J. & Joseph, S. (2009). Biochar for Environmental Management: Science and Technology. Earthscan, London.

Lehmann, J., & Joseph, S. (2015). Biochar for Environmental Management: Science, Technology and Implementation (2nd ed.). Routledge. https://doi.org/10.4324/9780203762264

Mohan, D., Sarswat, A., Ok, Y. S., & Pittman, C. U., Jr. (2014). Organic and inorganic contaminants removal from water with biochar, a renewable, low cost and sustainable adsorbent: A critical review. *Bioresource Technology, 160*(Supplement C), 191–202. https://doi.org/10.1016/j.biotech.2014.01.120

Nematian, M., Keske, C., & Ng'ombe, J. N. (2021). A techno-economic analysis of biochar production and the bioeconomy for orchard biomass. *Waste Management, 135*, 467–477.

Nepal, J., Ahmad, W., Munsif, F., Khan, A., & Zou, Z. (2023). Advances and prospects of biochar in improving soil fertility, biochemical quality, and environmental applications. *Frontiers in Environmental Science, 11*, 1114752.

Pang, S. (2019). Advances in thermochemical conversion of woody biomass to energy, fuels and chemicals. *Biotechnology Advances, 37*, 589–597.

Pingree, M. R., Kardol, P., Nilsson, M. C., Wardle, D. A., Maaroufi, N. I., & Gundale, M. J. (2022). No evidence that conifer biochar impacts soil functioning by serving as microbial refugia in boreal soils. *GCB Bioenergy, 14*(8), 972–988.

Rahmanian, M., & Khadem, A. (2024). The effects of biochar on soil extra and intracellular enzymes activity. *Biomass Conversion and Biorefinery, 14*, 21993–22005.

Ruthiraan, M., Abdullah, E.C., Mubarak, N.M., & Noraini, M.N. (2017). A promising route of magnetic based materials for removal of cadmium and methylene blue from waste water, *Journal of Environmental Chemical Engineering, 5*, 1447–1455, https://doi.org/10.1016/j.jece.2017.02.038.

Spears, S. (2018). *What is biochare?* The Regeneration International. https://regenerationinternational.org/2018/05/16/what-is-biochar/

Shareef, T., & Zhao, B. (2017) Review Paper: The Fundamentals of Biochar as a Soil Amendment Tool and Management in Agriculture Scope: An Overview for Farmers and Gardeners. *Journal of Agricultural Chemistry and Environment, 6*, 38-61. Biochar from Animal Waste and Its Applications https://doi.org/10.4236/jacen.2017.61003

Tan, Z., Khakbaz, P., Chen, Y., Lombardo, J., Yoon, J. M., Shanks, J. V., et al. (2017). Engineering Escherichia coli membrane phospholipid head distribution improves tolerance and production of biorenewables. *Metabolic Engineering, 44*, 1–12.

Tisserant, A., & Cherubini, F. (2019). Potentials, limitations, co-benefits, and trade-offs of biochar applications to soils for climate change mitigation. *Land, 8*(12), 179.

Xiong, W., Zeng, Z., Li, X., Zeng, G., Xiao, R., Yang, Z., et al. (2019). Ni-doped MIL-53 (Fe) nanoparticles for optimized doxycycline removal by using response surface methodology from aqueous solution. *Chemosphere, 232*, 186–194.

Yaashikaa, P. R., Kumar, P. S., Varjani, S., & Saravanan, A. J. B. R. (2020). A critical review on the biochar production techniques, characterization, stability and applications for circular bioeconomy. *Biotechnology Reports, 28*, e00570.

Yoder, J., Galinato, S., Granatstein, D., & Garcia-Pérez, M. (2011). Economic tradeoff between biochar and bio-oil production via pyrolysis. *Biomass and Bioenergy, 35*(5), 1851–1862.

Zhang, K., Khan, Z., Yu, Q., Qu, Z., Liu, J., Luo, T., et al. (2022). Biochar coating is a sustainable and economical approach to promote seed coating technology, seed germination, plant performance, and soil health. *Plants, 11*(21), 2864.

Zhou, Y., Qin, S., Verma, S., Sar, T., Sarsaiya, S., Ravindran, B., et al. (2021). Production and beneficial impact of biochar for environmental application: A comprehensive review. *Bioresource Technology, 337*, 125451.

An Overview of Methods and Techniques for the Preparation and Characterization of Biochar

Md. Parvez Kabir, Akbar Hossain, Annika Jahan Aonti, Bulbul Ahmed, and Md. Israfil Haq

Abstract

The concept of "biochar" is relatively new but is gaining popularity worldwide. The term itself is a combination of "bio," meaning "biomass," and "char," referring to "charcoal." It has great potential as a method not only for removing waste but also for preventing management from fouling our future too much more than we already have in terms of soil health, carbon sequestration, and greenhouse gas (GHG) emissions, at least by producing biochar instead. Identifying soil and crop management techniques that increase the net accumulation of carbon in soils and reduce the rate at which microbes breakdown organic materials may mitigate some of the negative impacts associated with climate change. In addition, biochar has a broad specific surface area, allowing for maximal cation exchange in the soil, which results in an increased nutrient mineralization cycle that can improve physiobiochemical properties. Furthermore, biochar serves as a carbon sink that can help restore ecological integrity and environmental sustainability. However, there is significant variability in biochar yield and quality from different sources of organic matter due to differences in production temperature as well as poorly designed oxygen control systems. The manufacturing process involves a range of techniques and procedures to enhance its characteristics and applications. This chapter provides a summary of the different methods used for biochar preparation with the characterization techniques adopted.

M. P. Kabir · A. Hossain (✉) · A. J. Aonti · B. Ahmed
Soil Science Division, Bangladesh Wheat and Maize Research Institute, Dinajpur, Bangladesh

M. I. Haq
Department of Agricultural Extension (DAE), Langadu, Rangamati, Bangladesh

S. Singh, S. Srivastava (eds.), *Biochar: A Precious Resource from Biological Waste*,
https://doi.org/10.1007/978-981-95-0425-1_4

Keywords

Biochar · Organic matter · Soil health · Carbon sequestration · Greenhouse gas · Global warming

1 Introduction

The development of urban infrastructures, including roads, industries, markets, residential areas, hospitals, and schools, is causing a gradual decline in farmland in several Asian countries, such as Bangladesh and India. This decline is occurring at a rate of approximately 1% per year, as reported in studies. The expected global requirement, from such projections, calls for an approximate 60% increase in agricultural productivity and will be necessary to meet the increased demands of food by a hungry world in 2050 (Rosenstock et al., 2016). However, the environment and the soil cannot be allowed to degenerate to achieve such a dramatic increase in agricultural yields. Greenhouse gas (GHG) emissions of CO_2 and CH_4 from a variety of agricultural operations affect climate change to a large extent. However, appropriate scientific integrated management techniques for soil fertility have the potential to store large pools of carbon within these soils, mitigating adverse climatic effects (Rahman, 2013).

Nearly all these issues (and more) can be cured or mitigated by the use of copious amounts of natural organic amendments. Large quantities of organic waste remain in farming states, and they are usually discarded at dumps or stacked before being put to use as a fuel. The combustion of these organic residues releases high amounts of greenhouse gases (GHGs), causing air pollution and depleting key nutrients such as sulfur and nitrogen via volatilization (Tipayarom & Oanh, 2007). The increasing replenishment of diverse organic residues in soil represents a new dimension for fostering land fertility recycling (Kamara et al., 2014). OM rapidly decomposes when it comes into contact with microbes, which generate heat and convert some of the carbon into CO_2 within a short time. This process contributes to the breakdown of soil and the formation of compost. This method of maintaining soil fertility is not sustainable in the long term because it requires the annual application of organic waste.

Biochar is a potential alternative for organic materials and is synthesized from different types of naturally occurring biomass at relatively high temperatures in the absence or presence of low oxygen (Singh et al., 2010; Lehmann & Joseph et al., 2024). Biochar can be rich in organic carbon and hence stable, making it capable of remaining untouched on the soil over extended periods of time; up to possibly centuries, it is dependent on the source material, quality, perhaps production procedure, or temperature at pyrolysis (Zimmerman, 2010). In contrast to other organic wastes, manures, or composts, they possess a highly resistant aromatic structure that delays their degradation and maintains soil properties for a longer period (Baldock & Smernik, 2002; Antal & Grønli, 2003). Compared with other conventional organic amendments, the effects of the addition of biochar to agroecosystems on soil health (physical, chemical, and biological properties) are thought to be longer in agricultural fields (Amonette & Joseph, 2012). This innovative technology has the ability to significantly increase crop productivity and soil quality in different types of soils by increasing nutrient availability as well as organic matter replacement, particularly in acidic soils with low levels or the absence of both nutrients (Biederman & Harpole, 2013; Jeffery et al., 2011).

Based on these findings, obtaining a comprehensive understanding of the essential procedures and techniques for the creation and characterization of biochar is time demanding. This chapter provides a comprehensive introduction to several processes used to create biochar, as well as the methodologies used to characterize and describe it.

2 Techniques for Biochar Preparation

The aim of all the techniques in biochar preparation is to walk through a series of processes, bringing biomass into a stable carbon form by pyrolysis. Biomass is subjected to pyrolysis, which involves heating in the absence of oxygen and results in biochar as one product (along with useful syngas or bio-oil). A flowchart of the key activities involved in biochar preparation is presented below.

2.1 Feedstock Selection

The initial stage of biochar preparation involves the careful selection of appropriate biomass feedstocks. The category encompasses organic waste (municipal solid waste, sewage sludge), forestry residues (wood chips and sawdust), agricultural leftovers such as crop residues or straw, and finally dedicated energy crops such as miscanthus and switchgrass, which are grown on a large scale (Lehmann & Joseph, 2024).

2.2 Drying

The percentage of moisture contained in the biomass feedstock must first be decreased so that conversion is efficient and that there are no energy losses. Drying can be carried out via mechanical processes or via natural drying techniques (Brewer et al., 2009).

2.3 Size Reduction

Biomass feedstocks are generally too large to be effectively pyrolyzed and must often be ground, chipped, etc., to obtain the appropriate particle size (Downie et al., 2012).

2.4 Pyrolysis

The leading process of developing biochar involves the heating of biomass in the range of 300–800 °C (oxygen deficit conditions), although higher temperatures can also be used. Pyrolysis can be performed via various reactor setups (batches, continuous flow reactors, or fluidized-bed reactors), among other methods (Bridgwater, 2012).

2.5 Cooling

The biochar is rapidly cooled after pyrolysis has completed to stop any additional reactions, and the final product is set. Cooling can be accomplished by using water or air-cooling systems (Mohan et al., 2006).

2.6 Product Harvesting and Separation

After complete cooling, the biochar is harvested or separated from other pyrolysis products (syngas and bio-oil). For the biochar to be pulled from this mixture, it likely may force further centrifuges or filters (Jirka & Tomlinson, 2014).

2.7 Characterization and Quality Control

Properties such as pH, pore structure, surface area, and elemental content greatly influence the effectiveness of a biochar for a variety of applications. Characterizing biochar and conducting quality control inspections ensure consistency and appropriateness for its intended application (Mukherjee et al., 2011).

2.8 Conditioning (Optional)

Biochar may be further treated, such as being modified or activated to improve its properties for some uses. Conditioning methods include surface modification, steam activation, chemical activation, etc. (Kambo & Dutta, 2015).

2.9 Storage and Packaging

Finally, the generated biochar is stored in appropriate containers until use. The quality of biochar must be ensured in time by adhering to the correct storage procedure, which includes keeping it dry and away from the air (Shackley et al., 2016).

The following flowcharts illustrate (Fig. 1) how these activities interact to produce biochar, an important carbon-rich output with a range of agricultural and

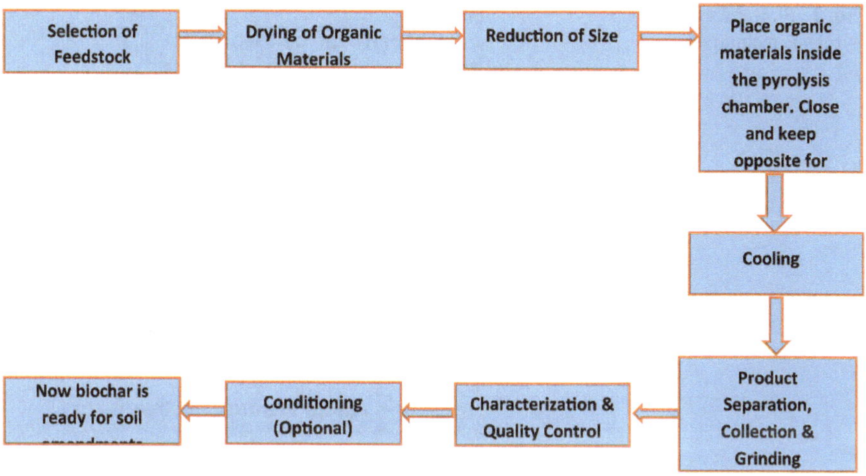

Fig. 1 Flowchart of biochar production

environmental uses, including soil treatment, carbon storage, and renewable energy generation:

Several sun-dried organic materials, such as rice straw, rice husk, wood grass, manure, and sewage sludge, can be used as raw materials for biochar production through the following techniques:

3 Methods of Biochar Preparation

3.1 Pyrolysis Process

The most commonly employed procedure in the production of biochar is pyrolysis. This is due to the pyrolysis of biomass in an oxygen-limiting environment, which generates three main products: syngas, bio-oil, and biochar. The biochar properties are largely dependent on the temperature and heating rate during pyrolysis. Slow pyrolysis at low temperatures (300–500 °C) typically results in biochar with greater carbon content and surface area, whereas fast pyrolysis of biomass substrates at high temperatures may produce less porous biochars but an increased yield of volatile material during breakdown to gases from the amount of biochar produced (Lehmann & Joseph, 2015). The processes needed to produce biochar by pyrolysis are as follows (Fig. 2):

3.1.1 Feedstock Selection and Preparation

Pyrolysis can utilize a range of organic materials as feedstocks, for example, rice straw/husk, wood chips and flakes, or crop residues such as animal manures, agricultural waste, etc. To increase the surface area and provide uniform heating, it is often dried and ground (Brewer et al., 2009).

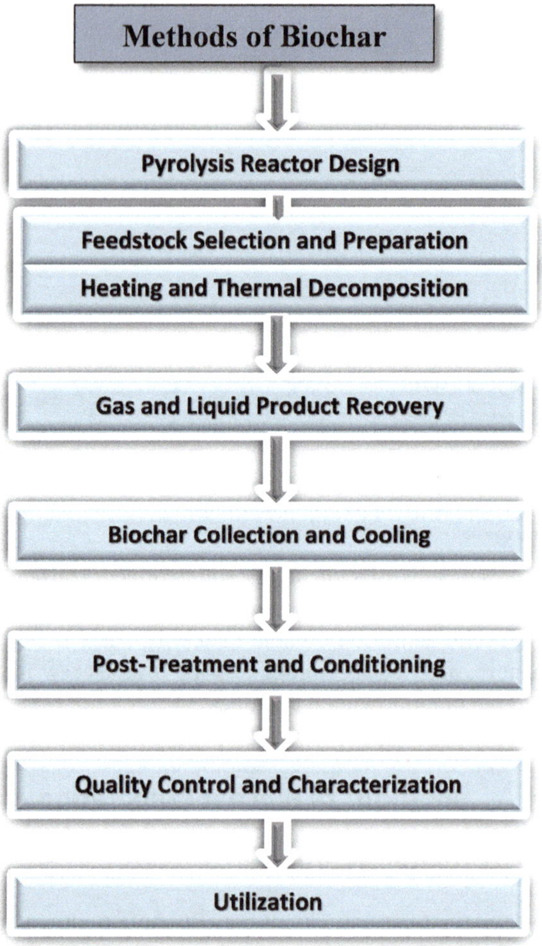

Fig. 2 Flowchart of the pyrolysis techniques

3.1.2 Design of the Pyrolysis Reactor

Different types and designs of pyrolysis reactors, such as rotary kilns, fixed-bed reactors, fluidized-bed reactors, and retorts, exist. The pretreatment of biomass to produce biochar is a critical process within the pyrolysis chain because it directly influences the biochar production cost, but for this reason, complex interaction factors make reactor selection difficult (Bridgwater, 2012).

3.1.3 Heating and Thermal Decomposition

Once the feedstock has been prepared, it is loaded into a pyrolysis reactor. Then the samples are heated to temperatures typically between 300 and 800 °C in an oxygen-deprived environment, during which organic materials thermally decompose, resulting in volatile gases, liquid bio-oil, and solid char (biochar) or carbon (Mohan et al., 2006).

3.1.3.1 Gas and Liquid Products

When biochar is produced through the pyrolysis process, some residual volatile gases and bio-oils in liquid form are also obtained. These byproducts can be collected and further processed for various applications, such as energy production, chemical synthesis, and value-added products (Jirka & Tomlinson, 2014).

3.1.3.2 Biochar Collection and Cooling

Once the decomposition of carbon is complete, the remaining solid biochar is released from the pyrolysis reactor. Depending on the type of reactor, biochar may have to be cooled further so that its temperature decreases and it becomes stable (Mukherjee et al., 2011).

3.1.3.3 Posttreatment and Conditioning

To enhance its physical and chemical properties, the generated biochar may undergo a series of posttreatments, such as sieving, grinding, or adding activators. This may require nutritional content modification, porosity, increased surface area, and particle size distribution for a particular application (Kambo & Dutta, 2015).

3.1.3.4 Quality Control and Characterization

This is an extended process of examining physical and chemical qualities, including particle size, surface area, pH, elemental composition, and nutrient content. Quality control measures ensure that the biochar meets specifications deemed necessary for its intended use (Shackley et al., 2016).

3.1.3.5 Utilization

Pyrolysis-produced biochar applications are diverse, comprising industrial processes, soil amendments, and carbon sequestration, and can also be used in livestock bedding as well as in environmental remediation. The porous and high carbon content of this biochar makes it very useful for improving soil fertility, water-holding capacity, and nutrient cycling (Lehmann & Joseph, 2015).

Thus, pyrolysis is a versatile method of converting organic materials into biochar for better soil health and enables renewable energy as well as proper waste management.

Numerous methods of heating plant biomass to produce biochar (mostly in the near absence of oxygen) have been suggested by many scientists. According to, the Department of Soil Science, Bangabandhu Sheikh Mujibur Rahman Agricultural University (BSMRAU), has set up two improved chambers with pyrolysis stoves for small-scale biochar production for research purposes only (Figs. 3 and 4).

The organic materials are placed in the outer chamber facing away to keep O_2 levels low. The gas flame burner is located in the middle and bottom of the chamber, where heat is subsequently increased to 400–500 °C after a gradual increase and remains constant for another 4–5 h. The biochar was first allowed to cool after pyrolysis, collected, and then powdered before being stored for use in agricultural fields (Fig. 4).

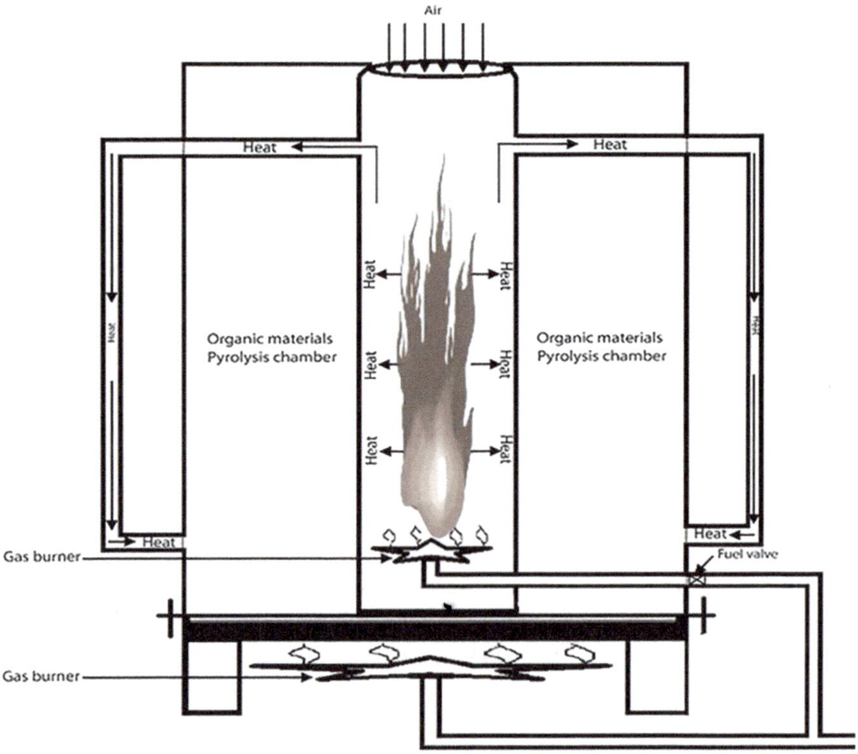

Fig. 3 Handy modified biochar preparation stove.

Fig. 4 Energy-efficient options for generating biochar via conversion into a furnace.

The temperature of pyrolysis and the material source are critical for determining the biochar yield (Table 1). Some of the key biomass materials used to produce biochar organically are forestry wastes, crop residues, and animal manures. The type of feedstock chosen determines how affordable biochar can be and to what yields are possible. The results revealed that rice husk, rice straw, cow dung, and chicken manure have the highest energy potential, whereas sawdust is the most feasible for accessing biomass, as it is readily available in Bangladesh for biogas production. In addition, several studies conducted by Laghari et al. (2016a, b), Zhao et al. (2013), and Kinney et al. (2012) reported that a number of organic materials, including wood, plant branches, magnolia leaves, and straw from various crops (Table 1), are suitable for biochar production.

Table 1 Effects of different biomasses on biochar production yield at different pyrolysis temperatures

Source of biomass	Pyrolysis temperature (°C)	Biochar production yield (%)	References
Rice straw	450–500	35.6	Islam et al. (2018)
Rice husk		46.1	
Mixed sawdust		53.3	
Poultry manure		49.1	
Cow dung		48.5	
Rice straw	300	50.1	Wu et al. (2012)
	400	42.8	
	500	39.0	
	600	36.3	
	700	36.5	
Canola straw	350	24.4	
Rice straw		33.3	
Soybean straw		32.5	
Pea straw		30.2	
Pea straw	400	55.0	Laghari et al. (2016a, b)
Sawdust	550	28.0	Zhao et al. (2013)
Magnolia leaves	300	61.6	Kinney et al. (2012)
Apple wood	600	25.0	
Spotted gum wood	400	51.0	

4 Gasification Process

Another thermochemical reaction used when creating biochar is gasification. Gasification, unlike pyrolysis, is the oxidation of partial biomass in a controlled environment {semi-O_2}, which leads to the production of a flammable gas mixture known as syngas ($CO + H_2$) at high temperatures (>500 °C).

This process produces biochar as a byproduct. Gasification is more energy efficient than pyrolysis but needs accurate control over working parameters to optimize the quantity and quality of biochar produced. According to You et al. (2017), gasification generally consists of four distinct steps: drying, pyrolysis (thermal-induced fragmentation by bond dissociation and dehydrogenation), partial oxidation, and reduction. Gasification for biochar synthesis involves many procedures (Fig. 5), which are discussed in the following subheadings:

4.1 Fixed-Bed Gasification

This procedure packs organic materials into the gasifier's fixed bed. Heat is provided from below the bed to induce processes such as pyrolysis, gasification, and combustion. Pyrolysis or incomplete combustion leads to the development of biochar, which is deposited in the top bed (Zhang et al., 2010).

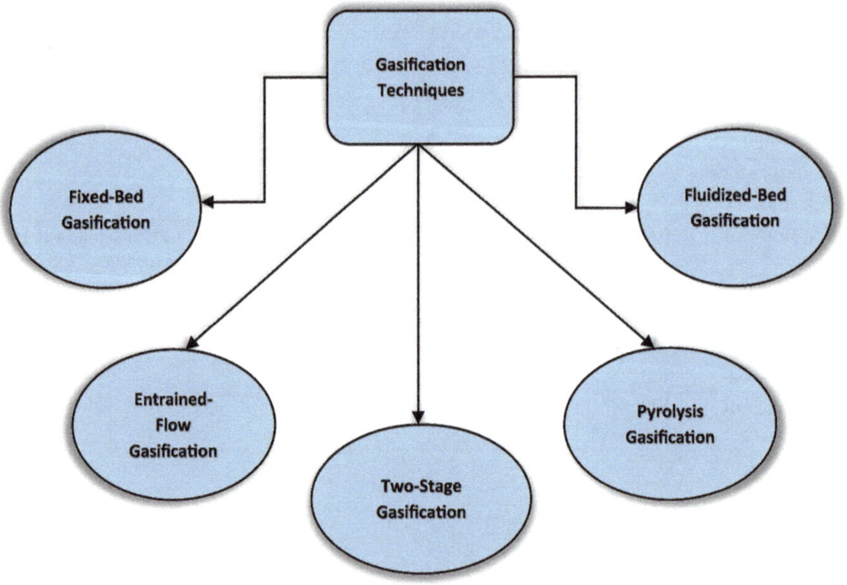

Fig. 5 Fundamental mechanisms of gasification

4.2 Fluidized-Bed Gasification

Organic substances are added to a bed of inactive material, such as sand or char particles, which are made to flow like a fluid by a stream of gas, commonly air or steam. This is due to the fluidized bed, which provides efficient mixing and heat transfer, enhancing pyrolysis as well as gasification reactions. Biochar is collected in beds as a solid substance (Gómez-Barea & Leckner, 2010).

4.3 Entrained-Flow Gasification

The organic materials are finely ground and entrained in a high-rate gas stream (steam, oxygen, or air) that is fired into the reactor. With high temperature and rapid mixing, the slurry fully pyrolyzes and gasses. The main gas product is typically synthesis gas or "syngas," and biochar is collected as a byproduct (You et al., 2017).

4.4 Pyrolysis Gasification

In this method, phenolic compounds are processed via a combination of both gasifying and pyrolyzing steps. Organic materials are heated to high temperatures without potentially contaminating steam or oxygen. This leads to reactions such as gasification, pyrolysis, and mild combustion. Biochar is not the only product; it produces syngas and some other byproducts (Mohan et al., 2006).

4.5 Two-Stage Gasification

Here, the process deploys two different stages or reactors. In the first stage, organic resources are pyrolyzed into volatile gases and biochar. Volatile gases are then added to the second stage, which runs at higher temperatures and leads to gasification processes that create more syngas. The biochar was collected as a solid residue from the first step (Lv et al., 2004).

Generally, gasifiers are classified as entrained-flow, fluidized-bed, or fixed-bed gasifiers on the basis of the contact mode of solids and syngas. The appropriate process is used to maximize benefits on the basis of feedstock characteristics, desired biochar properties, energy needs, etc. To optimize the efficient and high-quality production of biochar, a suitable gasification technology must be selected (see Table 2).

Table 2 Comments on gasification parameters (gasifier type, temperature, and type of gasifying agent), total carbon yield (%), ash content (%), and inorganic elements found in biochar under various conditions

Feedstock	Gasifier types	Temperature (°C)	Gasifying agent	Ash (wt.%) [b]	C (wt.%)	Inorganic elements (mg/g)	References
Pinewood	Fluidized bed	600–710	Steam and/or N2	18.69	60.04	SiO_2 (73.15), Al_2O_3 (3.2), TiO_2 (0.03), Fe_2O_3 (12.54), CaO (3.30), MgO (83.38), Na_2O (0.17), K_2O (2.36), P_2O_5 (0.99), SO_3 (0.09), Cl (0.02), CO_2 (1.96)	Shen et al. (2016)
White oak	Fluidized bed	600–710	Steam and/or N2	34.90	59.49	SiO_2 (141.33), Al_2O_3 (5.97), TiO_2 (0.07), Fe_2O_3 (23.55), CaO (7.94), MgO (149.99), Na_2O (0.17), K_2O (0.20), P_2O_5 (0.44), SO_3 (0.42), Cl (0.03), CO_2 (4.37)	Shen et al. (2016)
Kentucky bluegrass seed mill screenings	Up draft	600–650	Air	46.3	45.7	NH_4-N (0.0149), NO_3-N (1.081), incubation-N (0.0248), mineralizable-N (0.0099), K (48.203), Ca (13.580), P (11.680), Mg (5.702), S (3.879), Cl (2.268), Fe (1.031), Mn (0.703), B (0.0272), Na (0.275), Zn (0.0895), Cu (0.0251), Al (0.869)	Griffith et al. (2013)

Switchgrass	Down draft	N.A.	N.A.	N.A.	64.80	P (0.27), Ca (2.81), K (0.73), Mg (0.45), Na (0.13), Fe (0.29), Zn (97.8)[a], Cu (8.8)[a], Mn (394.5)[a], Ni (4.6)[a], Al (1408.0)[a]	Bhandari et al. (2014)
Switchgrass	Fluidized bed	N.A.	N.A.	N.A.	1.38	P (0.04), Ca (0.13), K (0.17), Mg (2.65), Na (0.03), Fe (0.78), Zn (79.4)[a], Cu (5.7)[a], Mn (130.1)[a], Ni (303.7)[a], Al (426.0)[a]	Bhandari et al. (2014)
Sewage sludge	Up draft	800–950	Air	73.17	89.85	Si (180), Al (63.7), P (53.3), Fe (41.9), K (20.6), Ca (4.89), Mg (11.1), Na (10.1), Mn (2.4), Zn (0.863), Cu (1.21), Cr (0.27), Pb (0.0617), Ni (0.237), Zr (0.0766)	Kim et al. (2016)
Corn stover	Fluidized bed	850	Steam	47.1	42.0	Fe (10)[a], Ca (14), Mg (9.2)[a], K (54) Si (14)	Cheah et al. (2014)
Winter wheat straw	Fluidized bed	750	Air	N. A.	46.8	K (72), Na (2.6)	Hansen et al. (2016)
Pine wood	Down draft	1200	Steam	N.A.	65.29	K (19), Na (2.0)	Hansen et al. (2016)

(continued)

Table 2 (continued)

Feedstock	Gasifier types	Temperature (°C)	Gasifying agent	Ash (wt.%) [b]	C (wt.%)	Inorganic elements (mg/g)	References
Pine wood	N.A.	600–900	N.A.	10.79	79.34	P (1.337), Na (0.48), K (9.36), Ca (20.52), Mg (2.1), Cu (0.012), Co (0.008), Cr (0.034), Ni (0.025), Pb (0.016), Zn (0.256), As (9×10^{-4}), Cd (1.38×10^{-3}), Sb (4.35×10^{-4}), $CaCO_3$ (33.4), C-CO_3 (4)	Marks et al. (2016)
Rice husk	Down draft	900–1100	N.A.	63	35	Al (0.092–0.543), Ba (0.019–0.048), B (0.00181–0.00538), Ca (0.609–1.94), Cu (0.0082–0.0153), Fe (0.066–0.107), Mg (0.162–0.658), Mn (0.135–0.47), K (0.595–2.418), Si (0.066–0.199), Na (0.076–0.65), Sr (0.00187–0.0091), Ti (0.00179–0.00525), Zn (0.0117–0.0442)	Shackley et al. (2012)
Pine wood	Fluidized bed	600–900	Air	N.A.	71	Ca (92.3), K (8.3), Na (0.8), P (0.8), Fe (0.08), Cd (1.2), Cr (26.0), Cu (224), Ni (10), Pb (9.1), Zn (982–1504)	Ojeda et al. (2015)

Pine wood	Up draft	1000	Air	N.A.	N.A.	Al (0.35×10^{-3}), Ca (12×10^{-3}), Co (0), K (2.0×10^{-3}), Mg (0.11×10^{-3}), Mn (0.02×10^{-3}), Na (0.18×10^{-3}), P (0.06×10^{-3}), Sc (0.02×10^{-3}), Si (2.31×10^{-3}), Ti (0.18×10^{-3}), Zn (0.07×10^{-3})	Huggins et al. (2015)
Poultry litter	Up draft	580–680	Air	54.8	33.1	P (29.3), K (87.8), Na (10.4), Ca (44.1), Cl (1.2), Mg (18.3), Fe (2.389), Mn (1.577), Zn (1.229), Cu (0.273), B (0.148), Al (1.121), Pb (0.87), Cd (1.03), Ni (21.6), Cr (16.1)	Taupe et al. (2016)

[a]The unit is ppm
[b]The values are determined on a dry basis

5 Hydrothermal Carbonization (HTC)

Hydrothermal carbonization (HTC) is a method in which high temperature and pressure are used in the aqueous phase to convert organic materials into biochar. It occurs within hours to a few days, but it is meant to mimic the geological time periods over which coal normally forms. The unique characteristics of HTC are different from those of conventional (pyrolysis)-generated biochar, such as higher nutrient retention capacity and superior energy density (Benavente & Fullana, 2015; Román et al., 2018). A flowchart of hydrothermal carbonization techniques is available in Fig. 6.

These are the stages involved in generating biochar via HTC:

Fig. 6 Flowchart of hydrothermal carbonization techniques

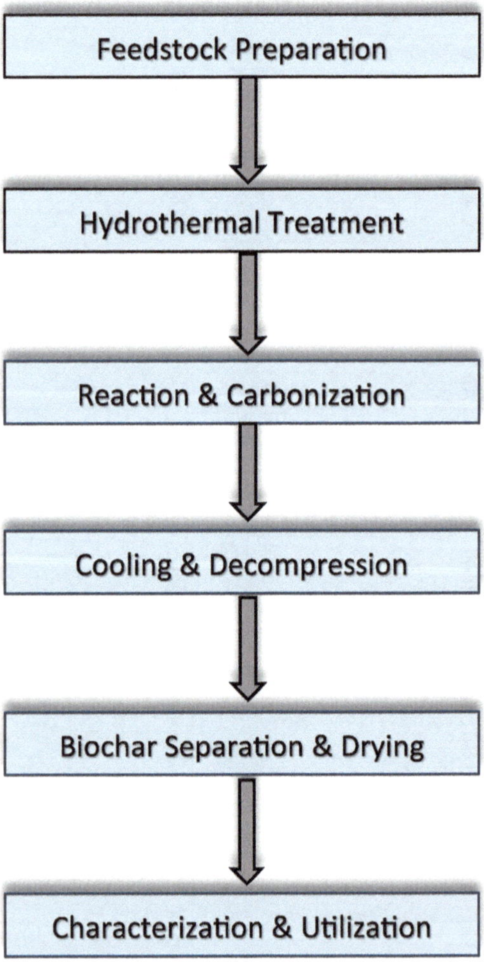

5.1 Feedstock Preparation

Organic matter such as biomass residues, agricultural waste, or sewage sludge are shredded and ground into smaller sizes to maximize the surface area and make them suitable for consistent processing (Nunoura et al., 2006).

5.2 Hydrothermal Treatment

In this step, once the organic matter is ready, it is placed into sealed reactor vessels under water pressure. To create hydrothermal conditions in the reactor, it is heated at temperatures typically varying from 180 to 250 °C under high pressure (approximately 10–25 bar) (Wade et al., 2006).

5.3 Reactions and Carbonization

Under high temperature and pressure in a water environment, organic components in reactors usually undergo many types of chemical reactions, such as hydrolysis, dehydration, or decarbonation. Therefore, organic material is converted into liquid and gaseous end products as well as solid carbonaceous materials (such as biochar) (Jin et al., 2016).

5.4 Cooling and Decompression

Once the hydrothermal treatment of the reactor is complete, the pressure begins to decrease. Hence, the material can be safely removed from the reactor (Chen et al., 2008).

5.5 Drying of the Biochar

The biochar is dried before it can be recovered, and in this process, slurries or solid mixtures that remain are separated to ensure that all the remaining moisture will vaporize easily. Depending upon its initial moisture content and the properties the biochar producer desires in their finished material, additional drying steps may be needed to decrease those levels further through increasing stability (Benavente & Fullana, 2015).

5.6 Method of Analysis

Analyses are performed to determine the physical and chemical properties of the biochar produced, such as its stability (via a CO_2 evolution assay), pore structure, and surface area, and N_2 adsorption is used to estimate the C sequestration potential and elemental composition. It can then be used in many cases to create energy, treat water, lock up carbon from the atmosphere, or act as a feedstock for new materials (Nunoura et al., 2006).

Hydrothermal carbonization can assist in the production of biochar, which offers several advantages, such as being able to treat an array of feedstocks at faster reaction rates and creating biochars that may represent high value for specific applications. Control and optimization are required to achieve all these expectations simultaneously, and several process parameters affect each other and thus interact critically in biochar production, including the resulting quality but not the long-term environmental impact.

Notably, in addition to the known methods of preparation, flash carbonization and torrefaction techniques have also been used for biochar production (Benavente & Fullana, 2015; Chen et al., 2008; Nunoura et al., 2006; Wade et al., 2006). In the flash carbonization process, feedstock is converted to both solids and gases under $1e^2$ MPa pressure at temperatures ranging from 300 °C to approximately 600 °C, with a residence time of approximately half an hour. Torrefaction in an inert atmosphere causes a reduction in moisture and oxygen contents, which results in the formation of hydrophobic solid products. It operates at a temperature of 200–300 °C, and solid products with a low oxygen-to-carbon ratio are generally produced from the torrefaction process (Jin et al., 2016). However, the partial oxidation of feedstock during torrefaction tends to severely limit the biochar adsorption capacity (Wade et al., 2006).

6 Characterization of the Biochar

Biochar is a carbon-rich material produced when biomass is heated in the absence of air (also known as pyrolysis). Its unique diverse and dynamic properties are valuable for a wide range of applications, particularly in environmental management or agriculture. Surface area analysis through the methods of scanning electron microscopy (SEM) (Fig. 7), Fourier transform infrared spectroscopy (FTIR), X-ray diffraction (XRD), and elemental analysis are other important characteristics that can be used to characterize biochar (Chia et al., 2015). These characteristics are essential for improving biochar production processes, predicting its performance in specific applications and exploiting its opportunities as a sustainable way to produce energy or improve soil quality while remediating environmental problems.

6.1 Physical Characteristics

When any organic material, such as wood chips, dung, or agricultural waste, is heated in the absence of oxygen, it remains rich in carbon substances from the process called biochar. It possesses many physical properties that can be useful in different applications, some of which are as follows:

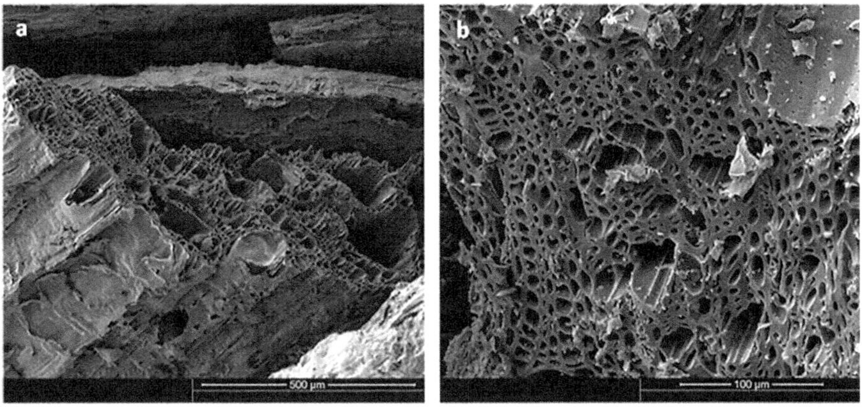

Fig. 7 SEM images confirming the macroporosity of biochar made via slow pyrolysis from wood. (Source: Chia et al. (2015))

6.1.1 Porosity

A useful physical characteristic of biochar is its net porosity, which is extremely important for different types of applications in addition to environmental remediation and agriculture. Biochar is typically highly porous in nature because it is produced through a pyrolysis method, resulting in the formation of an intricate network of pores (Chia et al., 2015). Pore sizes in different ranges:

6.1.1.1 Micropores

The pores with diameters of less than two nanometers are called micropores. They contribute greatly to the total surface area of biochar. Micropores are required to carry out adsorption activities such as retaining pollutants, organic molecules, and nutrients (Ahmad et al., 2014).

6.1.1.2 Mesopores

Pores with diameters between 2 and 50 nm. The air and water move freely between the biochar matrix, allowing mesopores to form. They also promote the material's ability to host a microbial ecology and retain nutrients (Lehmann & Joseph et al., 2024

6.1.1.3 Macropores

These pores are larger than 50 nm in diameter. Greater macroporosity caused the soil structure to be more friable; therefore, the possibility of flooding could also decrease since biochar has a greater capacity for water infiltration and drainage (Jeffery et al., 2011).

Various factors, such as feedstock selection, pyrolysis conditions (temperature, heating rate, and residence time) and posttreatment techniques, control the specific porosity of biochar. Therefore, it is possible that the biochars produced at higher temperatures generally contain a larger portion of microporosity and less than meso- or macroporosity than those produced at lower pyrolysis temperatures (Mohan et al., 2014). Porosity is a key requirement in the application of biochar. Below, we cover some applications in a concise manner.

- *Soil amendments*:
  Biochar, which has a highly porous structure, improves plant growth by improving the water, nutrient, and agrochemical retention properties of soils.
- *Water purification*:
  Owing to its porosity, biochar is an excellent water treatment solution because it is able to extract impurities and contaminants from the liquid.
- *Climate change mitigation and carbon sequestration*:
  The selection of biochar has broad potential to mitigate climate change by storing carbon owing to the abundant pore size network supplying physical sites for adsorption.

  Overall, the porosity of biochar influences its performance in a range of applications and is therefore a potentially useful and versatile material for use in projects involving carbon management, environmental remediation, or sustainable agriculture.

6.1.2 Surface Area

The porous structure of biochar results in a high specific surface area per unit mass. Owing to this enormous surface area, functional groups such as carboxyls, hydroxys, and phenolic groups are present to chemically interact with other molecules. This is due to their broad surface area, which provides many sites for adsorption, nutrient recirculation, and interaction with soil bacteria. Surface area measurements are commonly denoted in measures of square meters per gram (m^2/g) or square meters per unit of biochar volume (m^2/m^3) (Chia et al., 2015). There are several ways to measure the surface area of biochar, including the following:

- *Brunauer-Emmett-Teller (BET) method*:
  The first adsorption method is based on gas molecules and is the Brunauer-Emmett-Teller (BET) technique. BET isotherms allow the determination of surface area by measuring how much gas is adsorbed on a material at different pressures (Brunauer et al., 1938).
- *Nitrogen adsorption method (Brunauer-Emmett-Teller (BET) nitrogen adsorption)*:
  Like the BET approach, the biochar surface is exposed to nitrogen gas at low temperatures, causing it to be adsorbed. This is followed by the calculation of the surface area via nitrogen adsorption data and the application of the BET equation (Sing, 1985).
- *Methylene green adsorption process*:
  This method uses methylene green, where the surface of the biochar is layered according to color. Its specific area and volumetric porosity are calculated by considering the quantity of dye retained on the biochar surface (Martin et al., 1999).

- *Pore volume measurement*:
  Gas adsorption and mercury intrusion porosimetry are two methods that can be used to indirectly quantify the pore volume and total surface area of biochar. If a certain pore shape is considered, the surface area and pore volume are coupled (Barrett et al., 1951). The surface area of biochar is highly dependent on numerous parameters, such as the raw material, pyrolysis temperature, and posttreatment procedures (Cheng et al., 2008). Biochar generally has a relatively large specific surface area because carbonization and pore formation are promoted at relatively high temperatures. However, the surface area does not fully characterize biochar behavior or performance since the pore size distribution, availability of functional groups, and degree of porosity can also contribute significantly to biochar properties, which are in turn related to and influence different applications. Biochar has a high surface area, which allows it to absorb minerals, organic compounds, and pollutants more efficiently, thus increasing its effectiveness in agricultural and environmental contexts. It helps maintain the fertility of soil, water quality, and environmental remediation.

6.1.3 Density

The density of biochar can influence a variety of factors, including the feedstock material, pyrolysis conditions, and posttreatment processes. Additionally, pyrolyzing its parent materials leaves few to no volatiles for biochar production; hence, biochar often has a low to moderate density relative to the organic material it is made of. Biochar density is expressed in grams per cubic centimeter (g/cm^3) or kilograms per cubic meter (kg/m^3) (Chia et al., 2015). The density of biochar ranges from ~0.1 to 0.8 g/cm^3 and depends on several factors, such as the following:

- *Feedstock materials*:
  The densities of different types of biomass feedstocks might affect the density of the final biochar. For example, woody biomass tends to produce biochar denser than grasses or agricultural residues do (Lehmann & Joseph et al., 2024).

- *Parameters for pyrolysis*:
  The density of biochar can be affected by the operating conditions during pyrolysis, such as the temperature range, heating rate, and residence time. In a normal gasification process, as the temperature increases, the volatile elements of only the gaseous parts are eliminated, resulting in biochar having a higher density than it would have during pyrolysis alone (Downie, 2011).

- *Posttreatment techniques*:
  The density of biochar is also affected by posttreatment techniques such as chemical modification or activation. When the biochar was mobilized by each activation process, there could be an increase in porosity and hence a decrease in overall density (Zhang et al., 2010).

- *Particle size and structure*:
  The particle size, aspect ratio, and meso-porosity of biochar influence its density as well. Bulk density may differ in part because smaller particles generally pack more closely than larger particles do (Mohan et al., 2014).

- *Moisture content*:

  The moisture content of biochar can also affect the bulk density. Drier biochar may have a greater density, but wetter material will appear to be denser (Patterson et al., 2013).

  Importantly, although the density of biochar has a significant effect, other features and performance may not be fully described. Porosity, surface area, and pore structure are three more physical traits that influence how biochar affects these applications of soil amendments for the remediation of carbon sequestration in the environment.

6.1.4 Color

Biochar may have different colors depending on factors such as the type of feedstock used, the pyrolysis conditions, and any posttreatment steps taken. The color of biochar is typically black to brown, but there may be wide variation even within these fairly large ranges as a grouping (Chia et al., 2015). These are some of the factors that play a role in determining biochar color:

- *Biochar feedstock material considerations*:

  The color of biochars is affected by the biomass feedstocks used in their production. For example, woody materials such as softwood or hardwood tend to produce biochar that has a darker black/brown color and is deeper. With biochar manufactured from waste agricultural products, such as rice husk and maize stover, the color could be a lighter shade of brown (O'Callaghan et al., 2015).

- *Pyrolysis conditions*:

  The initial color of biochar is also influenced by the pyrolysis temperature and time. Biochar darkens when it is heated to higher temperatures with a longer pyrolysis time. This is because the black color results from more widespread carbonization and aromatic carbon structure synthesis at high temperatures (Laird et al., 2010).

- *Inorganic mineral content*:

  The presence of ash or inorganic minerals in feedstock can influence biochar color. Biochar can appear in various colors due to certain minerals, such as red from iron oxides and yellow from silica (Novak et al., 2009).

- *Posttreatment processes*:

  Biochar may change color after posttreatment processes such as chemical modification or activation. The biochar can be activated by further heating in the presence of oxidizing chemicals, resulting in a darker color. Similarly, several studies have shown that with time, certain chemical processes or the impregnation of biochar may change its color (Jien & Wang, 2013).

- *Surface oxidation*:

  The color of biochar varies over time as a result of oxidation or exposure to certain elements. The surface of the biochar might discolor or take on a patina due to oxidation processes (Lehmann & Joseph et al., 2024).

The color of biochar offers partial insight into its makeup and how it is produced. However, it should not be solely dependent upon determining its characteristics or overall excellence. The porosity, surface area, and chemical composition are important characteristics to consider when assessing biochar for various uses, including improving soil quality, storing carbon, and cleaning the environment.

6.1.5 Particle Size and Structure

The two major physical properties of biochar that are critical to its performance and application are its size and structure. These qualities can be summarized as follows:

6.1.5.1 Particle Size

Biochar can be used to make a wide range of particle sizes and materials, from small powders to larger granules or pellets (Chia et al., 2015). There are several issues related to the particle size distribution of biochar that can influence either its use or performance.

- *Fine particles*

  Fine biochar particles have a relatively large surface area per unit mass, which can increase the reactivity and adsorption capacity of such fractions. They are also used in applications where there is a need for close proximity to the soil, water, or other materials, such as when they are applied to soils (e.g., through incorporation into the soil), water, or sludge (Yang et al., 2006).

- *Coarse particles*

  While their surface area may be lower, coarse biochar particles can help with soil aeration and structure. It is often used in agricultural applications to improve soil structure and water retention (Van Zwieten et al., 2010).

- *Particle size distribution*

  In soil and other matrices, an appropriate particle size distribution (with a combination of fine and coarse particles) balances the reactivity with structural benefits (Jeffery et al., 2011).

6.1.5.2 Particle Structure

The structure of biochar particles refers to their internal arrangement and pore characteristics, which influence properties such as porosity and permeability. Common structures include:

- *Permeable structure*:

  The semipermeable structure of biochar is an interconnected network of pores. Properties such as porosity provide available microbial habitat, adsorption sites, and nutrient retention sites (Mohan et al., 2014). The pore size distribution (micropores, mesopores, and macropores) varies depending on the feedstock material as well as the pyrolysis conditions.

- *Agglomerates*:

  Agglomerates are small packs of biochar particles that cluster and combine to create larger aggregates (Lehmann & Joseph et al., 2024). Finally, agglomerates can result in alterations in biochar handling and application physically, but they may also degrade after some time because of environmental factors.

- *Pellets or briquettes*:

  Biochar can be shaped into pellets and briquettes for easier handling and application (Downie, 2011). For agricultural uses, this consistency in size and shape may allow pelletized biochar to be applied more uniformly.

  The utility of biochar for numerous applications, including soil enhancement and remediation, water filtration, or greenhouse gas mitigation, may vary on the basis of physical integrity in environmental scenarios. This knowledge facilitates effective optimization of the biochar production process and application method so that the desired outcomes can be achieved.

6.1.6 Hydrophobicity/Hydrophilicity

The ability of biochar to repel or attract water is its hydrophobicity or hydrophilicity. This is impacted by the feedstock material, pyrolysis conditions, and surface chemistry combined with the pore structure of these biochar products. How these variables affect the hydrophobic or hydrophilic properties of biochar is as follows:

- *Feedstock material*:

  The hydrophobicity or hydrophilicity of the biochar may be related to the feedstock because different types of biomasses have distinct physical and chemical characteristics (Lehmann & Joseph et al., 2024). For example, biochars produced from lignin-rich materials, such as hardwoods or wood chips, may be more hydrophobic since they contain hydrophobic aromatic compounds, whereas their counterparts with higher cellulose/hemicellulose/higher ash contents (e.g., grasses and agricultural residues) are likely to be more hydrophilic.

- *Pyrolysis conditions*:

  Hydrophobic/hydrophilic biochar can be generated on the basis of the pyrolysis temperature and heating rate. Although more hydrophilic surface functional groups can be produced on biochars at lower pyrolysis temperatures, high temperatures probably create aromatic carbon structures that are much more hydrophobic (Mohan et al., 2014).

- *Surface chemistry*:

  The presence of specific functional groups on the biochar surface can affect how it interacts with water molecules (Masiello, 2004). For example, biochar with hydrophilic functional groups, such as hydroxyl (−OH) or carboxyl (−COOH) groups, is more hygroscopic and can form hydrogen bonds with water molecules. However, the influence of biochars with aromatic carbon structures and phenolic or carboxyl hydroxides was apparent to be hydrophobic in that their nonpolar nature created strong interactions between these groups.

- *Pore structure*:

  The pore structure includes the size distribution and connectivity of pores, which affects the hydrophobic/hydrophilic properties of biochar. Biochars with predominantly microporous structures may exhibit greater hydrophobicity due to restricted water accessibility, whereas biochars with more meso- and macropores may be more hydrophilic (Basso et al., 2013).

- *Surface area*:

  Biochar surfaces, which are much larger than other surfaces (as applied in this study), may have more contact with water molecules, which can influence their hydrophobicity and hydrophilicity (Cheng et al., 2008). To understand the variety of biochar used, its hydrophobicity and hydrophobicity should be elucidated. Biochar with hydrophilic characteristics can increase the soil moisture level and retention when it is applied to the soil. Similarly, hydrophobic biochars may also more effectively capture dissolved-phase contaminants because they are mostly adsorbed on the outer surface and tend not to enter micropores. The function of biochar can be controlled via different surface modifications and carbon feedstocks. Therefore, the hydrophobic/hydrophilic nature of biochar is multifactorial. Understanding these factors may help maximize biochar efficiency for its numerous applications.

Finally, all the physical properties combine to improve biochar functionality in applications such as environmental remediation, water treatment, carbon sequestration, and soil improvement.

6.2 Chemical Characteristics

However, the chemical properties of biochar can be heavily influenced if it is a material of origin as well as by the methodology used to manufacture and process it. The composition of biochar is generally similar to that of coal, which means that its main component is carbon, followed by varying amounts of nitrogen, oxygen, and hydrogen, as well as minerals. In addition, hydroxyl (–OH), carboxyl (–COOH), and phenolic functional groups can be found in biochar. Properties such as pH and CEC contribute to its reactivity and ability to interact with soil and other environmental components. In the next part, there are several key chemical characteristics.

6.2.1 Carbon Percentage

Biochar has 50–90% carbon by weight depending on the type of feedstock, pyrolysis temperature, and time. Higher pyrolysis temperatures often result in biochar with higher carbon contents (Table 3). This carbon-rich composition is what makes biochar an effective tool for carbon sequestration and soil amendment, as it provides a stable form of organic carbon that can persist in soils for hundreds to thousands of years, thereby helping to mitigate climate change and improve soil fertility and structure (Lehmann & Joseph et al., 2024).

Table 3 Chemical properties of different types of biochar at different temperatures

Source of biochar	Pyrolysis temperature (°C)	pH	Organic carbon (C) (%)	CEC (Cmol/kg)	Total N (%)	P (%)	References
Rice straw	450–500	8.96	46.45	32.12	0.800	0.070	Islam et al. (2018)
Rice husk		7.75	44.50	34.67	0.673	0.065	
Mixed sawdust		7.92	53.10	28.90	0.584	7.310	
Poultry manure		7.57	42.70	45.17	0.091	0.311	
Cow dung		8.12	39.80	47.80	0.132	0.143	
Rice straw	300	9.91	72.1	56.9	1.550	0.005	Wu et al. (2012)
	400	9.96	77.2	61.6	0.174	0.006	
	500	10.45	82.8	32.0	1.770	0.003	
	600	10.84	87.1	23.9	1.520	0.002	
	700	10.87	85.2	23.1	1.610	0.018	
Canola straw	350	8.00	61.7	180.0	0.170	0.220	Yuan and Xu (2011)
Rice straw		7.69	42.5	152.0	1.650	0.330	
Soybean straw		9.02	54.1	98.0	3.620	0.720	
Pea straw		Unavailable	Unavailable	Unavailable	Unavailable	Unavailable	
Pine sawdust	400	6.35	51.7	27.5	0.930	12.000	Laghari et al. (2016a, b)
	700	9.08	73.6	24.9	0.600	21.000	
Pig manure	350	9.65	39.1	49.0	0.290	0.230	Zhao et al. (2013)
Wheat straw	350	8.69	58.9	87.2	0.004	0.294	
Cow manure	500	10.20	43.7	149.0	0.064	0.100	
Sawdust	500	10.50	75.8	41.7	0.006	Unavailable	

6.2.2 Functional Groups

Most compounds that are attached to the carbon structure of biochar contain various functional groups. Temperature pyrolysis is the key for determining which functional groups are present in biochar. Biochar produced at low temperatures (below 400 °C) tends to retain more oxygen-containing functional groups, such as hydroxyl (–OH), carbonyl (C=O), and carboxyl (–COOH) groups. These functional groups contribute to the polarity, hydrophilicity, and surface reactivity of biochar. Low-temperature biochar typically has a relatively large surface area due to the preservation of these functional groups (Zhang et al., 2010).

At temperatures ranging from 400 to 600 °C, biochar undergoes substantial modifications that result in a decrease in the number of oxygen-containing functional groups. However, aromatic structures are subsequently formed, and functional groups such as phenolic (–OH), ketone (C=O), or ether (–O–) groups arise. The decrease in the number of oxygen-containing functional groups and increase in aromatic structures resulting from the pyrolysis procedure improved the stability and carbon content of the biochar (Mohan et al., 2014).

Owing to high temperatures (>600 °C), the majority of functional oxygen-containing groups are removed, resulting in a significant alteration in chemistry (organic matter composition). In addition, biochars produced under proper conditions have abundant graphitized carbon and a low content of less polar functional groups (Ahmad et al., 2014). At high temperatures, biochar is highly stable, has low reactivity, and has an increased carbon content.

The nature and profile of these functional groups play important roles in the surface chemistry, reactivity, and interaction of biochar with other materials. Understanding how functional groups in biochar are arranged and distributed is important for predicting its behavior or performance, regardless of whether it is used as a medium-soil amendment, for water treatment, or for carbon sequestration.

6.2.3 pH

The pH of biochar varies, depending on the type of organic material used for pyrolysis (charring), as well as its pyrolytic conditions and any posttreatment processes. Most biochar samples at temperatures ranging from 450 to 500 °C have neutral to slightly alkaline pH values (usually in the range of pH = 7–10). The pH of the surface of biochar can potentially be affected by acidic or alkaline functional groups. For example, biochars derived from high-lignin feedstocks could have higher pH values because of the presence of alkaline ash residues. Conversely, biochars derived from cellulose- or hemicellulose-rich feedstocks can result in a lower pH due to the various numbers of acidic functional groups generated during pyrolysis (Lehmann et al., 2024). The biochar pH is also independent of the biomass source, but it depends on the pyrolysis temperature (for all feedstocks). Specifically, Table 3 shows the pH of the biochar as a function of temperature. Biochar is a soil amendment, and the pH of biochar influences that of soils. Similarly, pH also plays an important role in the availability of nutrients; thus, before its use for agriculture, pH must be analyzed.

6.2.4 Cation Exchange Capacity (CEC)

The capacity of biochar to absorb, retain, and release positively charged ions is regarded as its cation exchange property, which has implications for soil fertility and nutrient availability. The CEC of biochar is influenced mostly by the type of biomass and not the pyrolysis temperature. This phenomenon is attributed to the higher temperatures during pyrolysis leading to the removal of some acidic functional groups from the biochar surface, as observed by Zhao et al. in 2013. During pyrolysis, organic residues such as Ca, Mg, K, Na, and P contribute to the formation of oxygen-containing functional groups on the biochar surface, thereby increasing the CEC (Agrafioti et al., 2013). For example, rice straw-derived biochar contains notable amounts of N (0.80%), P (0.07%), S (0.005%), Ca (16.80 Cmol/kg), Mg (4.86 Cmol/kg), K (8.36 Cmol/kg), Na (2.10 Cmol/kg), and organic carbon (44.5%). Song and Guo (2012) reported that biochar usually has relatively high levels of CEC and that the pyrolysis temperature has little effect on the CEC of biochar. Specifically, biochar made from crop straw has greater levels of CEC than does biochar made from manure; these increased levels of CEC (Table 3) indicate that biochar may significantly enhance soil quality as well as plant growth as a result of enhancing nutrient retention and mitigating leaching.

6.2.5 Ash Content

Compared with high-temperature operations, organic materials may not fully carbonize and become ash; as a result, low-temperature pyrolysis techniques can produce biochars with higher ash contents. In addition, postprocessing steps such as washing can help reduce the ash content of biochar (or acid leaching). The biochar ash concentration is an important consideration in applications such as soil remediation and agriculture (Zhang et al., 2010). A high ash content can affect the pH and nutrient composition of biochar, possibly limiting its potential as a soil amendment or carbon sequestration technique. Therefore, to increase the functionality of biochar across diverse applications, it is necessary to understand and adjust its ash content.

From the above discussion, it is clear that the chemical properties of biochar should be modified to achieve optimal efficacy and tailor its production conditions (production feedstock). By understanding more about the chemical makeup of biochar, researchers and practitioners might be able to further fine-tune production settings to generate particular features that are best suited for desired outcomes. For example, by measuring the carbon concentration of a biochar, producers may adjust pyrolysis settings to reach an optimal degree of carbonization. The reactivity and incorporation of various environmental matrices depend on the distribution and type of surface chemistry available on biochar. By detailing these functional groups, researchers can predict how biochar will perform in soil ecosystems, for example, whether it is a good nutrient retainer or remediation agent. In addition, biochar pH and cation exchange capacity (CEC) are important parameters for enhancing soil handling. Modifying these traits through customized production techniques can increase nutrient retention, pH levels, and soil fertility, which are key to sustainable agricultural practices. In addition, the stability of biochar, which affects its longevity and carbon storage potential in different applications, is a function of either chemical composition or introduction conditions. Understanding the key drivers governing biochar stability is critically important when designing durable carbon sequestration strategies that can effectively mitigate climatic alterations.

6.3 Biological Properties

These properties are called the biological properties of biochar and include all the effects that occur when biochar is used on environmental flora. The effects of biochar on soil biological processes could be positive, negative, or neutral depending on its origin (e.g., feedstock type).

6.3.1 Microbial Habitat

Properties such as the presence of porous media, good surface area for nitrification, and moisture and nutrient retention provide the best home material to microorganisms. These properties facilitate the colonization and activation of microorganisms, as they generate small pressure cells in the inner biochar matrix. Owing to its porous nature, biochar has a range of niches and shelters where microorganisms can reside, protecting it from environmental stress, such as desiccation, extreme temperatures, or predation (Dong & Lin, 2018). This elaborate network of pores also provides space for nutrients, air, and water to circulate, which ultimately supports the growth (population) of microorganisms already present even in large lignite deposits (Karhu et al., 2011).

Biochar can also capture and retain organic matter and some microbial wastes to serve as a supply of carbon, which is needed by the formation of new colonies (Venkatesh et al., 2018). The biochar matrix enhances microbial diversity and abundance due to competition for resources between microbes, in addition to setting a scene with defined roles of various bacterial components. Consequently, the microbial habitat of biochar promotes a repertoire of dynamic microbial communities that are essential for soil health, nutrient cycling, and ecosystem function. To maximize the potential benefits of biochar in sustainable land, agriculture, and environmental cleanup practices, it is essential that we understand the role with which it acts as a home for microbes.

6.3.2 Diversity of Microbes

The use of biochar can significantly increase microbial diversity in soil ecosystems. Its porous structure and open space provide an opportunity for various ranges of microorganisms to colonize this niche, resulting in inhabited homes or small ecosystems. Some of the main influences of biochar on microbial diversity are as follows:

- *More diverse habitats*:

  Biochar is porous, enabling the creation of microhabitats with different moisture levels, oxygen statuses, and nutrient structures. Different niches foster the colonization of several species with very specific targets for certain kinds of microorganisms to adapt to and be present in those niches (Kookana et al., 2011).

- *Enhanced microbial community structure*:
  When biochar is applied, beneficial microorganisms that promote plant health and nutrient cycling (e.g., nitrogen-fixing bacteria), mycorrhizal fungi, or other soil microbes flourish. Additionally, biochar may have the potential to suppress the growth of pathogenic microbes, leading to a shift in community balance toward one that is more beneficial (Tong et al., 2014).

- *Stimulation of microbial activity*:
  Biochar can improve soil conditions for microbial growth and serve as a substrate that attracts a source of carbon (and energy). In this sense, higher microbial activity alters the soil structure by increasing the decomposition of organic matter and nitrogen cycling in the soil (Xie et al., 2014).

- *Long-term effects on microbial diversity*:
  Biochar can provide a stable home for microorganisms over longer timescales. The effect of biochar on microbial species diversity may be affected by weathering and its interaction with soil organic matter when applied to agricultural soils in the long term, potentially influencing ecosystem function downstream (Bailey et al., 2011).

Thus, biochar has the potential to dramatically alter microbial community diversity in soil ecosystems (and thereby plant productivity and sustainability) through direct nutrient-cycling interactions. Understanding these effects is necessary to optimize the positive impacts of biochar on agricultural and environmental management practices.

6.3.3 Biochar in SOM Dynamics

Biochar impacts soil organic matter (SOM) dynamics in multiple ways. Some are discussed below:

- *Biotic stability of organic carbon*:
  Biochar contains recalcitrant carbon, which is resistant to microbial degradation. They possibly add biochar to soil that acts as a stable "container" of organic carbon where SOC is slow in turning over and quite recalcitrant or largely stabilized for the remainder. They act as sinks that remove carbon from the atmosphere, play an important role in mitigating climate change, and help store soil carbon (Lammirato et al., 2011).

- *Soil organic matter protection*:
  Biochar acts as a shield that prevents microorganisms from degrading SOM. By providing a refuge for organic matter within its porous structure, biochar could serve to stabilize SOM pools in the longer term and increase soil carbon stocks by decreasing the accessibility of decaying biomass to decomposers (Zhao et al., 2013).

- *Soil aggregation*:
  Biochar induces microbes to secrete exudates that act as glues, which leads to soil aggregation. Sometimes, microbes serve as a physical shield to SOM by wrapping their biomass in aggregates. This protection reduces the susceptibility of SOM to erosion and decomposition, contributing to the preservation of soil carbon and soil fertility (Masto et al., 2013).
- *Enhanced microbial activity*:
  Biochar provides support for biosignificant soil microbial populations and the process of SOM turnover. Even though biochar is resistant to degradation by microorganisms, it can improve the recycling of soil organic matter (SOM) as a substrate for microbial growth. They could have decomposed more of the SOM in the vicinity of biochar, which might lead to spatially localized enhanced availability or rates for nutrient cycling and labile soil organic matter (SOM) turnover (Kumar et al., 2013).
- *Alterations in SOM composition*:
  Biochar amendments can alter the chemical composition of SOM by influencing microbial processes and organic matter transformations. This may result in changes in SOM quality, such as shifts in the relative abundance of different organic compounds or alterations in the stability of SOM fractions (Xie et al., 2014).

A diverse community of soil microorganisms can reside in biochar. In addition, more microorganisms may be retained in high-surface areas, hydrophobic soils amended with biochar, leading to greater activity and diversity. As microorganism adhesion to solids increases with increasing hydrophobicity of the surfaces, it is plausible that biochar may possess high reactivity toward bacteria. As a technique for providing life in nutrient-poor soils, biochar is quite effective. Carbonized biomass for soil amendment has been shown to be useful as a plant fertilizer. According to Venkatesh et al. (2018), ideal biochar has the potential to turn on soil microbes, leading to nutrient release and fertilization (Lehmann et al., 2018), as well as increasing the decadal pool of SOC. Biochar affects soil physics by reducing the bulk density and increasing the pore size and surface area, which are all advantageous for aerobic activities such as methane oxidation (Karhu et al., 2011; Sun et al., 2017; Lammirato et al., 2011). Kookana et al. (2011) and Tong et al. (2014) noted that biochar could serve as a residence for soilborne microorganisms in addition to their protection from natural enemies. Enzyme activity from biochar-amended soils is poorly described in the literature, and Bailey et al. (2011) reported inconsistent results, likely due to differences in parent material characteristics; the variation in soil types for enzyme types remains incomprehensible, which necessitates further systematic research on these fields. Masto et al. (2013) reported that a maximum dose of *Eichornia* biomass biochar (20 g kg^{-1}) resulted in the greatest increases in alkaline phosphatase (22.8%), acid phosphatase (32%), and dehydrogenase (21%) activities. A similar study by Kumar et al. (2013) revealed that 20 g kg^{-1} biochar from *P. hysterophorus* (L.) yielded the highest daily DHA, with a peak of 1071 mg TPF kg^{-1} 24 h^{-1}. Therefore, the biological properties of biochar have become a key factor in its ability to promote soil amendment, sustainable agriculture, and environmentally friendly management practices. Understanding these properties is essential for optimizing the use of biochar in various applications and ensuring its beneficial effects on soil health and ecosystem functioning.

7 Conclusion

Rising atmospheric CO_2 levels are a global concern of the twenty-first century, and long-term storage of carbon in soil is considered one of the solutions to this problem. Biochar has been increasingly discussed as a carbon storage strategy in soil since high-productivity Amazon soils were discovered to be partially created by burning biomass, called Terra preta. Biochar application can also increase the efficiency of water and fertilizer use, decrease N_2O emissions, and increase soil fertility, leading to better agricultural performance. Moreover, the potential for sequestering carbon also differs because various feedstocks and methods are used to produce different biochars. As a result, to address certain stakeholder requirements, identifying need-based biochar production methodologies is crucial. The feedstock and/or pyrolysis temperature affect the characteristics, C content, nutritional quality, and liming potential of biochar (outputs). For that reason, it is economical to characterize biochar according to its properties so that users can select the biochar that is best suited for soil application.

References

Agrafioti, E., Kotsopoulos, S., & Kalderis, D. (2013). Impact of pyrolysis temperature on biochar properties and its application in the remediation of contaminated soil. *Journal of Environmental Management, 128*, 130–136.

Ahmad, M., Rajapaksha, A. U., Lim, J. E., Zhang, M., Bolan, N., Mohan, D., & Ok, Y. S. (2014). Biochar as a sorbent for contaminant management in soil and water: A review. *Chemosphere, 99*, 19–33.

Amonette, J. E., & Joseph, S. (2012). Characteristics of biochar: Microchemical properties. In *Biochar for environmental management* (pp. 65–84). Routledge.

Antal, M. J., & Grønli, M. (2003). The art, science, and technology of charcoal production. *Industrial & Engineering Chemistry Research, 42*(8), 1619–1640.

Bailey, V. L., Fansler, S. J., Smith, J. L., & Bolton, H., Jr. (2011). Reconciling apparent variability in effects of biochar amendment on soil enzyme activities by assay optimization. *Soil Biology and Biochemistry, 43*(2), 296–301.

Baldock, J. A., & Smernik, R. J. (2002). Chemical composition and bioavailability of thermally altered Pinus resinosa (Red pine) wood. *Organic Geochemistry, 33*(9), 1093–1109.

Barrett, E. P., Joyner, L. G., & Halenda, P. P. (1951). The determination of pore volume and area distributions in porous substances. I. Computations from nitrogen isotherms. *Journal of the American Chemical Society, 73*(1), 373–380.

Basso, A. S., Caires, E. F., Gava, J. L., et al. (2013). Biochar influences soil water retention and nutrient availability. *Journal of Soil Science and Environmental Management, 4*(2), 57–65.

Benavente, J., & Fullana, A. (2015). Biochar: A tool for sustainable agriculture and environmental management. *Environmental Technology & Innovation, 4*, 1–14.

Bhandari, P. N., Kumar, A., Bellmer, D. D., & Huhnke, R. L. (2014). Synthesis and evaluation of biochar-derived catalysts for removal of toluene (model tar) from biomass-generated producer gas. *Renewable Energy, 66*, 346–353.

Biederman, L. A., & Harpole, W. S. (2013). Biochar and its effects on plant productivity and nutrient cycling: A meta-analysis. *GCB Bioenergy, 5*(2), 202–214.

Brewer, C. E., Schmidt-Rohr, K., Satrio, J. A., & Brown, R. C. (2009). Characterization of biochar from fast pyrolysis and gasification systems. *Environmental Progress & Sustainable Energy, 28*(3), 386–396.

Bridgwater, A. V. (2012). Review of fast pyrolysis of biomass and product upgrading. *Biomass and Bioenergy, 38*, 68–94.

Brunauer, S., Emmett, P. H., & Teller, E. (1938). Adsorption of gases in multimolecular layers. *Journal of the American Chemical Society, 60*(2), 309–319.

Cheah, S., Malone, S. C., & Feik, C. J. (2014). Speciation of sulfur in biochar produced from pyrolysis and gasification of oak and corn stover. *Environmental Science & Technology, 48*(15), 8474–8480.

Chen, K. S., Wang, H. K., Peng, Y. P., Wang, W. C., Chen, C. H., & Lai, C. H. (2008). Effects of open burning of rice straw on concentrations of atmospheric polycyclic aromatic hydrocarbons in Central Taiwan. *Journal of the Air & Waste Management Association, 58*(10), 1318–1327.

Cheng, C. H., Lehmann, J., Thies, J. E., & Burton, S. D. (2008). Stability of black carbon in soils across a climatic gradient. *Journal of Geophysical Research: Biogeosciences, 113*(G2), G02027.

Chia, C. H., Downie, A., & Munroe, P. (2015). Characteristics of biochar: Physical and structural properties. In *Biochar for environmental management* (pp. 89–109). Routledge.

Dong, X., & Lin, Q. (2018). Biochar effect on soil physical properties: A review. *Chinese Journal of Eco-Agriculture, 26*(12), 1846–1854.

Downie, A. (2011). *Biochar production and use: environmental risks and rewards*. Doctoral Thesis, School of Materials Science and Engineering, The University of New South Wales. pp. 309. https://doi.org/10.26190/unsworks/15191

Downie, A., Crosky, A., & Munroe, P. (2012). Physical properties of biochar. In *Biochar for environmental management* (pp. 45–64). Routledge.

Gómez-Barea, A., & Leckner, B. (2010). Modelling of biomass gasification in fluidized bed. *Progress in Energy and Combustion Science, 36*(4), 444–509.

Griffith, S. M., Banowetz, G. M., & Gady, D. (2013). Chemical characterization of chars developed from thermochemical treatment of Kentucky bluegrass seed screenings. *Chemosphere, 92*(10), 1275–1279.

Hansen, V., Müller-Stöver, D., Munkholm, L. J., Peltre, C., Hauggaard-Nielsen, H., & Jensen, L. S. (2016). The effect of straw and wood gasification biochar on carbon sequestration, selected soil fertility indicators and functional groups in soil: An incubation study. *Geoderma, 269*, 99–107.

Huggins, T. M., Pietron, J. J., Wang, H., Ren, Z. J., & Biffinger, J. C. (2015). Graphitic biochar as a cathode electrocatalyst support for microbial fuel cells. *Bioresource Technology, 195*, 147–153.

Islam, S. J. M., Mannan, M. A., Khaliq, Q. A., & Rahman, M. M. (2018). Growth and yield response of maize to rice husk biochar. *Australian Journal of Crop Science, 12*(12), 1813–1819.

Jeffery, S., Verheijen, F. G., van der Velde, M., & Bastos, A. C. (2011). A quantitative review of the effects of biochar application to soils on crop productivity using meta-analysis. *Agriculture, Ecosystems & Environment, 144*(1), 175–187.

Jien, S. H., & Wang, C. S. (2013). Effects of biochar on soil properties and erosion potential in a highly weathered soil. *CATENA, 110*, 225–233.

Jin, W., Zhang, X., & Wang, H. (2016). Production and characterization of biochar from agricultural residues by hydrothermal carbonization. *Biomass and Bioenergy, 85*, 72–80.

Jirka, S., & Tomlinson, T. (2014). *State of the biochar industry*. International Biochar Initiative.

Kamara, A., Mansaray, M. M., Kamara, A., & Sawyerr, P. A. (2014). Effects of biochar derived from maize stover and rice straw on the early growth of their seedlings. *American Journal of Agriculture and Forestry, 2*(5), 232–236.

Kambo, H. S., & Dutta, A. (2015). A comparative review of biochar and hydrochar in terms of production, physico-chemical properties and applications. *Renewable and Sustainable Energy Reviews, 45*, 359–378.

Karhu, K., Mattila, T., Bergström, I., & Regina, K. (2011). Biochar addition to agricultural soil increased CH4 uptake and water holding capacity – Results from a short-term pilot field study. *Agriculture, Ecosystems & Environment, 140*(1–2), 309–313.

Kim, M., Lee, Y., Park, J., Ryu, C., & Ohm, T. I. (2016). Partial oxidation of sewage sludge briquettes in an updraft fixed bed. *Waste Management, 49*, 204–211.

Kinney, T. J., Masiello, C. A., Dugan, B., Hockaday, W. C., Dean, M. R., Zygourakis, K., & Barnes, R. T. (2012). Hydrologic properties of biochars produced at different temperatures. *Biomass and Bioenergy, 41*, 34–43.

Kookana, R. S., Sarmah, A. K., Van Zwieten, L., Krull, E., & Singh, B. (2011). Biochar application to soil: Agronomic and environmental benefits and unintended consequences. *Advances in Agronomy, 112*, 103–143.

Kumar, S., Masto, R. E., Ram, L. C., Sarkar, P., George, J., & Selvi, V. A. (2013). Biochar preparation from Parthenium hysterophorus and its potential use in soil application. *Ecological Engineering, 55*, 67–72.

Laghari, M., Hu, Z., Mirjat, M. S., Xiao, B., Tagar, A. A., & Hu, M. (2016a). Fast pyrolysis biochar from sawdust improves the quality of desert soils and enhances plant growth. *Journal of the Science of Food and Agriculture, 96*(1), 199–206.

Laghari, M., Naidu, R., Xiao, B., Hu, Z., Mirjat, M. S., Hu, M., & Fazal, S. (2016b). Recent developments in biochar as an effective tool for agricultural soil management: A review. *Journal of the Science of Food and Agriculture, 96*(15), 4840–4849.

Laird, D., Fleming, P., Wang, B., Horton, R., & Karlen, D. (2010). Biochar impact on nutrient leaching from a Midwestern agricultural soil. *Geoderma, 158*(3–4), 436–442.

Lammirato, C., Miltner, A., & Kaestner, M. (2011). Effects of wood char and activated carbon on the hydrolysis of cellobiose by β-glucosidase from *Aspergillus niger*. *Soil Biology and Biochemistry, 43*(9), 1936–1942.

Lehmann, J., Rillig, M. C., Thies, J., Masiello, C. A., Hockaday, W. C. & Crowley, D. (2011). Biochar effects on soil biota–a review. *Soil biology and biochemistry, 43*(9), 1812–1836. https://doi.org/10.1016/j.soilbio.2011.04.022

Lehmann, J., Abiven, S., Azzi, E., Fang, Y., Singh, B.P., Sohi, S., Sundberg, C., Woolf, D. & Zimmerman, A. R. (2024). Persistence of biochar: Mechanisms, measurements, predictions. In Biochar for environmental management (pp. 277–311). Routledge.

Lehmann, K., Schaefer, S., Babin, D., Köhne, J. M., Schlüter, S., Smalla, K., Vogel, H. J. & Totsche, K.U. (2018). Selective transport and retention of organic matter and bacteria shapes initial pedogenesis in artificial soil-a two-layer column study. *Geoderma, 325*, 37–48. https://doi.org/10.1016/j.geoderma.2018.03.016

Lehmann, J. & Joseph, S. (2015). Biochar for environmental management: an introduction. In Biochar for environmental management (pp. 1-13). Routledge. Doi: 10.4324/9781003297673-1

Lehmann, J., & Joseph, S. (Eds.). (2024). *Biochar for environmental management: Science, technology and implementation*. Taylor & Francis.

Lehmann, J., Rillig, M. C., Thies, J., Masiello, C. A., Hockaday, W. C., & Crowley, D. (2011). Biochar effects on soil biota – A review. *Soil Biology and Biochemistry, 43*(9), 1812–1836. https://doi.org/10.1016/j.soilbio.2011.04.022

Lv, P. M., Xiong, Z. H., Chang, J., Wu, C. Z., Chen, Y., & Zhu, J. X. (2004). An experimental study on biomass air-steam gasification in a fluidized bed. *Bioresource Technology, 95*(1), 95–101.

Marks, E. A., Mattana, S., Alcañiz, J. M., Pérez-Herrero, E., & Domene, X. (2016). Gasifier biochar effects on nutrient availability, organic matter mineralization, and soil fauna activity in a multiyear Mediterranean trial. *Agriculture, Ecosystems & Environment, 215*, 30–39.

Martin, J. P., Gerard, J. M., Leal, A., Rojas, J., Pinto, S., & Valderrama, J. (1999). The use of methylene green for evaluating the surface area and porosity of adsorbents. *Journal of Chemical Technology & Biotechnology, 74*(3), 279–286.

Masiello, C. A. (2004). New directions in black carbon organic geochemistry. *Marine Chemistry, 92*(1–4), 201–213.

Masto, R. E., Ansari, M. A., George, J., Selvi, V. A., & Ram, L. C. (2013). Coapplication of biochar and lignite fly ash on soil nutrients and biological parameters at different crop growth stages of *Zea mays*. *Ecological Engineering, 58*, 314–322.

Mohan, D., Pittman, C. U., Jr., & Steele, P. H. (2006). Pyrolysis of wood/biomass for bio-oil: A critical review. *Energy & Fuels, 20*(3), 848–889.

Mohan, D., Sarswat, A., Ok, Y. S., & Pittman, C. U., Jr. (2014). Organic and inorganic contaminants removal from water with biochar, a renewable, low cost and sustainable adsorbent – A critical review. *Bioresource Technology, 160*, 191–202.

Mukherjee, A., Zimmerman, A. R., & Harris, W. (2011). Surface chemistry variations among a series of laboratory-produced biochars. *Geoderma, 163*(3–4), 247–255.

Novak, J. M., Lima, I., Xing, B., Gaskin, J. W., Steiner, C., Das, K. C., & Schomberg, H. (2009). Characterization of designer biochar produced at different temperatures and their effects on a loamy sand. *Annals of Environmental Science, 3*, 195–206.

Nunoura, T., Nakazawa, T., & Kato, M. (2006). Flash carbonization of biomass: A review on process and applications. *Fuel Processing Technology, 87*(6), 509–517.

O'Callaghan, M., John, R., & McCann, M. (2015). Color change of biochar in the soil and its implications for soil management. *Journal of Soil Science and Environmental Management, 6*(1), 1–8.

Ojeda, G., Mattana, S., Àvila, A., Alcañiz, J. M., Volkmann, M., & Bachmann, J. (2015). Is soil–water functions affected by biochar application? *Geoderma, 249*, 1–11.

Patterson, J. C., Wright, K. J., & Rees, D. E. (2013). Biochar density and its implications for soil amendment and carbon sequestration. *Environmental Science & Technology, 47*(12), 6511–6520.

Rahman, M. M. (2013). Nutrient-use and carbon-sequestration efficiencies in soils from different organic wastes in rice and tomato cultivation. *Communications in Soil Science and Plant Analysis, 44*(9), 1457–1471.

Rahman, F., Rahman, M. M., Rahman, G. M., Saleque, M. A., Hossain, A. S., & Miah, M. G. (2016). Effect of organic and inorganic fertilizers and rice straw on carbon sequestration and soil fertility under a rice–rice cropping pattern. *Carbon Management, 7*(1–2), 41–53.

Román, S., Libra, J., Berge, N., Sabio, E., Ro, K., Li, L., & Bae, S. (2018). Hydrothermal carbonization: Modelling, final properties design and applications: A review. *Energies, 11*(1), 216.

Rosenstock, T. S., Lamanna, C., Chesterman, S., Bell, P., Arslan, A., Richards, M. B., & Zhou, W. (2016). *The scientific basis of climate-smart agriculture: A systematic review protocol* (CCAFS Working Paper No. 138). CGIAR Research Program on Climate Change, Agriculture and Food Security (CCAFS).

Shackley, S., Carter, S., Knowles, T., Middelink, E., Haefele, S., Sohi, S., & Haszeldine, S. (2012). Sustainable gasification–biochar systems? A case-study of rice-husk gasification in Cambodia, Part I: Context, chemical properties, environmental and health and safety issues. *Energy Policy, 42*, 49–58.

Shackley, S., Ruysschaert, G., Zwart, K., & Glaser, B. (2016). *Biochar in European soils and agriculture. Science and practice*. Taylor & Francis.

Shen, Y., Linville, J. L., Ignacio-de Leon, P. A. A., Schoene, R. P., & Urgun-Demirtas, M. (2016). Towards a sustainable paradigm of waste-to-energy process: Enhanced anaerobic digestion of sludge with woody biochar. *Journal of Cleaner Production, 135*, 1054–1064.

Sing, K. S. (1985). Reporting physisorption data for gas/solid systems with special reference to the determination of surface area and porosity (Recommendations 1984). *Pure and Applied Chemistry, 57*(4), 603–619.

Singh, B., Singh, B. P., & Cowie, A. L. (2010). Characterization and evaluation of biochars for their application as a soil amendment. *Soil Research, 48*(7), 516–525.

Song, W. & Guo, M. (2012). Quality variations of poultry litter biochar generated at different pyrolysis temperatures. *Journal of Analytical and Applied Pyrolysis, 94*, 138–145. https://doi.org/10.1016/j.jaap.2011.11.018

Sun, J., He, F., Pan, Y., & Zhang, Z. (2017). Effects of pyrolysis temperature and residence time on physicochemical properties of different biochar types. *Acta Agriculturae Scandinavica, Section B—Soil & Plant Science, 67*(1), 12–22.
Taupe, N. C., Lynch, D., Wnetrzak, R., Kwapinska, M., Kwapinski, W., & Leahy, J. J. (2016). Updraft gasification of poultry litter at farm-scale – A case study. *Waste Management, 50*, 324–333.
Tipayarom, D., & Oanh, N. K. (2007). Effects from open rice straw burning emission on air quality in the Bangkok Metropolitan Region. *Science Asia, 33*(3), 339–345.
Tong, H., Hu, M., Li, F. B., Liu, C. S., & Chen, M. J. (2014). Biochar enhances the microbial and chemical transformation of pentachlorophenol in paddy soil. *Soil Biology and Biochemistry, 70*, 142–150.
Van Zwieten, L., Kimber, S., Morris, S., Chan, K. Y., Downie, A., Rust, J., & Cowie, A. (2010). Effects of biochar from slow pyrolysis of papermill waste on agronomic performance and soil fertility. *Plant and Soil, 327*, 235–246.
Venkatesh, G., Gopinath, K. A., Reddy, K. S., Reddy, B. S., Prasad, J. V. N. S., Rao, G. R., & Venkateswarlu, B. (2018). *Biochar production and its use in rainfed agriculture: Experiences from CRIDA*. National Initiative on Climate Resilient Agriculture, ICAR – Central Research Institute for Dryland Agriculture.
Wade, T., Zhang, J., & Goodall, J. (2006). Torrefaction of biomass as a precursor to biochar production: Impact on yield and quality. *Energy, 31*(4), 479–487.
Wu, W., Yang, M., Feng, Q., McGrouther, K., Wang, H., Lu, H., & Chen, Y. (2012). Chemical characterization of rice straw-derived biochar for soil amendment. *Biomass and Bioenergy, 47*, 268–276.
Xie, J., et al. (2014). Effects of biochar on soil microbial activity and community structure in a reclaimed mined land. *Soil Biology and Biochemistry, 78*, 275–283.
Yang, X., Xu, J., Zhang, Z., et al. (2006). Effects of biochar on soil properties and plant growth: A review. *Journal of Soil and Water Conservation, 61*(2), 84–92.
You, S., Ok, Y. S., Chen, S. S., Tsang, D. C., Kwon, E. E., Lee, J., & Wang, C. H. (2017). A critical review on sustainable biochar system through gasification: Energy and environmental applications. *Bioresource Technology, 246*, 242–253.
Zhang, L., Xu, C. C., & Champagne, P. (2010). Overview of recent advances in the thermochemical conversion of biomass. *Energy Conversion and Management, 51*(5), 969–982.
Zhao, L., Cao, X., Mašek, O., & Zimmerman, A. (2013). Heterogeneity of biochar properties as a function of feedstock source and production temperature. *Journal of Hazardous Materials, 256*, 1–9.
Zimmerman, A. R. (2010). Abiotic and microbial oxidation of laboratory-produced black carbon (biochar). *Environmental Science & Technology, 44*(4), 1295–1301.

Household Kitchen Waste: A Resource That Needs to Be Tapped

Urvashi Lama, Archita Dey, Anuradha Mahato, Debadrita Das, Sharmistha Majumder, Swetanjana Ghosh, Antara Das, Barun Kumar Thakur, Markkandan Ganesan, and Tarit Roychowdhury

Abstract

Household kitchen waste (HKW) presents a significant challenge due to its volume and decomposition tendencies, contributing to environmental pollution and greenhouse gas emissions. By converting this waste into biochar through controlled pyrolysis, not only can its detrimental impacts be mitigated, but its latent potential as a resource can also be unlocked. This chapter offers a succinct examination of the issue surrounding HKW and introduces the transformative concept of harnessing this waste as a valuable resource through biochar production. It further underscores the potential of biochar as a sustainable solution to the dual challenges of HKW management and environmental stewardship. By transforming waste into a valuable resource, kitchen waste biochar (KWB) not only addresses waste disposal issues but also contributes positively to agricultural productivity, soil health, and climate change mitigation. This chapter aims to explore these facets in greater detail, highlighting biochar's role as a cornerstone

U. Lama
School of Environmental Studies, Jadavpur University, Kolkata, India

Department of Botany, Sovarani Memorial College, Howrah, India

A. Dey · D. Das · S. Majumder · S. Ghosh · T. Roychowdhury (✉)
School of Environmental Studies, Jadavpur University, Kolkata, India

A. Mahato · M. Ganesan
Plant Cell Engineering Laboratory, Department of Life Sciences, Presidency University, Kolkata, India

A. Das
School of Environmental Studies, Jadavpur University, Kolkata, India

Department of Economics, FLAME University, Pune, Maharashtra, India

B. K. Thakur
Department of Economics, FLAME University, Pune, Maharashtra, India

S. Singh, S. Srivastava (eds.), *Biochar: A Precious Resource from Biological Waste*,
https://doi.org/10.1007/978-981-95-0425-1_5

in the sustainable management of kitchen waste (KW) and environmental conservation efforts.

Keywords

Household kitchen waste · Kitchen waste biochar · Waste management · Mitigation · Sustainable solution

1 Introduction

1.1 Brief Overview of the Problem of Household Kitchen Waste

Kitchen waste (KW) is a type of municipal solid waste which includes any leftover food scraps, peels, or organic materials generated during food preparation and consumption (Gui et al., 2021). Nowadays, KW is a critical environmental issue that is increasing significantly day by day with the development of the economic situation (Mokhtari et al., 2019; Singh, 2019; Zhang et al., 2021). In India, a huge amount of KW is generated, i.e., approx. 65 mt yr.$^{-1}$, which consists of greater than 50% of the total solid waste (Dora et al., 2021). As per the report of Food and Agriculture Organization (FAO) (2019), food wastage is 1.3 billion tons annually, which will increase to 2 billion tons by 2025 with an 11% annual growth rate, generating many societal, economic, and environmental issues (Dong et al., 2023). Along with that, very harmful greenhouse gases (i.e., methane and carbon dioxide) are produced by their disposal in landfills, which contributes to climate change and global warming (Kaur et al., 2020). Besides, it has been reported that KW is one of the major resources of the heavy metals such as Hg, Ni, Cd, Cu, Zn, and Pb in the soil. The heavy metals escaped into the soil by the process of anaerobic digestion and incineration treatment of kitchen waste (Long et al., 2011; Ke et al., 2017). Hence, to manage this KW problem over the world, affordable and eco-friendly technologies are required. Mostly, incineration and landfilling techniques are used for treating municipal solid waste (Bashir et al., 2018; Chen et al., 2019). However, transportation and incineration technology for KW are not suitable due to its high-water content (Shen et al., 2020; Wang et al., 2019). Besides, landfilling process also emits methane into the atmosphere and leached into the soil (De Clercq et al., 2016).

1.2 Utilizing Kitchen Waste Biochar as a Resource

The problem of KW can be solved by the development of its sustainable disposal and treatment technology. Household kitchen waste can be used as a resource to reduce greenhouse gas emissions and energy consumption which can enhance sustainable development (Pandey et al., 2009; Wainaina et al., 2020). Mostly, kitchen waste is used as a source of biochar production, which has several technical advantages like wide sources of raw materials, large-scale promotion, affordability, and bioecological safety. Kitchen waste is a good source for biochar production, which has several technical advantages like wide sources of raw materials, large-scale promotion, affordability, and bioecological safety (Kamau et al., 2019). To enhance the kitchen waste treatment, there are mainly three types of fundamental techniques—(i) thermal conversion (incineration, pyrolysis, and gasification), (ii) biochemical conversion (compost, vermicompost, and anaerobic digestion), and (iii) landfilling (Kalyani & Pandey, 2014) as represented in Fig. 1.

In recent years, biochar formation from waste materials has played a significant role in reducing environmental hazards and acting as a waste management resource (Downie et al., 2012; Ahmad et al., 2017). Biochar is derived from the organic matter of KW via pyrolysis methods in the absence of oxygen, which makes it highly aromatic in structure and a very stable solid product (Cao & Harris, 2010). The organic matter, present in KW, plays a crucial role in producing biochar. Biochar is highly porous and contains numerous functional groups like carbonyl, phenolic hydroxyl, carboxyl, pyrone, anhydride and pyrone groups (Li et al., 2011; Cha et al., 2016; Tripathi et al., 2016). Characterized by its high surface area and ability to retain water and nutrients, biochar serves multiple beneficial purposes in agriculture. Apart from these, biochar shows chemical binding properties, linked with nitrogen-containing functional groups -NH_2 and -NH and atmospheric CO_2 to form carbamates (Chatterjee et al., 2018; Shafawi et al., 2021). The functional groups found in biochar contribute in lowering CO_2 and heavy metal levels in the soil and transforming it into a material that is more environmentally friendly (Genuino et al., 2018).

Currently, biochar is applied in several fields. It has been used as an effective tool in agriculture, soil amendment (Ahmad et al., 2017), environmental remediation (Xu et al., 2021), and renewable energy production (Kumar et al., 2021). It enhances soil fertility by improving water retention, nutrient availability, and microbial

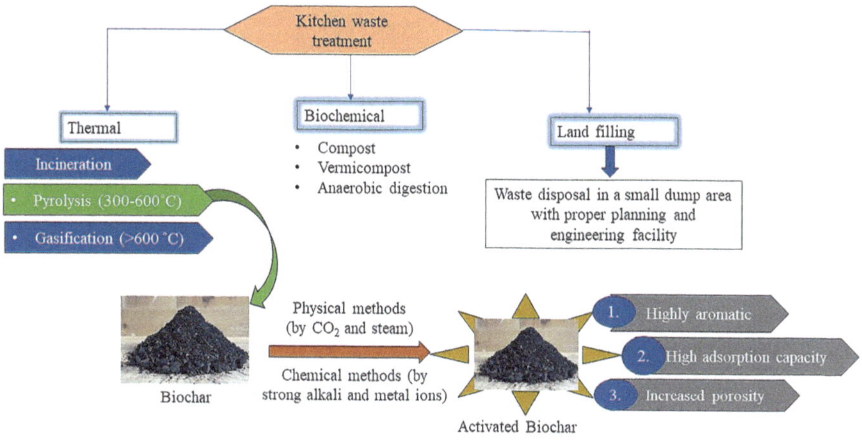

Fig. 1 Some kitchen waste treatment processes

activity, thereby boosting crop productivity. Moreover, biochar acts as a long-term carbon sink, sequestering carbon dioxide from the atmosphere and mitigating climate change impacts. The applications of biochar extend beyond agriculture to include soil remediation, where it helps to mitigate pollutants and restore degraded soils. Its introduction into contaminated soils can immobilize heavy metals and other harmful substances, rendering them less harmful to the environment and human health.

2 Understanding Household Kitchen Waste

2.1 Household Kitchen Waste Around the World

Uneaten food and food preparation residues (rice, vegetables, fruits, meats, etc.) are the major content of KW within a household (Hafid et al., 2011). Huge amount of KW is produced annually. Based on the references of Kumar et al. (2021), KW generation (million tons/year) recorded in several countries (Girotto et al., 2015; Bellemare et al., 2017; Joshi & Visvanathan, 2019; Shoar et al., 2019; Wang et al., 2019; Tian et al., 2020) is shown in Fig. 2.

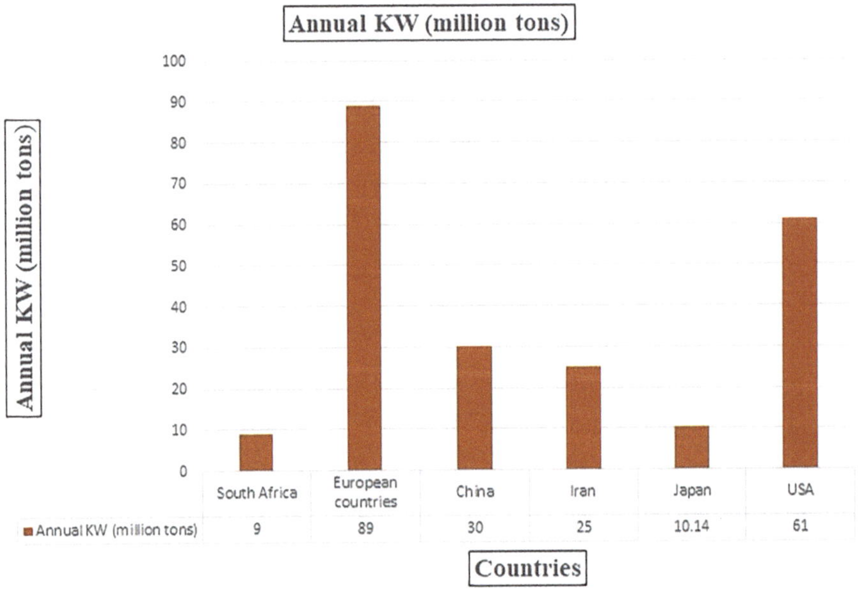

Fig. 2 Annual kitchen waste generation (million/tons) in several countries (Kumar et al., 2021)

2.2 Quantity and Composition of Kitchen Waste

Kitchen waste within household can be of different types depending on its source and nature. In India, municipal solid waste has been categorized into 40–60% compostable, 30–50% inert waste, and 10–30% recyclable (Joshi & Ahmed, 2016). Kitchen waste may be broadly categorized into following types, like organic waste (food scraps, leaf and small branches, etc.), non-organic waste (plastics, aluminum foil, glass jars, etc.), recyclable waste (glass bottles, plastic containers and metal cans, etc.), and hazardous waste (batteries, expired medicines and empty chemical bottles, etc.). To predict the specific composition of KW is very difficult due to the heterogenous distribution of organic materials (Cesaro & Belgiorno, 2014). Vavouraki et al. (2013) characterized the KW which consists mainly of carbohydrate (60%), protein (20%), and lipid (10%). Besides, KW is highly rich in moisture and humidity with high calorific value. The pH value and moisture content of the KW showed a range from pH 4 to 6 and 70 to 80% (w/w), respectively (Silva et al., 2018). Total moisture content and other valuable materials are altered with the changing of pH value of kitchen waste (Kiran & Prasad, 2019).

Kitchen waste typically exhibits a substantial moisture content ranging from 75.09% to 86.90%. It also contains a high mass percentage of total carbon (TC), reaching up to 65.1% (Chen et al., 2019). Raw materials rich in carbon are favorable for biomass production. Additionally, KW includes significant fractions of nitrogen (N) and sulfur (S). Total nitrogen (TN) levels in range from 1.82% to 4.27%, while S content in KW analyzed by Li et al. (2016) was found to be 0.29%. As a result, gases emitted from KW are also notably enriched with N and S elements due to their high concentrations in KW (Yu et al., 2022). Unlike agricultural and forestry waste, KW typically exhibits higher moisture content, necessitating more energy-intensive drying processes (Sindhu et al., 2019). The combination of high moisture, organic matter, and mineral salts makes KW highly perishable. Nevertheless, as a biomass raw material rich in N and S, KW holds promise for producing high-performance catalytic materials without the need for these additional elements (Xing et al., 2020).

2.3 Environmental and Social Impacts of Kitchen Waste Disposal

The disposal of KW exerts significant environmental and social impacts, intricately linked to the escalating global waste generation crisis. Disposing of kitchen waste in the environment poses significant hazards due to the offensive odors and the emission of toxic gases, which result from the extensive microbiome present in the waste materials (Moon et al., 2009). Environmentally, the mismanagement of kitchen waste contributes to greenhouse gas emissions, primarily methane, from anaerobic decomposition in landfills, exacerbating climate change (EPA, 2020). Leachate from improperly disposed kitchen waste contaminates soil and groundwater, posing substantial risks to ecosystems and human health (Kumar et al., 2021). Furthermore, the inefficient recycling and recovery of kitchen waste materials lead to the depletion of natural resources, undermining sustainability efforts. Socially, the improper disposal of kitchen waste has far-reaching implications. In urban areas, inadequate waste management systems result in the accumulation of kitchen waste in public spaces, fostering unsanitary conditions that can lead to the proliferation of disease vectors and adverse public health outcomes (World Health Organization, 2018). Communities, particularly in low-income regions, bear the consequences of these health impacts, highlighting an inequity in waste management practices (Kwenda et al., 2022). Additionally, the odor and unsightly appearance of accumulated waste degrade the quality of life and reduce property values, further disadvantaging already marginalized communities.

3 Benefits and Applications of Kitchen Waste

3.1 As an Adsorption Material

The utilization of KW for developing materials with superior adsorption capabilities has garnered significant attention in recent research (Han et al., 2021). Several studies have specifically focused on converting KW into effective adsorption materials to address environmental pollution challenges. For instance, biochar derived from KW effectively reduced the toxicity risk of heavy metals like Cd (II) in water, achieving notable adsorption capacities of 49.0 mg g^{-1}, under pyrolysis conditions, particularly at 500 °C (Xu et al., 2021). The enhanced porosity observed in the biochar surface post-pyrolysis contributed significantly to its adsorption performance. Furthermore, efforts have been made to develop magnetic KW biochar, facilitating easy separation and potential reuse in water treatment applications. Zhang XinWang et al. (2019) employed high-temperature pyrolysis and coprecipitation techniques to synthesize KW for both natural and magnetic KWB. The adsorbent exhibited a removal efficiency of 97.94% for methylene blue from water, with the magnetic KW biochar demonstrating a methylene blue removal capacity and adsorption amount of 9.2 mg g^{-1}. The natural KW biochar demonstrated a removal efficiency of 79.40% and an adsorption capacity of 7.6 mg g^{-1} for methylene blue (Zhang XinWang et al., 2019). Kitchen waste biochar obtained from pyrolysis at 500 °C was employed successfully to remediate arsenic (As) contaminated soil, demonstrating effective removal with an adsorption capacity of up to 11.3 mg g^{-1}. The mechanism underlying KWB's adsorption of As in soil was analyzed, revealing processes such as electrostatic attraction, physisorption, ionic exchange, and hydrogen bonding (Kumar et al., 2021). KWB significantly decreased the extractable levels of Pb (22.61%) and Cd (71.01%), surpassing the reductions achieved by cornstalks (Pb-18.54% and Cd-64.35%) and peanut shells (Pb-3.28% and Cd-60.25%). This inhibition effectively limited the absorption of Pb and Cd by *Ipomoea aquatica* throughout its growth cycle (Xu et al., 2020).

Overall, biochar derived from KW presents a sustainable solution for mitigating both water and soil pollution, showcasing its potential as a versatile adsorbent material in environmental remediation strategies. These advancements underscore its role in transforming KW into a valuable resource for addressing contemporary environmental challenges.

3.2 As a Soil Amendment for Fertility Enhancement

Biochar, a carbon-rich material derived from biomass pyrolysis, has garnered attention for its potential to enhance soil quality and sequester carbon dioxide from the atmosphere (Lehmann et al., 2011). Biochar application significantly increases soil carbon storage, thereby improving soil fertility. This is attributed to biochar's ability to serve as a carbon source, fostering microbial activity (Steinbeiss et al., 2009). Biochar derived from empty fruit bunches at a pyrolysis temperature of 400 °C, yielded 50.60% with a carbon content of 31.89% and a pH of 10.88 (Aziz et al., 2020). This study underscores biochar's versatility in influencing soil pH and carbon content, particularly beneficial for acidic soils. Yao et al. (2018) investigated the use of sugarcane bagasse-derived biochar as a soil amendment, highlighting its efficacy in nutrient retention. Such findings are pivotal as they underscore biochar's role in sustainable agriculture by improving soil nutrient availability and water retention. In summation, the amalgamated findings from these studies highlight the manifold advantages of biochar in agricultural contexts, encompassing enhancements in soil fertility, stimulation of microbial activity, carbon sequestration, and bolstered nutrient retention. These outcomes substantiate the imperative for continued investigation and widespread adoption of biochar as a pivotal strategy to confront contemporary agricultural challenges and advance efforts in climate change mitigation.

3.3 As an Energy Material

In the production process of KW-derived fuel, various valuable gases, liquids, and solids can be generated. Among these, particular attention has been given to gases and liquids such as biodiesel, methane, hydrogen, and fuel ethanol (Priyadarshi & Paul, 2018). Conversely, solid products from KW have received less emphasis as an energy resource. Nevertheless, KWB shows promising potential as a viable energy source. Jin utilized hydrothermal carbonization to convert KW into biochar, yielding a biochar with a calorific value as high as 30.18 MJ kg^{-1}, meeting the standards for "clean coal" and suitable for use as a coal energy fuel (Jin et al., 2014). The processed solids of KW have been identified as a high-quality energy fuel (Li et al., 2013). Additionally, the research demonstrated that biochar produced from KW retained a significant portion of the carbon, yielding a product comparable to high-value coal, which is both easily storable and transportable.

3.4 For Carbon Sequestration and Greenhouse Gas Mitigation

Moreover, biochar's application aligns with global efforts to mitigate climate change by sequestering carbon in soil over prolonged periods. This dual benefit of enhancing soil fertility while mitigating greenhouse gas emissions positions biochar as a promising tool in sustainable agricultural practices.

An overall application of KWB from KW for different purposes is represented in Fig. 3.

3.5 Emerging Resource Utilization Methods of Kitchen Waste: A New Perspective

Carbon quantum dots (CQDs) are nanoscale carbon-based materials known for their distinctive optical and electronic characteristics (Huang et al., 2021). These materials are typically smaller than 10 nanometers and possess a considerable surface area. CQDs have garnered considerable interest across diverse fields owing to their exceptional photoluminescence and chemical durability. They hold promise for applications in environmental monitoring (Wu et al., 2021a, b). Kitchen waste can serve as a viable raw material for producing carbon quantum dots. CQDs can be categorized into three main types: carbon nanodots, graphene quantum dots, and polymer dots (Fan et al., 2020). *Citrus sinensis* and *Citrus limon* peels were used as raw materials to produce biochar under controlled conditions of 180 °C for 2 hours in separate porcelain crucibles. Subsequently, the resulting biochar was processed into carbon quantum dots (CQDs), which exhibited the ability to detect Fe^{3+} ions. A

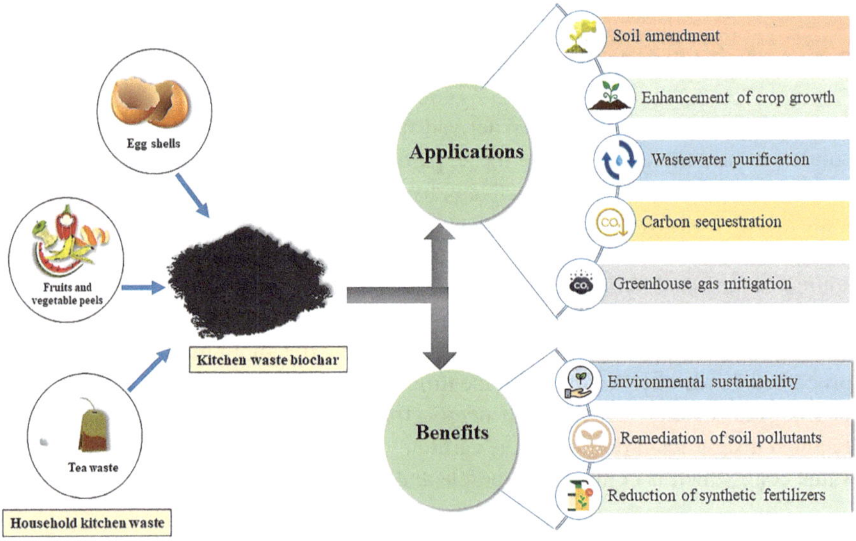

Fig. 3 Applications and benefits of kitchen waste biochar (KWB)

Table 1 Application of biochar from different wastes derived from household

Types of biochar	Temperature used (°C)	Usage	Reference
Peanut hull biochar	300–600	Soil amendment	Sathe et al. (2021)
Empty fruit bunch biochar	400–600	Soil amendment	Aziz et al. (2020)
Crab shell biochar	300–800	Adsorption	Dai et al. (2017)
Citrus sinensis peel biochar	180	Prepared into carbon quantum dots (CQDs)	Chatzimitakos et al. (2017)
Citrus limon peel biochar	180	Prepared into CQDs for detection of heavy metals from wastewater	Chatzimitakos et al. (2017)
Corn cob biochar	300–650	Adsorption	Eduah et al. (2020)
Peanut hulls biochar	400	Adsorption	Xu et al. (2021)
Black tea biochar	700, 900	Electrode materials	Gao et al. (2021)
Bean dregs biochar	400–900	Activated oxidant	Yu et al. (2022)
KWB	400	Adsorption of Pb and Cd	Xu et al. (2021)
KWB	180	Energy material	Li et al. (2013)
KWB	400–600	Energy material	Jin et al. (2014)

strong linear correlation was observed within the concentration ranges of 0.01–1.0 μM and 0.6–23.5 μM, indicating that Fe^{3+} could be accurately measured within these specified ranges (Chatzimitakos et al., 2017). Utilizing KW as a raw material for electrodes has introduced a novel approach for its utilization. Specifically, black tea has been explored as a type of KW for potassium ion storage electrodes. Gao et al. (2021) employed black tea to produce biochar, which serves as an environmentally friendly, cost-effective carbon electrode. This electrode demonstrated exceptional performance and maintained stability through repeated use. Table 1 highlights different uses of biochar obtained from different materials of kitchen waste.

4 Challenges and Considerations

The food waste from the household is highly susceptible to natural degradation due to its elevated moisture content. Additionally, its heterogeneous composition, fluctuating sugar levels, microbial contamination, and substantial production costs pose significant challenges to its commercialization into useful products. To address these issues, effective management strategies are essential.

Ensuring quality-controlled rapid logistics, proper handling, storage, and preservation, along with regulatory support and economic incentives, are vital for maintaining the quality of food waste throughout the supply chain. Nevertheless, the heterogeneous composition, high moisture content, variable sugar levels, microbial contamination, biodegradability, and high production costs of food waste pose obstacles to large-scale commercialization. Other impediments include a shortage of skilled labor, inadequate infrastructure, limited logistical resources, and the lack of a robust food waste supply chain.

5 Future Directions and Recommendations/Discussion

The findings revealed that increasing the pyrolysis temperature to 300–600 °C elevated the pH and ash content of biochar while reducing its yield, H/C, and N/C ratios. Notably, pyrolysis may enhance the remediation efficacy for Cd (II) and diminish the risk of heavy metals in KWB. Pyrolysis of KWB at 500 °C exhibited significant potential as an environmentally friendly, cost-effective, and efficient absorbent for the remediation of Cd (III)-contaminated water, considering environmental risk, remediation performance, and production cost. The pyrolyzed KWB when applied were proved to be a good phytoremediator in heavy metal stress and reduced methane emissions from ruminants and help to sequester carbon (Fig. 4). The production of biochar through pyrolysis provides an eco-friendly solution for treating KW. Pyrolyzed food waste-derived biochars can enhance soil water and nutrient retention and can be further processed into higher-value products such as activated carbons. The yield and characteristics of biochars are influenced by the process conditions and the specific type of food waste utilized.

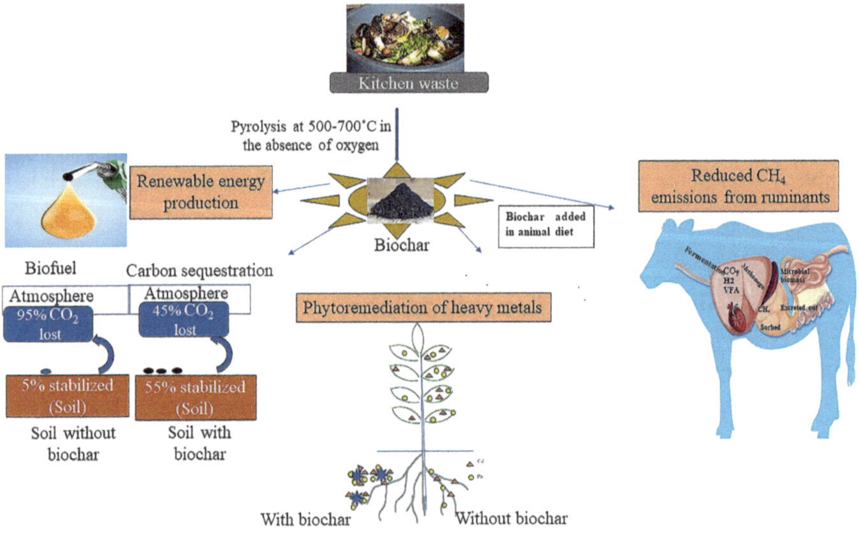

Fig. 4 Role of kitchen waste biochar (KWB) in environment

Recent applications have highlighted the benefits of employing food waste pyrolysis for generating value-added products, particularly biochars. However, several research areas warrant further investigation: (a) integrating the gas byproduct into the pyrolysis process to enhance both sustainability and economic viability, (b) identifying optimal feedstocks to achieve desired product qualities, (c) exploring the co-pyrolysis of food waste with various materials to improve yield and minimize unwanted byproducts, (d) evaluating microwave-assisted technology for converting mixed food waste into valuable products, and (e) optimizing process design and conditions (e.g., temperature, pressure, residence time) through modeling and experiments to produce biochars with specific characteristics for applications such as soil amendment and contaminant adsorption. Overcoming the technical challenges of pyrolyzing food waste to create value-added products will expand its use in waste management and the production of diverse materials, either for direct application or further processing (Elkhalifa et al., 2019). Efforts to mitigate these impacts require a multifaceted strategy that combines technological innovations with strong policy frameworks. Advanced composting and anaerobic digestion technologies present sustainable solutions by transforming kitchen waste into valuable bioproducts, such as biogas and organic fertilizers, thereby reducing environmental pollution and creating economic benefits (Sindhu et al., 2019). Legislative measures mandating segregation at the source, along with public awareness campaigns, are essential for promoting responsible waste disposal practices (Malinowsky et al., 2021). Together, these strategies can turn kitchen waste from an environmental burden into a valuable resource, aligning with global sustainability goals and promoting social equity.

6 Conclusion

As the global population continues to grow, it is expected to rise from 7.7 billion to 8 billion by 2030 (Nanda & Berruti, 2021). If this trend continues, the environment will face increased hazards from pollutants, kitchen waste, energy shortages, and other issues. Pan et al. (2021) reported that annual kitchen waste production is approximately 30 million tons. Improper disposal of this waste without treatment can lead to severe environmental pollution (Shi et al., 2020). To address this environmental issue, the organization should prioritize corporate social responsibility by focusing on the recycling of kitchen waste, waste disposal management, and utilizing it as a sustainable renewable resource. Consequently, proper disposal and treatment methods for kitchen waste are crucial for mitigating environmental pollution. Using kitchen waste as a biochar resource provides a dual environmental advantage: It reduces the impact of kitchen waste on the environment while turning it into a valuable resource. This conversion addresses environmental challenges and helps resolve further issues, effectively achieving the objective of using waste to manage waste (Kumar et al., 2021).

References

Ahmad, M., Lee, S. S., Lee, S. E., Al-Wabel, M. I., Tsang, D. C., & Ok, Y. S. (2017). Biochar-induced changes in soil properties affected immobilization/mobilization of metals/metalloids in contaminated soils. *Journal of Soils and Sediments, 17*, 717–730.

Aziz, S., Ali, M. I., Farooq, U., Jamal, A., Liu, F. J., He, H., Guo, H., Urynowicz, M., & Huang, Z. (2020). Enhanced bioremediation of diesel range hydrocarbons in soil using biochar made from organic wastes. *Environmental Monitoring and Assessment, 192*, 1–14.

Bashir, M. A., Khalid, M., Naveed, M., Ahmad, R., & Gao, B. (2018). Influence of feedstock and pyrolytic temperature of biochar on physico-chemical characteristics and sorption of chromium in tannery polluted soil. *International Journal of Agriculture and Biology, 20*, 2823–2834.

Bellemare, M. F., Çakir, M., Peterson, H. H., Novak, L., & Rudi, J. (2017). On the measurement of food waste. *American Journal of Agricultural Economics, 99*(5), 1148–1158.

Cao, X., & Harris, W. (2010). Properties of dairy-manure-derived biochar pertinent to its potential use in remediation. *Bioresource Technology, 101*(14), 5222–5228.

Cesaro, A., & Belgiorno, V. (2014). Pretreatment methods to improve anaerobic biodegradability of organic municipal solid waste fractions. *Chemical Engineering Journal, 240*, 24–37.

Cha, J. S., Park, S. H., Jung, S. C., Ryu, C., Jeon, J. K., Shin, M. C., & Park, Y. K. (2016). Production and utilization of biochar: A review. *Journal of Industrial and Engineering Chemistry, 40*, 1–15.

Chatterjee, R., Sajjadi, B., Mattern, D. L., Chen, W. Y., Zubatiuk, T., Leszczynska, D., Leszczynski, J., Egiebor, N. O., & Hammer, N. (2018). Ultrasound cavitation intensified amine functionalization: a feasible strategy for enhancing CO2 capture capacity of biochar. *Fuel, 225*, 287–298.

Chatzimitakos, T., Kasouni, A., Sygellou, L., Avgeropoulos, A., Troganis, A., & Stalikas, C. (2017). Two of a kind but different: Luminescent carbon quantum dots from Citrus peels for iron and tartrazine sensing and cell imaging. *Talanta, 175*, 305–312.

Chen, Y., Zhu, R., Jiang, Q., Sun, T., Li, M., Shi, J., Chai, H., Gu, L., Ai, H., & He, Q. (2019). Effects of green waste participation on the co-digestion of residual sludge and kitchen waste: A preliminary study. *Science of the Total Environment, 671*, 838–849.

Dai, L., Tan, F., Li, H., Zhu, N., He, M., Zhu, Q., Hu, G., Wang, L., & Zhao, J. (2017). Calcium-rich biochar from the pyrolysis of crab shell for phosphorus removal. *Journal of Environmental Management, 198*, 70–74.

De Clercq, D., Wen, Z., Fan, F., & Caicedo, L. (2016). Biomethane production potential from restaurant food waste in megacities and project level-bottlenecks: a case study in Beijing. *Renewable and Sustainable Energy Reviews, 59*, 1676–1685.

Dong, Z., Rene, E. R., Zhang, P., Hu, Q., & Ma, W. (2023). Design and preparation of carbon material catalyst modified with metal framework and sulfonate for biochar generation from low-temperature directional pyrolysis of kitchen waste: Mechanism and performance. *Bioresource Technology, 371*, 128616.

Dora, M., Biswas, S., Choudhary, S., Nayak, R., & Irani, Z. (2021). A system-wide interdisciplinary conceptual framework for food loss and waste mitigation strategies in the supply chain. *Industrial Marketing Management, 93*, 492–508.

Downie, A., Crosky, A., & Munroe, P. (2012). Physical properties of biochar. In *Biochar for environmental management* (pp. 45–64). Routledge.

Eduah, J. O., Nartey, E. K., Abekoe, M. K., & Andersen, M. N. (2020). Changes in phosphorus fractions in three tropical soils amended with corn cob and rice husk biochars. *Communications in Soil Science and Plant Analysis, 51*(10), 1331–1340.

Elkhalifa, S., Al-Ansari, T., Mackey, H. R., & McKay, G. (2019). Food waste to biochars through pyrolysis: A review. *Resources, Conservation and Recycling, 144*, 310–320.

EPA. (2020). *Greenhouse Gas Emissions, Overview of Greenhouse Gases*. Available online at: https://www.epa.gov/ghgemissions/overview-greenhouse-gases. Accessed 20 Nov 2020.

Fan, H., Zhang, M., Bhandari, B., & Yang, C. H. (2020). Food waste as a carbon source in carbon quantum dots technology and their applications in food safety detection. *Trends in Food Science & Technology, 95*, 86–96.

FAO. (2019). *The state of food and agriculture 2019: Moving forward on food loss and waste reduction*. UN.

Gao, Y., Ru, Q., Zheng, M., Pan, Z., Lei, T., Zhang, J., & An, Q. (2021). Recovery of kitchen biowaste from spent black tea as hierarchical biomorphic carbon electrodes for ultra-long lifespan potassium-ion storage. *Applied Surface Science, 555*, 149675.

Genuino, D. A. D., de Luna, M. D. G., & Capareda, S. C. (2018). Improving the surface properties of municipal solid waste-derived pyrolysis biochar by chemical and thermal activation: Optimization of process parameters and environmental application. *Waste Management, 72*, 255–264.

Girotto, F., Alibardi, L., & Cossu, R. (2015). Food waste generation and industrial uses: A review. *Waste Management, 45*, 32–41.

Gui, X., Li, Z., & Wang, Z. (2021). Kitchen waste hydrolysate enhances sewage treatment efficiency with different biological process compared with glucose. *Bioresource Technology, 341*, 125904.

Hafid, H. S., Rahman, A. A., Abd-Aziz, S., & Hassan, M. A. (2011). Enhancement of organic acids production from model kitchen waste via anaerobic digestion. *African Journal of Biotechnology, 10*(65), 14507–14515.

Han, H., Hou, J., Zhang, D., Zhong, Y., Chen, C., & Zhang, Q. (2021). External carbon source facilitates indirect Cr (VI) bioreduction process by anaerobic sludge produced from kitchen waste. *Sustainability, 13*(9), 4806.

Huang, X., Zhang, H., Zhao, J., Jiang, D., & Zhan, Q. (2021). Carbon quantum dot (CQD)-modified Bi3O4Br nanosheets possessing excellent photocatalytic activity under simulated sunlight. *Materials Science in Semiconductor Processing, 122*, 105489.

Jin, T., Yan, J., Michael, B., Moritz, M., Cao, X., & Lu, H. (2014). Experimental study on making bio-coal from food waste by hydrothermal carbonization. *Renewable Energy Resources, 32*(7).

Joshi, R., & Ahmed, S. (2016). Status and challenges of municipal solid waste management in India: A review. *Cogent Environmental Science, 2*(1), 1139434.

Joshi, P., & Visvanathan, C. (2019). Sustainable management practices of food waste in Asia: Technological and policy drivers. *Journal of Environmental Management, 247*, 538–550.

Kalyani, K. A., & Pandey, K. K. (2014). Waste to energy status in India: A short review. *Renewable and Sustainable Energy Reviews, 31*, 113–120.

Kamau, S., Karanja, N. K., Ayuke, F. O., & Lehmann, J. (2019). Short-term influence of biochar and fertilizer-biochar blends on soil nutrients, fauna and maize growth. *Biology and Fertility of Soils, 55*, 661–673.

Kaur, G., Johnravindar, D., & Wong, J. W. (2020). Enhanced volatile fatty acid degradation and methane production efficiency by biochar addition in food waste-sludge co-digestion: a step towards increased organic loading efficiency in co-digestion. *Bioresource Technology, 308*, 123250.

Ke, C., Ma, X., Tang, Y., Zheng, W., & Wu, Z. (2017). The volatilization of heavy metals during co-combustion of food waste and polyvinyl chloride in air and carbon dioxide/oxygen atmosphere. *Bioresource Technology, 244*, 1024–1030.

Kiran, B. R., & Prasad, M. N. V. (2019). Biochar and rice husk ash assisted phytoremediation potentials of Ricinus communis L. for lead-spiked soils. *Ecotoxicology and Environmental Safety, 183*, 109574.

Kumar, A., Bhattacharya, T., Shaikh, W. A., Roy, A., Mukherjee, S., & Kumar, M. (2021). Performance evaluation of crop residue and kitchen waste-derived biochar for eco-efficient removal of arsenic from soils of the Indo-Gangetic plain: A step towards sustainable pollution management. *Environmental Research, 200*, 111758.

Kwenda, P. R., Lagerwall, G., Eker, S., & Van Ruijven, B. (2022). A mini-review on the causes of poor household solid waste management in low-income developing countries: a case study of Urban Harare City, Zimbabwe. *The Journal of Solid Waste Technology and Management, 48*(3), 375–388.

Li, L. I., Liu, Y., Lu, Y. C., Liang, Z., Zhang, P., & Sun, H. (2011). Review on environmental effects and applications of biochar. *Environmental Chemistry, 30*(8), 1411–1421.

Li, L., Diederick, R., Flora, J. R., & Berge, N. D. (2013). Hydrothermal carbonization of food waste and associated packaging materials for energy source generation. *Waste Management, 33*(11), 2478–2492.
Li, Y., Jin, Y., & Li, J. (2016). Influence of thermal hydrolysis on composition characteristics of fatty acids in kitchen waste. *Energy, 102*, 139–147.
Long, Y. Y., Shen, D. S., Wang, H. T., Lu, W. J., & Zhao, Y. (2011). Heavy metal source analysis in municipal solid waste (MSW): case study on Cu and Zn. *Journal of Hazardous Materials, 186*(2–3), 1082–1087.
Malinowsky, C., Nadaleti, W., Debiasi, L. R., Moreira, A. J. G., Bayard, R., & de Castilhos Junior, A. B. (2021). Start-up phase optimization of two-phase anaerobic digestion of food waste: Effects of organic loading rate and hydraulic retention time. *Journal of Environmental Management, 296*, 113064.
Mokhtari, M., Heshmati Rafsanjani, A. A., & Shariatmadari, N. (2019). The effect of aging on the compressibility behavior and the physical properties of municipal solid wastes: a case study of Kahrizak landfill, Tehran. *Environmental Earth Sciences, 78*, 1–14.
Moon, H. C., Song, I. S., Kim, J. C., Shirai, Y., Lee, D. H., Kim, J. K., Chung, S. O., Kim, D. H., Oh, K. K., & Cho, Y. S. (2009). Enzymatic hydrolysis of food waste and ethanol fermentation. *International Journal of Energy Research, 33*(2), 164–172.
Nanda, S., & Berruti, F. (2021). A technical review of bioenergy and resource recovery from municipal solid waste. *Journal of Hazardous Materials, 403*, 123970.
Pan, F. D., Liu, S., Xu, Q. M., Chen, X. Y., & Cheng, J. S. (2021). Bioconversion of kitchen waste to surfactin via simultaneous enzymolysis and fermentation using mixed-culture of enzyme-producing fungi and Bacillus amyloliquefaciens HM618. *Biochemical Engineering Journal, 172*, 108036.
Pandey, M. C., Ashok Kumar, A. K., Patel, M., Palod, J., Singh, D. V., & Mondal, B. C. (2009). Effective utilization of kitchen waste using probiotics in pig under intensive system.
Priyadarshi, D., & Paul, K. K. (2018). Single phase blend: An advanced microwave process for improved quality low-cost biodiesel production from kitchen food waste. *Biochemical Engineering Journal, 137*, 273–283.
Sathe, P. S., Adivarekar, R. V., & Pandit, A. B. (2021). Valorization of peanut shell biochar for soil amendment. *Journal of Plant Nutrition, 45*(4), 503–521.
Shafawi, A. N., Mohamed, A. R., Lahijani, P., & Mohammadi, M. (2021). Recent advances in developing engineered biochar for CO2 capture: An insight into the biochar modification approaches. *Journal of Environmental Chemical Engineering, 9*(6), 106869.
Shen, Y., Song, S., Thian, B. W. Y., Fong, S. L., Ee, A. W. L., Arora, S., Ghosh, S., Li, S. F. Y., Tan, H. T. W., Dai, Y., & Wang, C. H. (2020). Impacts of biochar concentration on the growth performance of a leafy vegetable in a tropical city and its global warming potential. *Journal of Cleaner Production, 264*, 121678.
Shi, Y., Deng, Y., Wang, G., & Xu, J. (2020). Stackelberg equilibrium-based eco-economic approach for sustainable development of kitchen waste disposal with subsidy policy: A case study from China. *Energy, 196*, 117071.
Shoar, F. H., Abdi, R., Najafi, B., & Ardabili, S. F. (2019). The effect of thermochemical pre-treatment on biogas production efficiency from kitchen waste using a novel lab scale digester. *Renewable Energy Focus, 28*, 140–152.
Silva, F. M., Mahler, C. F., Oliveira, L. B., & Bassin, J. P. (2018). Hydrogen and methane production in a two-stage anaerobic digestion system by co-digestion of food waste, sewage sludge and glycerol. *Waste Management, 76*, 339–349.
Sindhu, R., Gnansounou, E., Rebello, S., Binod, P., Varjani, S., Thakur, I. S., Nair, R. B., & Pandey, A. (2019). Conversion of food and kitchen waste to value-added products. *Journal of Environmental Management, 241*, 619–630.
Singh, A. (2019). Managing the uncertainty problems of municipal solid waste disposal. *Journal of Environmental Management, 240*, 259–265.
Steinbeiss, S., Gleixner, G., & Antonietti, M. (2009). Effect of biochar amendment on soil carbon balance and soil microbial activity. *Soil Biology and Biochemistry, 41*(6), 1301–1310.

Tian, L., Zhang, L., Liu, Y., He, Y., Zhu, Y., Sun, R., Yi, S., & Xiang, J. (2020). Clean production of ethyl levulinate from kitchen waste. *Journal of Cleaner Production, 268*, 122296.

Tripathi, M., Sahu, J. N., & Ganesan, P. (2016). Effect of process parameters on production of biochar from biomass waste through pyrolysis: A review. *Renewable and Sustainable Energy Reviews, 55*, 467–481.

Vavouraki, A. I., Angelis, E. M., & Kornaros, M. (2013). Optimization of thermo-chemical hydrolysis of kitchen wastes. *Waste Management, 33*(3), 740–745.

Wainaina, S., Kisworini, A. D., Fanani, M., Wikandari, R., Millati, R., Niklasson, C., & Taherzadeh, M. J. (2020). Utilization of food waste-derived volatile fatty acids for production of edible Rhizopus oligosporus fungal biomass. *Bioresource Technology, 310*, 123444.

Wang, H., Xu, J., & Sheng, L. (2019). Study on the comprehensive utilization of city kitchen waste as a resource in China. *Energy, 173*, 263–277.

World Health Organization. (2018). Essential public health functions, health systems and health security: developing conceptual clarity and a WHO roadmap for action.

Wu, M., Hu, J., Shen, F., Huang, M., Zhao, L., Tian, D., Zhang, Y., Liu, Y., Zeng, Y., & Deng, S. (2021a). Conceptually integrating a multi-product strategy for the valorization of kitchen waste towards a more sustainable management. *Journal of Cleaner Production, 306*, 127292.

Wu, P., Zhou, C., Li, Y., Zhang, M., Tao, P., Liu, Q., & Cui, W. (2021b). Flower-like FeOOH hybridized with carbon quantum dots for efficient photo-Fenton degradation of organic pollutants. *Applied Surface Science, 540*, 148362.

Xing, B., Dong, J., Yang, G., Jiang, N., Liu, X., & Yuan, J. (2020). An insight into N, S-codoped activated carbon for the catalytic persulfate oxidation of organic pollutions in water: Effect of surface functionalization. *Applied Catalysis A: General, 602*, 117714.

Xu, C., Zhao, J., Yang, W., He, L., Wei, W., Tan, X., Wang, J., & Lin, A. (2020). Evaluation of biochar pyrolyzed from kitchen waste, corn straw, and peanut hulls on immobilization of Pb and Cd in contaminated soil. *Environmental Pollution, 261*, 114133.

Xu, C., Tan, X., Zhao, J., Cao, J., Ren, M., Xiao, Y., & Lin, A. (2021). Optimization of biochar production based on environmental risk and remediation performance: Take kitchen waste for example. *Journal of Hazardous Materials, 416*, 125785.

Yao, Y., Zhang, Y., Gao, B., Chen, R., & Wu, F. (2018). Removal of sulfamethoxazole (SMX) and sulfapyridine (SPY) from aqueous solutions by biochars derived from anaerobically digested bagasse. *Environmental Science and Pollution Research, 25*, 25659–25667.

Yu, D., Yu, Y., Tang, J., Li, X., Ke, C., & Yao, Z. (2022). Application fields of kitchen waste biochar and its prospects as catalytic material: A review. *Science of the Total Environment, 810*, 152171.

Zhang XinWang, Z. X., Li BaoPeng, L. B., Shen JianGuo, S. J., Zang ZhiLin, Z. Z., Ma XiaoHan, M. X., & Xin YanJun, X. Y. (2019). Research on the adsorption performance of kitchen waste biochar for methylene blue.

Zhang, X., Wang, J., Bao, S., & Zhang, X. (2021). Impact of internal conditions on the gas flow path in semi-aerobic landfill reactors. *Science of the Total Environment, 770*, 144673.

Lehmann, J., Rillig, M. C., Thies, J., Masiello, C. A., Hockaday, W. C., & Crowley, D. (2011). Biochar effects on soil biota–a review. *Soil biology and biochemistry, 43*(9), 1812–1836.

Agricultural Waste Biomass Management and Biochar Production

Ajay Shinde, Sankarshan Patwardhan, Rahul Kshirsagar, Farhad Jivani, Sarika Loni, Manju Tanwar, Abhishek Guldhe, and Penna Suprasanna

Abstract

Agricultural production globally generates substantial solid waste, leading to environmental pollution, economic losses, and health issues when poorly managed. Developing eco-friendly, cost-effective waste management technologies is essential, among which, an effective solution is the convertion of agricultural waste into value-added products like biochar. Biochar, a porous, carbon-rich material produced from organic sources like crop residues through pyrolysis, offers significant environmental benefits. It enhances soil fertility, aids in pollution control, and sequesters carbon, making it valuable in sustainable fuels, soil remediation, and bioremediation. This chapter explores biochar production from agricultural waste, focusing on pyrolysis processes and biochar's applications. Some properties of biochar, such as its high carbon content and oxygen-rich functional groups, make it ideal for improving soil health and addressing environmental challenges. Converting agricultural waste into biochar provides a sustainable solution for waste management while offering numerous environmental benefits.

Keywords

Agriculture · Waste · Biomass · Biochar · Pyrolysis · Fertilizer

A. Shinde · S. Patwardhan · R. Kshirsagar · F. Jivani · A. Guldhe · P. Suprasanna (✉)
Amity Institute of Biotechnology, Amity University Maharashtra, Mumbai, India

S. Loni
Amity Institute of Biotechnology, Amity University Maharashtra, Mumbai, India

Organic Recycling Systems Ltd., Mumbai, India

M. Tanwar
Organic Recycling Systems Ltd., Mumbai, India

S. Singh, S. Srivastava (eds.), *Biochar: A Precious Resource from Biological Waste*,
https://doi.org/10.1007/978-981-95-0425-1_6

1 Introduction

Agricultural waste varies based on its source and availability. Agricultural waste is rich in complex carbohydrates (mainly cellulose, hemicellulose, and lignin), proteins, fibers, polyphenols, and bioactive compounds, poses environmental risks, but also serves as valuable raw material for agricultural, food, and pharmaceutical products (Kumari et al. 2025). Effective pretreatment and advanced conversion technologies can harness these biochemical properties, improving the cost-effectiveness of bioproduct development (Bala et al., 2023). The carbon in lignocellulosic materials is converted into stable, aromatic structures upon carbonization. High lignin content in biomass favors char formation, while cellulose promotes tar formation, and high oxygen content increases liquid yield. Thermal degradation of carbohydrates, proteins, and lipids affects biochar properties differently due to variations in stability, biodegradability, and thermodegradability of carbon pools and polymer structures (Bong et al., 2020).

India generates a substantial amount of solid waste each year, primarily consisting of agricultural waste, estimated between 350 and 990 Mt/y. As the second largest producer of agricultural waste globally, following China, India produces over 130 million tons of paddy straw annually, with approximately half utilized as fodder and the remainder discarded (Singh & Sidhu, 2014). Burning rice residue in northwestern India significantly impacts air quality and public health. Mismanagement of crop residues results in the release of greenhouse gases (GHGs) such as carbon dioxide (CO_2), nitrous oxide (N_2O), and methane (CH_4), which threaten both human health and the environment (Kaab et al., 2019; Searchinger et al., 2008; Shyamsundar et al., 2019).

Biochar is a porous carbon-rich sorbent typically produced from biological materials such as crop residues through thermochemical conversion (pyrolysis) under limited oxygen conditions. It is often derived from plant and agricultural biomass and contains oxygen-rich functional and aromatic groups (Jatav et al., 2021). This material is derived from the pyrolysis of organic sources such as wood and plants. Pyrolysis occurs at temperatures up to 700 °C in an environment with little or no oxygen, excluding fossil fuel products. Biochar produced under these conditions has a notably high carbon content, resulting from thermal incineration in an oxygen-limited atmosphere (Bhattacharya et al., 2024). Biochar is produced from feedstocks like woody residues, crop straw, animal manure, sewage sludge, and food waste through pyrolysis at 350 °C to over 750 °C. Its properties depend on the feedstock, temperature, pyrolysis duration, and treatments. Wood-based biochars have the highest surface area, straw-based have the highest cation exchange capacity (CEC), and manure-based have the highest nitrogen and phosphorus. Biochars made at temperatures above 500 °C are more stable in soil, with higher ash content and pH (Joseph et al., 2021). Incorporating biochar into cement-based materials can enable carbon sequestration and potentially enhance physical and mechanical properties even at low dosages. Additionally, the pore structure formed during biochar's pyrolysis process can improve the thermal performance of the cement. Biochar also serves as an internal curing agent during cement hydration, supporting strength development (Akinyemi & Adesina, 2020). One of the many uses of biochar is phosphorus sorption. Phosphorus sorption by biochar is affected by its structure, surface area, functional groups, and chemical makeup. The presence of metal cations on the biochar surface enhances this process by enabling chemical precipitation. To improve biochar's sorption capacity, various modification methods can be employed, considering the biochar type, preparation techniques, pyrolysis temperature, chemical reagents, pollutant characteristics, and aqueous conditions such as pH and wastewater composition (Nobaharan et al., 2021).

2 Agricultural Waste Biomass

Agricultural waste biomass refers to organic matter derived from plant and animal sources generated during agricultural activities. It encompasses a wide range of materials, including crop residues (stalks, leaves, husks), animal manure, food processing byproducts, and forest residues (Riseh et al., 2024). This abundant resource holds significant potential for sustainable development and environmental benefits. Globally, agriculture generates substantial quantities of waste biomass. The Food and Agriculture Organization of the United Nations (FAO) estimates that approximately 30% of food produced is lost or wasted annually (Lahiri et al., 2023). This waste, coupled with crop residues and animal manure, represents a massive untapped resource. However, improper management of agricultural waste poses severe environmental challenges. Open burning and uncontrolled disposal contribute to air pollution, greenhouse gas emissions, and soil degradation (Conly et al., 2007). Additionally, the accumulation of waste in agricultural lands can lead to nutrient imbalances, water contamination, and public health risks. Despite these challenges, agricultural waste biomass presents a promising opportunity to address environmental and economic concerns. Its utilization can contribute to mitigating climate change by reducing greenhouse gas emissions and sequestering carbon. Furthermore, converting waste biomass into biofuels, bio-based products, and bioenergy can create new income streams for farmers and rural communities, stimulating economic growth. Moreover, the use of agricultural waste as a feedstock for bioenergy production can reduce dependence on fossil fuels and enhance energy security (Gontard et al., 2018). Agricultural waste biomass is a heterogeneous material with varying properties depending on its origin. It can be broadly classified into three primary categories: crop residues, animal manure, and food processing waste. Crop residues are the byproducts remaining after harvesting agricultural crops. They consist of various plant parts such as stalks, leaves, stems, husks, and cobs. Examples include rice straw, wheat straw, corn stover, sugarcane bagasse, and cotton stalks. These residues are predominantly lignocellulosic in nature, composed of cellulose, hemicellulose, and lignin. Their composition and physical properties vary depending on the crop type and cultivation practices (Prakash Sharma et al., 2018). Animal manure is generated from livestock, including cattle, pigs, poultry, and other farm animals. It comprises a complex mixture of organic matter, inorganic nutrients, microorganisms, and water. The composition of animal manure differs based on the animal species, diet, and management practices. It is a valuable source of nutrients for soil fertilization but also poses environmental challenges if not managed properly due to its high moisture content and potential for nutrient runoff (Rayne & Aula, 2020). Food processing waste arises from the transformation of raw agricultural products into food products. It includes a wide range of materials such as fruit and vegetable peels, seeds, and cores; dairy byproducts like whey and lactose; and residues from meat processing. The composition of food processing waste is highly variable, depending on the specific food product and processing method (Joshi, 2020). This category often contains high levels of organic matter and water, making it susceptible to spoilage and microbial contamination (Karanth et al., 2023). Understanding the classification of agricultural waste biomass is essential for developing effective management and utilization strategies. Each category possesses distinct characteristics and requires tailored approaches for conversion into valuable products or energy sources (Duque-Acevedo et al., 2022).

2.1 Physicochemical Properties of Agricultural Biomass

The suitability of agricultural waste biomass for various applications hinges on its physical, chemical, and biochemical properties. These characteristics determine the efficiency and feasibility of conversion processes and the quality of resulting products. Properties such as particle size, bulk density, moisture content, and ash content significantly impact biomass handling, storage, transportation, and preprocessing. For instance, large particle sizes may hinder efficient grinding for biofuel production, while high moisture content increases drying energy requirements. Ash content can influence equipment wear and fouling in combustion processes (Ibitoye et al., 2023). Chemical properties encompass the elemental composition and organic constituents of the biomass. Proximate and ultimate analysis provide insights into energy content, combustion behavior, and potential pollutants. The presence of cellulose, hemicellulose, and lignin in lignocellulosic biomass affects its digestibility for biofuels and the efficiency of pretreatment processes. Additionally, the nutrient content of biomass, particularly nitrogen, phosphorus, and potassium, is crucial for its application as a soil amendment (Kachroo et al., 2024). Biochemical properties related to the composition of organic compounds within the biomass, including proteins, lipids, and carbohydrates. These components influence the suitability of biomass for specific conversion pathways. For example, high lipid content makes biomass suitable for biodiesel production, while high sugar content favors bioethanol production. The presence of certain compounds may also require specific pretreatment steps to enhance conversion efficiency (Bušić et al., 2018). Understanding the interplay between these properties and the desired application is essential for optimizing biomass utilization. For example, biomass with high energy content and low moisture content is suitable for combustion, while biomass rich in cellulose and hemicellulose is more suitable for biofuel production. By carefully characterizing biomass and selecting appropriate conversion technologies, it is possible to maximize the value of this renewable resource.

Agricultural waste biomass is a potential feedstock for energy production (Alam et al., 2022). Its energy content is primarily determined by its composition, particularly the levels of cellulose, hemicellulose, and lignin. This biomass can be converted into various energy forms like biofuels through processes like ethanol production from sugars or biodiesel from oil-rich wastes. Biogas is generated from anaerobic digestion of organic matter, producing methane for energy. Direct combustion is burning biomass for heat or electricity generation. The energy potential of agricultural waste is substantial, offering a renewable alternative to fossil fuels. However, factors like biomass availability, conversion technology efficiency, and economic feasibility influence its practical realization. Proper characterization is crucial to optimize energy recovery and minimize environmental impacts (Saini et al., 2015).

3 Socioeconomic and Environmental Impact of Agriculture Waste

Over the past 50 years, agriculture sector has produced 23.7 million tons of food every day. Along with the food production, it produces a large amount of biomass which is deemed to be waste as it cannot be used for food production (Puglia et al., 2021). The Federal Service for Supervision of Natural Resources estimates that 46,096.7 thousand tons of production and consumption wastes are generated annually from agriculture and fisheries (Rudoy et al., 2024). Surprisingly large number of agricultural solid wastes are being produced at an average annual increase of 7.5% largely owing to ignorance. A significant contributor to agricultural waste is food spoilage. About 40% of food is wasted annually in the United States alone with a projected cost of approximately 162 billion dollars as estimated by Natural Resources Defense Council (Adejumo & Adebiyi, 2021).

3.1 Socioeconomic Impact

Agricultural waste includes pesticides, fertilizer runoff from fields, harvest waste, manure, salt, and silt drained from fields which are generated during various agricultural activity. Pesticides, like insecticides and herbicides along with fertilizers, are substantial contributors to pollution generated by agricultural waste. With the excessive use of chemical fertilizers, human diseases are caused through the contamination of food and water (Puglia et al., 2021). Approximately 2% of pesticides are said to stay in the containers after use, which some careless and ignorant users may dump in ponds or on open fields resulting in food poisoning, ultimately causing death (Adejumo & Adebiyi, 2021). Agricultural waste poses significant health and environmental risks, particularly to rural women working as farm laborers and those living nearby. Dioxins, produced in large quantities during biomass burning due to chlorinated pesticides, fungicides, and herbicides like pentachlorophenol (PCP), are a major concern. Exposure to dioxins can lead to severe health issues, including neurological effects on fetuses, early menopause, prolonged menstruation, thyroid hormone imbalances, cancer, and chloracne, particularly affecting women (Ugwuoke et al., 2018). Apart from this, the availability of land per capita is also reduced because of the various environmental pressures (Puglia et al., 2021).

The agricultural waste in its organic nature may be hazard free to the nature and could have the potential for economy. Making use of agro waste's economic potential is important as it fosters environmental preservation, societal and social satisfaction, and profitability (*Agricultural Waste Diversity and Sustainability Issues*, 2021). Since organic compounds make up about half of the waste, they act as crucial source of raw materials for the creation of organic fertilizers creating a window of possibility of developing an industry (Rudoy et al., 2024). Globalization and competition for sustainability drive positive ecosystem changes. Numerous applications of agricultural wastes are being noted, including the generation of biogas, filler in asphalt mixtures, and bricks. More investigation and modernization of the current methods are needed to convert this waste further (Harshwardhan & Upadhyay, 2017). The recycling of agricultural solid wastes has several advantages like lowering greenhouse gas emission and reducing their use as fossil fuels. It also greatly promotes the growth of new green markets, jobs, bioenergy production, and the bioconversion of agricultural solid wastes into animal feeds (Adejumo & Adebiyi, 2021). But the recycled goods have insufficient economic value to encourage individuals or farms to implement an agricultural waste management plan (Gupta et al., 2023). Processing agricultural waste offers farmers financial advantages including lower production costs and higher farmer income. Close cooperation among farmers, improvements in knowledge and attitude, and the provision of assistance among farmers are some of the social benefits that farmers receive (Yazid et al., 2020).

3.2 Impact on Air Quality

The agricultural wastes turned out in the farms and fields have the potential to influence the global air quality. Farmers use various inorganic fertilizers, insecticides, and pesticides primarily to increase yield and protect their crops from insects and pests. These constitute different toxic chemicals (chlorofluorocarbons and hydrocarbons) that are absorbed by the crops (Núñez-Delgado, 2024). Climate change can be influenced by the improper handling of agricultural solid wastes. About 14%–21% of the World's Greenhouse Gases (GHG) is released by agricultural activities with high carbon footprint (Adejumo & Adebiyi, 2021; Puglia et al., 2021). Many gasses and metals get released into the air when these crops are burned in an open field. It is estimated that 100,000 molecules of ozone get destroyed before one nascent chlorine (Cl) molecule is eliminated from the stratosphere.

Anaerobic conditions caused by inappropriate disposal of agricultural waste also emit carbynes, deadly greenhouse gasses that play a major role in the global warming. Nitrous oxides are produced from both nitrification and denitrification intensifying global warming. The process of making animal feed also results in higher emissions of trace gases and organic acids into the atmosphere (Núñez-Delgado,

2024; Wang et al., 2024). According to research, a significant reduction in greenhouse gas emission is necessary to prevent global warming of at least 35.6 °F, which might pose a threat to the world's population (Adejumo & Adebiyi, 2021). Swine manure, an essential plant nutrient, usually emits H_2S that contains foul smell. People with exposure to these gases may face respiratory diseases, and there is a possibility of death with the inhalation of gases released from liquid manure (Núñez-Delgado, 2024).

3.3 Impact on Water

Pollution can be direct or indirect at distinct times when waste is collected, treated, or disposed. Most often, the pollution arises from a single source or multiple sources. Sometimes, it arises by variety of events and undefined sources like farm fertilizers, forest debris, industrial sewage, and agricultural food processing which are later dumped into domestic water sources. This results in the contamination of water, making it unsafe for human use along with destroying the aquatic life forms (*Agricultural Waste Diversity and Sustainability Issues*, 2021). Researchers are concerned about water pollution from agricultural waste, as 80% of municipal wastewater is discharged untreated. Fruit, vegetable processing, and juice industries release significant sewage into streams, while agrochemicals and saline irrigation drainage contaminate groundwater, harming aquatic and human ecosystems (Núñez-Delgado, 2024). The organic contaminants released into water from agricultural and agro-industrial waste are highly dangerous and can cause serious health disorders like dysfunction of liver, brain, kidney, and reproductive system along with side effects like mutagenicity, carcinogenicity, and embryotoxicity (Zhou et al., 2015). A major water pollution source globally is the leaching of nutrients like phosphate, nitrate, and heavy metals from agricultural fields, leading to eutrophication and hypoxia due to increased Biological Oxygen Demand (BOD). Insecticides and pesticides harm the food chain, and in the European Union, 38% of water bodies are impacted by agricultural waste (Núñez-Delgado, 2024). The careless disposal of solid agricultural wastes has the potential to obstruct streams, causing floods that could cause property and human casualties (Adejumo & Adebiyi, 2021).

3.4 Impact on Soil

Unsustainable farming practices degrade soil by altering its physiochemical and biological properties through compaction, salinization, and erosion. The use of sewage sludge or wastewater from agro-industrial waste further reduces soil quality and microbial activity, negatively impacting its physical and chemical properties. Additionally, leachate from waste decomposition poses a threat to the ecosystem (Núñez-Delgado, 2024). Plastic waste like pesticide containers, mulching materials, and shade nets are nonbiodegradable and can cause significant damage to the soil if left untreated (Maraveas, 2020). The soil quality is reduced on farms by burning the crop waste. Vital nutrients and beneficial bacteria are at the risk of elimination when the straw is burned. Majority of the waste gets reduced, and the soil fertility can be increased with the proper handling of the agricultural waste (Núñez-Delgado, 2024). During the open field incineration, there is a risk of fire-related accidents because of the wind (Gupta et al., 2023). Improper disposal of the pesticides results in the reduction of soil fertility as well. The seeping of pesticides into the soil promotes the formation of more toxic compounds causing fatal damage to cattle, human, and aquatic ecosystem thereby disturbing the food chain (Rudoy et al., 2024).

Agriculture puts a lot of strain on the environment, and finding means to tackle these effects in short term is important. The large inputs of synthetic materials, greenhouse gas emissions, and the downstream issue of wastewater management are the primary challenges that need to be addressed (Puglia et al., 2021).

4 Agriculture Waste Management and Value-Added Products

The negligent release of agricultural waste into the environment has a negative impact on ecosystems, making the disposal of agricultural waste one of the main environmental issues. According to most reports, agro-industrial waste that has been underutilized and untreated is burned, dumped, or used in a landfill (Bala et al., 2023). Directly dumping agricultural waste increases greenhouse gas emissions and leads to water, soil, and air pollution, affecting human health. Effective waste management is crucial for mitigating pollution and transforming waste into valuable products, promoting environmental sustainability and soil health.

Agricultural waste, including crop residues, processing byproducts, and animal manure, poses challenges but also offers opportunities for resource recovery and environmental sustainability. Effective management minimizes environmental impact and maximizes resource recovery, with a focus on converting biomass into energy. Utilizing agricultural waste as an alternative energy source is a promising way to meet global energy demands (Misra et al., 2023).

4.1 Anaerobic Digestion Technology: For Agriculture Waste to Biogas Production

Anaerobic digestion (AD), also known as bio methanation, is a waste-to-energy technology that converts organic waste into energy. This natural process occurs in the absence of oxygen (anoxic conditions), breaking down organic matter and producing methane gas and other byproducts

Numerous microorganisms, including acetogens and methanogens, play a significant role in the complex process of anaerobic digestion, facilitating the conversion of biomass into biogas. This orchestration of microorganisms leads to a series of chemical processes documented by NNFCC in 2012. Anaerobic digestion involves four pivotal stages (Prasanna Kumar et al., 2024; Usman et al., 2021).

Anaerobic digestion involves four stages: Hydrolysis breaks biomass into simpler components; acidogenesis converts these into volatile fatty acids and gases; acetogenesis refines them into acetic acid, carbon dioxide, and hydrogen; and methanogenesis produces methane from acetate and hydrogen. This process transforms organic waste, including manure and crop residues, into biogas. The gas is ideally used for heating applications such as grain drying, water heating, and broiler operation (Mamo & Zewide, 2022). Biogas can be further utilized for electricity generation.

4.2 Fermentation Technology: For Agricultural Waste to Bioethanol and Biohydrogen Production

Wheat, rice, corn, and sugarcane are the four agricultural products that are grown most widely and yield the highest total amounts worldwide; that is, they produce the most agricultural lignocellulose waste. Because lignocellulose wastes are inexpensive, plentiful, and do not interfere with food production, they are potential substrates for the generation of bioethanol (Harinikumar et al., 2020). The majority of bioethanol produced globally comes from corn and sugarcane. In order to solve the environmental and economic concerns affecting our world, several countries are working with their lawmakers to boost the use of bioethanol and decrease their reliance on fossil fuels by finding renewable energy sources (Hamdi et al., 2024). Fermentation technology, which produces biohydrogen from biological waste, is less expensive, offers environmentally beneficial energy balance techniques, and opens up a way to use enormous amounts of agricultural waste sustainably (Zheng et al., 2022). Lignocellulosic sources like rice, corn, and wheat residues have been studied for biohydrogen production, with yields ranging from 12 to 7019 mL H_2/L. This variation is influenced by the lignocellulosic source, pretreatment method, inoculum, and fermentation parameters. Acid hydrolysis, often followed by enzymatic hydrolysis, is the most common method for breaking down the complex biomass structure to improve hydrogen production (Soares et al., 2020).

4.3 Composting Technology: For Agriculture Waste to Organic Fertilizer Production

Composting is the biological process which uses a variety of microorganisms, such as actinomycetes, bacteria, and fungus, to convert organic waste from plants and animals into a fertile matrix in the presence of oxygen. An assortment of microorganisms added to solid waste can turn it into compost (Dróżdż et al., 2020). The biological process of composting involves four stages: mesophilic, thermophilic, cooling, and compost maturation. These steps might occur simultaneously rather than sequentially (Waqas et al., 2023). The duration of these four stages is dependent on the water quantity, air circulation, microbial composition, and inceptive framework of the combination. The mesophilic stage involves actinomycetes, fungi, and bacteria metabolizing carbon-rich substrates at temperatures between 15 and 40 °C. This stage is followed by the thermophilic phase, where temperatures rise to about 40 °C, favoring thermophilic bacteria like bacillus, and then slightly drop. The cooling phase follows, marked by a slight temperature decrease after substrate reduction. Finally, during the maturation phase, compost quality improves as fungi degrade lignin and cellulose, reducing the oxygen uptake rate and lowering the temperature to around 25 °C. The process concludes with a pH decrease and a C/N ratio of 15 to 20, indicating compost maturity. The increase in agricultural waste exacerbates concerns related to aesthetics, health, and the environment; dumping of waste in landfill can cause groundwater pollution also. Thus, it is imperative to conduct research on safe disposal techniques. Composting is a useful way to intensify and preserve agricultural goods in the form of organic fertilizer. Organic fertilizer produced from agriculture waste has a wide application, especially in sustainable agriculture and waste management. Organic fertilizer enhances the soil fertility, and this leads to better root development and plant growth. Organic matter in this fertilizer improves the soil ability to retain the water.

4.4 Vermicomposting Technology: For Agriculture Waste to Biofertilizer Production

Vermicomposting technology involves the use of earthworms which convert organic waste into fertilizer. Earthworms were referred to as the "intestine of earth" by Aristotle and the "unheralded soldiers of mankind" by Charles Darwin due to their ability to digest a wide range of organic materials. Earthworms rely on organic waste, which helps to produce soil, degrades cellulose, and accumulates humus, which affects the soil health. Earthworms consume a small amount of substrate and excrete the partially digestated organic waste (Gupta et al., 2019). This is because the earthworm's intestines are home to a variety of microorganisms, hydrolytic enzymes, and hormones that aid in the quick breakdown of partially digested material, turning complex organic waste into vermicompost in a much shorter amount of time about 1–2 months (Gupta et al., 2019). Earthworms utilize a multistep process to break down waste materials to produce vermicompost. Earthworms grind their organic stuff into a fine powder inside their gizzard. Subsequently, the breakdown of the materials occurs inside the gut by using hydrolytic enzymes such urease, lipase, amylase, cellulase, and chitinase (Mallappa Munnoli et al., n.d.). Decomposed materials, known as vermicompost, are produced by treating agricultural wastes like floral, banana, bagasse, and agro-industrial residues with various earthworm species. Vermicomposting, a highly regarded biofertilizer, enhances plant growth and crop yield by supplying essential nutrients and promoting nutrient mineralization. It improves soil health, reduces heavy metals, and helps prevent groundwater and soil pollution. Additionally, wastewater residues can be treated using vermicompost.

4.5 Pyrolysis Technology: For Agriculture Waste to Biochar Production

Pyrolysis technology is an innovative and effective technology for the agriculture waste management. This technology involves the thermal decomposition of organic waste in the absence of oxygen, producing value-added product in the form of biochar (Kavan Kumar & Panwar, 2024).

In the twentieth century, pyrolysis techniques were used to commercially produce a variety of fuels, solvents, chemicals, and other products from biomass feedstock. Nowadays, the agricultural waste including sugarcane bagasse, corn stover, cornstalks, rice and wheat straw, and coconut shells attracts the attention of researchers as feedstock for pyrolysis. The composition of agriculture biomass feedstocks varies, containing varying proportions of lignin, cellulose, hemicellulose, and other fundamental components like proteins, lipids, and starch (Li et al., 2023). Pyrolysis technique is a thermal process which converts biomass waste into value-added products in the form of biochar, bio-oil, and syngas which are further utilized as an energy source. Pyrolysis technique is classified into slow pyrolysis and fast pyrolysis based on the heating rate (Kappler & Alberto Mendes Moraes CMORAES, n.d.; Li et al., 2023). Biochar is a carbon-rich product which enhances the soil fertility by increasing the soil capacity to retain water and nutrients. Along with that, it also helps to improve the soil structure. Biochar has an alkaline pH, so it helps to neutralize acidic soil so it makes plants more suitable for nutrient uptake and plant growth. Biochar can provide the habitat for beneficial microbial flora which boosts soil health and plant productivity (Fig. 1).

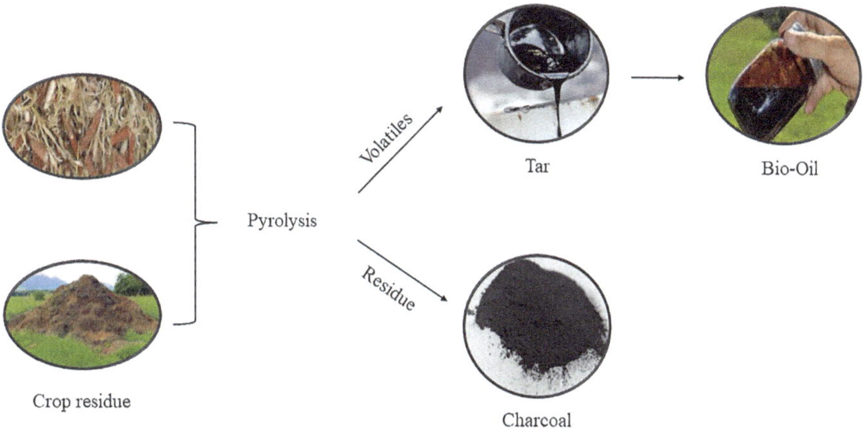

Fig. 1 Value added product from crop residues by pyrolysis

5 Fundamentals of Biochar Production

The thermochemical activities of biomass in the absence of oxygen or with low oxygen result in the production of biochar. Subclasses of the thermochemical processes used to produce biochar are quick pyrolysis, slow pyrolysis, gasification, hydrothermal carbonization (HTC), torrefaction, and flash carbonization. These are distinguished based on the temperature, heating rate, pressure, and residence time (Zhang et al., 2018). Most biochar is produced in traditional earthen charcoal furnaces. However, a variety of inexpensive portable biochar production devices are now available (Das et al., 2021). Carbonization of biomass decomposes it, but a large amount of carbon will be still present in it. The physical and chemical properties are changed (Weber & Quicker, 2018). Three different types of material sewage sludge, plant residue, and animal litter can be converted into biochar under low or no oxygen pyrolysis temperatures, which are typically lower than 700 °C. The creation of biochar has been done to address waste as well as to turn waste into a profitable resource (Zhang et al., 2018). Biochar is the carbon-rich byproduct obtained from the pyrolyzed biomass. It has properties like large surface area, high porosity, high adsorption capacity, surplus functional groups. Because of these properties, heavy metals can be removed from polluted water and soil using biochar as a substrate (Kumar & Bhattacharya, 2021).

5.1 Various Techniques for Biochar Production (Fig. 2)

5.1.1 Pyrolysis

Pyrolysis is a thermochemical conversion process that takes place in an oxygen-starved environment and involves a complicated series of competitive and simultaneous reactions (Vikram et al., 2021). Pyrolysis yields a gas mixture mostly made up of CO_2, CO, H_2, and CH_4, as well as solid or carbonized products and liquid products (tars, bio-oils, and water) (Uddin et al., 2018). There are many kinds of biomass that can be utilized for pyrolysis, such as wood, sewage, and agricultural waste (Fahmy et al., 2020). The primary advantage of the pyrolysis process over gasification is the ease of storage and transportation of the liquid fuel that is produced (Dhyani & Bhaskar, 2018). Pre-pyrolysis, main pyrolysis, and generation of carbonaceous products are the three stages of the biochar production process. The first stage (up to 200 °C), the evaporation of light volatiles and moisture, takes place breaking bonds forming COOH, hydroperoxide, and CO groups. The second stage involves the rapid decomposition and volatilization of hemicelluloses and cellulose between 200 and 500 °C. Degradation of lignin and other organic materials with stronger chemical bonds occurs at the last stage (over 500 °C) (Tomczyk et al., 2020). Fast pyrolysis aims to turn biomass into liquid to the maximum extent. This reduces the volume of energy in the biomass feedstock, which lowers transportation expenses and storage space requirements (Venderbosch, 2019). Currently, rapid pyrolysis is the recommended method because of its quick reaction rate and comparatively greater bio-oil yields (Hu & Gholizadeh, 2019). The optimal working of

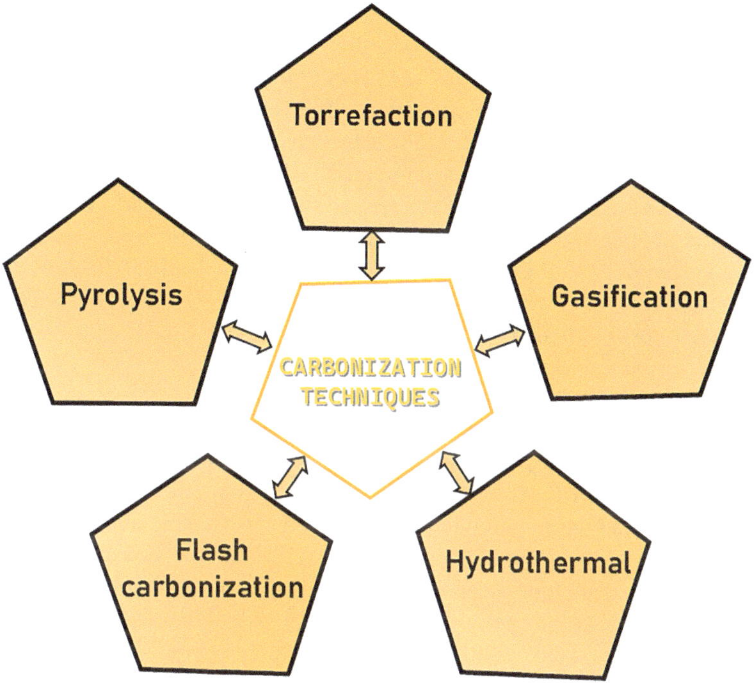

Fig. 2 Various carbonization techniques for biochar production

fast pyrolysis is temperature control (around 500 °C), high heating and very high heat transfer rates, short hot vapor residence period (less than 2 s), quick char removal, and quick cooling of vapors (Gómez et al., 2018). To investigate the thermal breakdown at or near fast pyrolysis settings, a range of lignocellulosic biomass feedstocks, including maize stover, pinewood, poplar, and model compounds including glucose, xylan, and monolignols, have been used (SriBala et al., 2019).

The most common method for producing biochar with gas as a byproduct is slow pyrolysis. This method is comparatively durable and easy to use, making it suitable for both farm-based and small-scale biochar manufacturing (Manyà et al., 2018). Slow pyrolysis is less strict on operating conditions. It uses longer residence durations of biomass (>1 mm) and lower temperatures around 350–450 °C. About 60% of biochar is the primary result in the slow pyrolysis phase (Cai et al., 2020). Under ideal circumstances, flash pyrolysis oil outputs can reach 60–75% of weight. The heating rate (10^3–10^4 °C/s), temperature, must be between 450 and 600 °C, and residence time must be between and less than 1 s for the method to be effective. High water content (often >15 wt.%) is a defining characteristic of flash pyrolysis oil (Ighalo et al., 2022). Slow and fast pyrolysis produces biochar and bio-oil, respectively, whereas intermediate pyrolysis simultaneously yields both (Kazawadi et al., 2021). The range of 300–500 °C is used for intermediate pyrolysis (Dhyani & Bhaskar, 2019). The heating rate is 1–10 °C/s and residence time is 0.5–20 s

(Kazawadi et al., 2021). Compared to fast pyrolysis, the bio-oil produced by intermediate pyrolysis has a reduced tar production and viscosity. Intermediate pyrolysis is more flexible and optimizable and can accept biomass of larger size compared to other pyrolysis (Dhyani & Bhaskar, 2019). The operational conditions of microwave pyrolysis ultimately determine the effectiveness. In microwave pyrolysis, the temperature rate, the initial moisture content, the addition of the microwave absorber and its concentration (such as metal oxides and sulfides, silicon carbide, and carbon-based materials), and the residence time are the factors that have the biggest effects on product recovery (Ethaib et al., 2020). The results obtained from traditional pyrolysis and low microwave heating are observed to be the same, while cellulose and hemicellulose are hydrolyzed at high heating rates (Robinson et al., 2022).

5.1.2 Hydrothermal Carbonization (HTC)

In hydrothermal carbonization, the feedstock is added to water at a moderate temperature (180–250 °C) and pressure (normally autogenous). Biomass undergoes a series of combination processes, including hydrolysis, dehydration, decarbonylation, decarboxylation, polymerization, and recondensation, to eventually become char. One of the primary benefits of HTC over other thermochemical processes is the absence of the need for expensive pre-drying procedures; also, residual moisture frequently serves both as a catalyst and a solvent in HTC (Román et al., 2018). HTC gives three main products: solids, liquids, and gas. The charcoal obtained in the solid form is the major product of HTC (González-Arias et al., 2022).

5.1.3 Gasification

Gasification is the process of breaking down biomass by reacting it with a regulated amount of oxidizing agent (oxygen, air, or/and steam) at high temperatures (>700 °C). This produces carbon dioxide, methane, hydrogen, and carbon monoxide (Jahromi et al., 2021). The process of gasification is regarded as one of the most effective ways to recover energy from biomass, primarily producing syngas (Tezer et al., 2022). The syngas can be directly used as a fuel to produce heat and power, or it can be used as a chemical feedstock. The phases of the gasification process are as follows: drying, pyrolysis, reduction (char gasification), oxidation (combustion), and cracking (Safarian et al., 2019).

5.1.4 Torrefaction and Flash Carbonization

Torrefaction is conducted in an inert atmosphere at a temperature of around 200–300 °C. The characteristics of biomass, such as its mass and energy density, ignitability, grindability, hydrophobicity, and reactivity during gasification and combustion, are greatly enhanced after torrefaction process (Niu et al., 2019). To enhance the fuel and physicochemical qualities of biomass, torrefaction can be used as a pretreatment method (Zadeh et al., 2020). Wet torrefaction is a sustainable subcritical water pretreatment process that transforms moist biomass into hydrochar solid fuel with improved fuel characteristics by eliminating the need for thermal drying (He et al., 2018). Flash carbonization is an easier and novel approach for the carbonization of biomass at a temperature of about 400 °C. With the help of flash heating, biomass precursor can be transformed into carbonaceous matter in just 5 to 10 min with the airflow. When activated, flash carbonized carbon provides greater yields of activated carbons with increased microporosity than activated carbons using hydrothermal carbonization or traditional pyrolysis (Hirst et al., 2018).

5.2 Factors Affecting Biochar Production

The physical and chemical nature of biochar can change mainly depending upon the raw material used and methods used during preparation. The important factors which affect the properties of biochar are feedstock of raw materials, pyrolysis temperature, and residence time (Haider et al., 2022).

5.2.1 Feedstock

The primary factors influencing the carbon concentration, ash content, and carbon sequestration potential of biochar are the type of feedstock used. Although non-lignocellulosic biomass, such as manure, sewage sludge, algae, etc., has also been studied for biochar production, lignocellulosic biomass, which includes agricultural and forest waste, is the most popular feedstock for biochar creation (Pandey et al., 2020). The feedstock can be of three primary types of biomasses containing cellulose, hemicellulose, and lignin each having different pyrolysis temperature which can affect the biomass yield (Panwar et al., 2019).

5.2.2 Temperature

The temperature of pyrolysis can have a major impact on the physicochemical characteristics of biochar, including its element composition, pH, and surface characteristics (Li et al., 2021). It has been found that higher operating temperatures have been observed to have lower oxygen levels and higher fixed carbon concentrations. The pyrolysis process carried out at high-temperature (750 °C) electrical conductivity significantly increased (Panwar et al., 2019). Increases in carbonized fractions, surface area, pH, and volatile matter were seen at higher pyrolysis temperatures, but decreases in CEC and surface functional group content were noted (Tomczyk et al., 2020).

5.2.3 Residence Time

There have been fewer research investigating the impact of residence time on the characteristics of biochar. Pyrolysis durations ranging from 30 min to more than 4 h have been used (Wang et al., 2020a). The pyrolysis temperature and heating rate are often related to the residence time. The yield of biochar drops as residence time increases at a constant pyrolysis temperature. Longer residence times increase char yield in slow pyrolysis situations by aiding in the repolymerization of components of char. According to studies, the biochar's volatile matter concentration dropped at a pyrolysis temperature, while its fixed carbon content rose as residence time increased (Pandey et al., 2020) (Fig. 3).

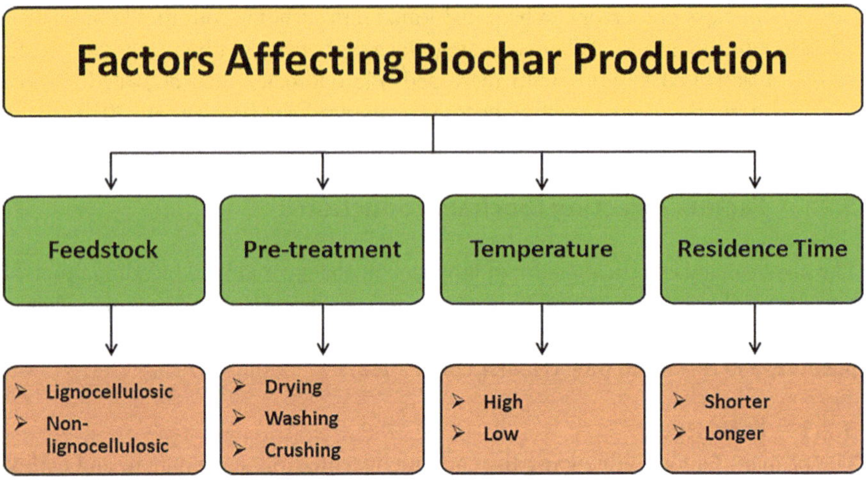

Fig. 3 Factors affecting on the biochar production

5.2.4 Pretreatment

The main methods for overcoming the drawbacks of biomass, such as its uneven shape, low bulk density, and high moisture content, are drying and pelleting. However, research suggests that there are four possible pretreatment techniques for biomass: drying, vacuum freezing drying (VFD), ultrasonic vibration-assisted pelleting (UV-AP), and pelleting (Meng et al., 2021). Washing biomass in water or acids (HNO_3, HF, and acetic acid) lowers the amount of alkaline and alkaline earth metals present, which increases the combustion property and decreases the ash content (Bhatnagar et al., 2022).

6 Biochar Production From Various Agricultural Waste Biomasses

One promising way to repurpose and add value to these agricultural residues is through the production of biochar (Kumari et al., 2020, Khawkomol et al., 2021). Due to its versatile application in a range of agricultural and industrial processes, biochar has gained significant interest in recent years (Amalina et al., 2022). In contrast to open burning of agricultural leftovers in fields, which can release toxic gases such as carbon monoxide, sulfur oxides, nitrogen oxides, and smoke particles carrying carcinogens, the process of creating biochar creates less air pollution (Khawkomol et al., 2021).

The characteristics of the biochar may depend on the kind of agricultural biomass used. For example, biochar made from manure has a greater potential for cation exchange than biochar made from woody materials (eucalyptus) (Tasim et al., 2019). Moreover, biochar made from switchgrass and corn stover at the same temperatures had less aromatic carbon and more ash content than biochar made from woody materials. Similarly, applying wood chip biochar produces greater saturated hydraulic conductivities than applying biochar derived from manure. As it is evident from the literature, the kind of feedstock used greatly influences the quality of the resulting biochar (Tasim et al., 2019). Although pyrolysis is the most widely used process for producing biochar, multiple approaches are used depending on the type of agricultural feedstock. Table 1 shows the production method and analysis of biochar from different agricultural feedstock.

6.1 Corncob

Corn is one of the most common and widely grown crops in the world. The world's corn production increased by 17.47 million tons, or 1.57%, globally in 2019/2020, to 1116.41 million metric tons, and is predicted to reach 1133.89 million metric tons in 2020/2021. The United States, China, and Brazil are the world's top three producers, making up over 65% of the total production (Wang et al., 2022). Corncob biochar is produced via slow pyrolysis after the corncob is cleaned, sun-dried to reduce moisture, and pretreated with o-phosphoric acid. Pyrolysis is conducted at 400, 500, and 600 °C, with proximate analysis showing 7.11% moisture, 1.43% ash, and 75.11% volatile matter, which contributes to bio-oil production. Carbon content increased from 46.69% in corncob to 80.22% in biochar, with a maximum biochar yield of 58.50% at 400 °C (Gupta et al., 2018). The findings demonstrated the great stability of corncob. They displayed the appropriate physical and chemical qualities for sustainable agricultural and environmental purposes (Wijitkosum, 2022).

Table 1 Different value-added products from agricultural waste biomass

Biomass	Value-added product	Process	Yield	Reference
Sugarcane tops	Bioethanol	Saccharification and fermentation	65.1%	Pereira et al. (2015)
Rice straw	Biogas	Hydrolysis, acidogenesis, acetogenesis, and methanogenesis	62.7 mL. g^{-1} VS	Olatunji et al. (2021)
Wheat straw	Bioethanol	Fermentation	15–22.5 g. L^{-1}	Cotta (2012)
Groundnut shell	Biogas	Acidogenesis, acetogenesis, and methanogenesis	147.6 L_N. kg^{-1}FM	Olatunji et al. (2021)
Rice hulls	Bioethanol	Lime pretreatment, enzymatic saccharification, and fermentation with recombinant *E. coli* strain FBR5	9.8–11 g. L^{-1}	Saha and Cotta (2008)
Corn stover	Biohydrogen	Acidic pretreatment and fermentation	3.0 g.mol^{-1}	DATAR et al. (2007)
Wheat straw	Biohydrogen	Chemical pretreatment and fertilization	3.8 mol. $glucose^{-1}$	Ivanova et al. (2009)

L_N.kg^{-1}FM: Liters of nitrogen gas per kilogram of fresh matter; mL.g^{-1} VS: milliliters of a substance per gram of volatile solids

Corncob biochar lacks surface functional groups for adsorption and has a high moisture content, making it unsuitable for use as fuel. Therefore, surface modification or pretreatment would be needed. It may also be applied as a soil amendment (Khawkomol et al., 2021).

6.2 Vegetable Waste

Approximately 630 million metric tons (MMT) of fruit and vegetables are wasted throughout agricultural production, transportation and packaging, postharvest handling, distribution activities, and consumption. Banana peels account for 35% of fruit and vegetable waste (220 MMT), followed by cabbage waste (50 MMT), cauliflower and brassica debris (2%; 13 MMT), potato peels (94 MMT), and tomatoes (20 MMT) (Pradhan et al., 2020). In one study, vegetable wastes were collected, cut into pieces, and pyrolyzed in batches at 300, 400, 500, and 600 °C to assess biochar yield and properties. Biochar yields ranged from 20 to 30%, with higher yields at lower temperatures. The decline in yield at higher temperatures is due to dehydration of hydroxyl groups and thermal degradation of lignocellulosic structures (Pradhan et al., 2020). Producing biochar from vegetable waste offers numerous applications, such as adsorption-based pollution removal from soil, water, and organic matrices, enhancing the characteristics of agricultural soils, altering the composting process, and serving as a feedstock to produce biogas (Molenda et al., 2018).

6.3 Rice Husk

Rice is a staple food in Asia, with countries like China, India, Indonesia, and others leading in production and consumption. Sustainable agricultural practices, particularly in soil fertility and waste management, are crucial in rice fields. Rice husk, a byproduct, is generated during the initial stage of rice milling when paddy rice is husked (Karam et al., 2022). Biochar was produced from granular and powdered rice husk using a modular pyrolizer. About 500–1000 g of biomass was pyrolyzed, taking an hour for burning and 2 h to convert fully into biochar. The yield ranged between 30% and 40%. Granular rice husk biochar had an ash content of 43.17% and 56% volatile matter (Sakib et al., 2021). The high ash content obtained from the production of rice husk biochar makes it appropriate as a soil supplement to improve soil fertility (Khawkomol et al., 2021). Rice husk biochar, rich in silica, enhances nutrient retention, adsorption, and plant nutrient absorption, improving plant structure and turgidity. Its ability to boost soil fertility and fertilizer efficiency makes it an ideal soil amendment (Karam et al., 2022).

6.4 Wood Residues

Crop residues are produced in large quantities all over the world, and they are not always recycled or disposed of appropriately. Brazil produces 50.8 million m^3 of lignocellulosic residue annually from the manufacturing of wood logs. These wastes formed from wood are high in lignin and cellulose content. Hence, these are perfectly suitable for slow pyrolysis (Domingues et al., 2017). Wood residue biochar was produced via slow pyrolysis at 400, 500, and 600 °C, yielding 22%. The process took around 2 h. Wood chips, with 77.3% volatile matter, produced biochar with high fixed carbon and low moisture and ash content, indicating good quality. This biochar effectively enhances soil quality, making it nutrient-rich and suitable for farming (Sakib et al., 2021). Wood biochar have the potential to reduce N_2O emissions in crop fields, but because they must be applied at high rates, composting and other more profitable processes should be the main uses of biochar to offset N_2O emissions rather than soil (Karam et al., 2022).

6.5 Coconut Husk and Coconut Shell

The coconut industry in Asia, Oceania, the Caribbean, the Americas, and Africa generates biomass from shells and husks. Coconut shell biochar is primarily produced via pyrolysis, with gasification as an alternative. Microwave-assisted pyrolysis at 400–800 °C yields 84–91% biochar. This carbon-rich biochar contains nitrogen, oxygen, hydrogen, and inorganic materials, according to proximate and final analysis (Ajien et al., 2023). Biochars derived from coconut husk and coconut shell residues can be used as fuel, soil amendment, or adsorbents due to their high heating value, high soluble N and P for plant growth, high acidic functional groups for cationic adsorption, and high iodine number for pollutant adsorption in pores (Khawkomol et al., 2021). Biochar and activated biochar are useful for various applications such as biodiesel production, capacitive deionization, soil conditioning, toxic remediation, and water treatment (Ajien et al., 2023) (Table 2).

7 Utilization of Biochar Produced From Agricultural Waste Biomass

Biochar, a low-cost alternative to activated carbon, is effective in removing organic and inorganic pollutants from water due to its high sorption capacity, large surface area, microporosity, and ion exchange capacity. Its environmental applications are influenced by the unique physicochemical properties, which depend on the biomass source and pyrolysis process used (Amalina et al., 2022).

Table 2 Biochar production through various processes

Feedstock	Process	Temperature	Yield	Application	References
Corncob	Slow pyrolysis (furnace)	400–600 °C	58.50% at 400 °C	Soil amendment	Gupta et al. (2018) and Wijitkosum (2022)
Corncob	Pyrolysis (drum kiln)	150–600 °C	32% at 480 °C	Soil amendment	Khawkomol et al. (2021)
Vegetable waste	Pyrolysis (batch)	300–600 °C	20–30% at 300 °C	Soil amendment, wastewater treatment, biogas feedstock	Molenda et al. (2018) and Pradhan et al. (2020)
Rice husk	Slow pyrolysis	400–600 °C	30–40% at 400 °C	Soil remediation and amendment	Khawkomol et al. (2021) and Sakib et al. (2021)
Wood residues	Slow pyrolysis	400–600 °C	22% at 400 °C	Soil amendment and composting	Karam et al. (2022) and Sakib et al. (2021)
Coconut husk and coconut shell	Microwave-assisted pyrolysis	200–800 °C	84–91% at 200 °C	Soil amendment and biofuels	Ajien et al. (2023)
Olive pit	Pyrolysis (batch)	300–600 °C	97.3% at 350 °C	Soil amendment	Campos et al. (2020)
Sawdust	Fast pyrolysis	500 °C	26–34%	Soil fertility	Haddad et al. (2021) and Khawkomol et al. (2021)
Coffee husk	Slow pyrolysis	350–750 °C	43.5% at 350 °C	Soil amendment	Domingues et al. (2017)
Sugarcane	Slow pyrolysis	350–750 °C	37.5% at 350 °C	Soil amendment	Domingues et al. (2017)

7.1 Biofertilizer Application of Biochar

In recent decades, inappropriate farming practices such as overuse of chemical fertilizers and monocropping have resulted in increasingly serious environmental problems, affecting the composition and function of the soil microbiome, as well as plant health and economic production (Sun et al., 2021). The use of inorganic fertilizers, which have several negative impacts, is a major component of strategies to boost agricultural output. Other challenges to agricultural land include erosion and declining soil fertility, which have an impact on the topsoil taken away by rainwater and create environmental issues for aquatic ecosystems. Due to its effects on land productivity and agricultural business, soil degradation poses a serious threat to farmers (Danapriatna et al., 2023). Therefore, it is essential to utilize biofertilizers effectively. Biofertilizers are inoculants containing active substances from living organisms that bind specific nutrients and increase the availability of other elements in the soil. Utilizing biochar as a biofertilizer can boost farmers revenue and promote sustainable agriculture (Danapriatna et al., 2023). Biofertilizers enhance soil biodiversity by expanding the microbial biome. In biofertilizer compounds, Azotobacter, Cyanobacteria, Rhizobium, Azospirillum, and Mycorrhizae are typically present (Mousavi et al., 2020). The nodule and nitrogen-fixing activity of legume plants was enhanced by inoculating them with rhizobium bacteria-based biofertilizer. Legumes can provide up to 90% of nitrogen when cultivated with the right strain of bacteria (Shikha et al., 2023).

Biochar, a low-cost pyrolytic derivative, offers numerous agricultural and environmental benefits, including enhancing soil quality and fertility. It helps retain soil carbon, potentially absorbing 1.4–2.9 tons of CO_2 per ton applied, while improving soil properties and plant nutrition (Ahmad et al., 2023). According to a study, adding 15 tons of biochar per hectare to the soil significantly increases the biological nitrogen fixation in common beans (Mousavi et al., 2020).

It is possible to reduce CO_2 emissions by soil carbon sequestration. Mineralization of organic matter in soil was more prominent in poor fertile soils than in excellent fertile soils and in soils with high carbon content than in low carbon content soil (Amalina et al., 2022). It has boosted crop yield by stimulating beneficial soil microbes such as mycorrhizal fungi, improving soil base saturation, increasing soil water-holding capacity, and retaining nutrients in the section of the soil column containing roots and enhancing nutrient usage efficiency (Ahmad et al., 2023). Remediation mechanisms are influenced by several factors, including pyrolysis temperature, which affects biochar's sorption efficiency. Higher temperatures result in biochar with larger surface areas and more developed pores, while lower temperatures yield biochar with more oxygen-containing functional groups and higher cation release, enhancing its effectiveness for inorganic contaminants. Sorption capacities are also influenced by the types of contaminants, biochar type, production conditions, and functional features (Nadarajah et al., 2024). Biochar's key qualities are derived from its carbon structures, which include both crystalline and amorphous carbon. The H/C ratio indicates the carbon structure of biochar as the conversion of hydrocarbons into aromatic rings, whereas the O/C ratio represents the polar functional groups and stability of biochar in soil (Wijitkosum, 2022). Applying biochar to soil enhances beneficial soil organisms and aeration, improving crop yield by developing a structured rhizosphere. Initially hydrophobic and low in functional groups, biochar becomes hydrophilic and gains large functional groups upon soil exposure, which increases its cation exchange capacity, nutrient retention, and water-holding capacity, all crucial for boosting crop yield (Nadarajah et al., 2024).

Biochar made from rice husk has been shown to improve the soil's characteristics and fertilization efficiency, allowing plants to absorb nutrients more effectively. The application of 10 g of biochar boosted the macronutrient content of the soil. In addition to carbon, biochar also includes additional elements like 10–15% moisture, 15–20% lignin, 10–15% cellulose, and 15–20% silica (Danapriatna et al., 2023). Inorganic pollutants, including heavy metals from sources like mining, metal finishing, and fertilizers, are resistant to decomposition. Biochar, a carbonaceous substance, has recently been used to absorb metals in soil. When applied to soils contaminated with metals such as Pb, Cu, Cd, and Zn, hardwood-derived biochar effectively moved Cu while immobilizing Cd and Zn (Nadarajah et al., 2024). Muhammad et al. studied the impact of wheat straw biochar on rice yield and soil parameters in Psammaquent and Plinthudult soils. Applying biochar at 3% by mass improved soil pH and total N and C content, reduced nutrient leaching, and significantly increased rice growth. Grain yields rose by 32% in Psammaquent soil and 41% in Plinthudult soil (Kalus et al., 2019). One study suggests that biochar's benefits depend on soil type and may require a tailored approach for different agricultural systems, potentially affecting soil microbial communities (Kalus et al., 2019). Biochar amendments significantly impact N_2O emissions and NH_3 volatilization from soils, either directly or indirectly. Recent studies show that combining biochar with other soil additives, such as urease inhibitors, can reduce N_2O emissions and NH_3 volatilization by approximately 35% and 80%, respectively. Additionally, combining 1% biochar with 1% vermicompost is more effective in reducing N_2O and NH_3 losses compared to using either substance alone (Sun et al., 2021).

Biochar with high cation exchange capacity (CEC) and water-holding capacity (WHC) can optimize plant nutrient intake, root development, and soil fertility, resulting in increased crop output. In studies involving coconut-based biochars, higher mineral concentration or mineral release rate suggests increased CEC of coconut shell and coconut husk biochar, which promotes higher nutrient uptake for plant growth (Ajien et al., 2023). Biochar, being porous and lightweight, enhances soil structure by increasing surface area and modifying the pore network. Its resistance to degradation improves long-term water-holding capacity and reduces soil bulk density. Studies show that biochar is more effective than manure at decreasing bulk density and increasing water retention (Arabi et al., 2018).

A recent study explored the use of plant growth-promoting bacteria (PGPB) with biochar derived from maize in a 3-year-old modified soil. This biochar, previously shown to have low decomposition and mitigate the loss of original soil organic matter, was found to retain its porous structure and support various cultivable bacterial strains after 3 years in a poplar SRC plantation. Notably, soils treated with biochar had higher abundances of Microbacterium and Pseudomonas, known as PGPB capable of producing plant growth hormones, suppressing ethylene production, and producing ion chelates (Bertola et al., 2019) (Table 3).

7.2 Wastewater Treatment Using Biochar

Water bodies around the world are deteriorating because of the constant release of numerous organic and inorganic pollutants from municipalities and industries, including pesticides, dyes, heavy metals, surfactants, medicines, and personal hygiene products. The unregulated discharge of these contaminants is cause of concern due to their suspected harmful effects on ecosystems (Enaime et al., 2020). Wastewater is a global issue that is produced because of home, industrial,

Table 3 Applications of biochar as biofertilizer

Feedstock	Role	Application	References
Biochar–rice husk	Absorb nutrients more effectively	Boosts the macronutrient content of the soil	Danapriatna et al. (2023)
Biochar–hardwood	Absorb metals in soil	Metals like Cd and Zn were immobilized in modified soils	Nadarajah et al. (2024)
Biochar–wheat straw	Boosting the pH and total N and C content	Boosted rice plant growth yield, including height, leaf length, and grain yield	Kalus et al. (2019)
Biochar and urease	Lowering N_2O emission and NH_3 volatilization	Reduce N_2O emission and NH_3 volatilization by roughly 35% and 80%	Sun et al. (2021)
Biochar–coconut shell	High cation exchange capacity (CEC) and water-holding capacity (WHC)	Higher mineral concentration promotes higher nutrient uptake for plant growth.	Ajien et al. (2023) and Arabi et al. (2018)
Biochar and PGPB	Stimulate production of plant growth-promoting bacteria	Production of plant growth hormones, suppressing ethylene production, and producing ions chelates	Bertola et al. (2019)

commercial, or agricultural activity (Xiang et al., 2020). Biological treatments, such as activated sludge and anaerobic digestion, do not effectively remove certain non-biodegradable contaminants and incur considerable expenses for sludge management. Another inclination has been the creation of different kinds of carbonaceous materials that can meet the sustainability standards needed for use in practical applications (Kamali et al., 2021). Biochar has a lot of potential for treating wastewater. Biochar's unique characteristics, like its large specific surface area and abundance of surface functional groups, make it a good adsorbent for the removal of a wide range of pollutants (Xiang et al., 2020). Chemical modification of biochar using acids, bases, or polymers can improve adsorption effectiveness by increasing surface area, modifying chemical functionality, and creating high-affinity adsorption sites (Obey et al., 2022). Compared to traditional soil and sand infiltration systems, the use of biochar for wastewater purification can improve treatment efficiency, lessen the spread of hazardous chemical contamination, and reduce the effects of climate change through carbon sequestration (Perez-Mercado et al., 2018). Biochar is highly efficient and cost-effective due to its low feedstock costs and simple processing processes. The operating and running costs of the wastewater treatment process are greatly decreased by the reactivation (regeneration) of biochar following the adsorption process to be used for multiple cycles (Alsawy et al., 2022).

Biochar filters may serve as a treatment device for wastewater and water supply during extreme weather. Biochar filters adsorption capacity can be extended through bio regeneration, potentially leading to better removal rates in long-term trials (Kaetzl et al., 2018). Ecological benefits of biochar amendment in biofilters include enhanced macrophyte growth and reduced greenhouse gas (GHG) emissions when nitrogen, phosphate, organic pollutants, heavy metals, and pathogens are removed

(Deng et al., 2021). A study conducted by M.M. Manyuchi showed that biochar with a surface area of 1000 m^2/g and a loading of 1 g/L was used to treat municipal wastewater. Standard methods measured changes in pH, total suspended solids (TSS), total Kjeldahl nitrogen (TKN), total phosphates (TP), and chemical oxygen demand (COD) before and after treatment. The biochar filter achieved reductions of 90% in COD, 89% in TSS, 64% in TKN, and 78% in TP (Manyuchi et al., 2018).

Biochar has also attracted significant attention in the removal of hazardous metals due to its environmentally benign nature, low replacement cost, and feasible application on a broad scale, when compared to polymeric and commercial adsorbents. The presence of oxygen- and nitrogen-containing functional groups on the biochar surface promotes adsorption via acid-base interactions and hydrogen bond formation (Weidner et al., 2022). Researchers have extensively examined the removal of heavy metals and metalloids, such as As, Cd, Pb, Hg, Cr, and Cu, using various biochars (Gopinath et al., 2021). The removal mechanism varies based on the target metal's valence state and solution pH. The following four pathways are suggested to be dominant in the removal of heavy metals from wastewater by biochar: (i) Heavy metals and the surface of biochar are attracted to each other electrostatically; (ii) heavy metals and alkali or alkaline earth metals exchange ions on the surface of biochar; (iii) complexation with the π electron-rich domain or surface functional groups; and (iv) coprecipitation to form insoluble compounds (Wang et al., 2020b).

Biochar formed from biomaterials effectively removes organic pollutants. Biochar sorbs organic pollutants by a variety of mechanisms, including pore-filling, hydrophobic effect, electrostatic interaction, and hydrogen bonding, which vary depending on the contaminants' and biochar's characteristics. A crucial mechanism for the sorption of organic molecules on biochar is pore-filling. The surface area of the micropores directly relates to the sorption capacity (Wang et al., 2020b). For treatment of organic pollutants, biochar-based photocatalysts can be employed. Biochar demonstrated significant benefits when mixed with other semiconductors. The microporous structure and extensive surface area of biochar-based photocatalysts allowed for effective diffusion of reactants to catalytic sites. Furthermore, it improved photodegradation for both organic and inorganic compounds (Cui et al., 2020).

Industrial Wastewater

Industrial wastewater originates from different sources such as mining, smelting, battery manufacture, chemical industries, leather manufacturing, dyes, and more. After cross-linking, biochar combined with chitosan can be cast into membranes, beads, and solutions. It is effective as an adsorbent for the adsorption of heavy metals from industrial wastes (Xiang et al., 2020). Emerging organic pollutants in industrial effluent, such as phenols and PAHs, have raised concerns. The biochar derived from sewage sludge displayed a significantly increased hydroquinone sorption capability. Biochar made from orange peels demonstrated excellent 1-naphthol and NAP sorption (Wang et al., 2020b). The use of layered double hydroxides (LDHs) in a variety of industries, including water treatment, has drawn a lot of attention as well. Composites made of biochar and LDH have shown great potential in the adsorption of heavy metals, dyes, anions, and pharmaceuticals from water (Zubair et al., 2021).

Municipal Wastewater Treatment

Municipal wastewater can be treated with biochar made from biowaste during the biofiltration process. Biochar can function as a biofilter because of its large porous surface area. Biochar can be used alone or in combination with biofilters and other technologies for treating municipal wastewater (Xiang et al., 2020). A study was conducted that aimed to create a low-cost wastewater treatment system that produces safe and nutrient-rich water, utilizing locally accessible resources and materials. Biochar made from rice husks was used in filters to accomplish this purpose (Kaetzl et al., 2019). Algal biochar can also be used for municipal wastewater treatment. Algal biochar has excellent biosorption characteristics and may effectively remove ammonium and other contaminants from wastewater investigated the sorption of phosphates, dissolved nitrogen, and phosphorous from municipal wastewater using biochar. The biochar used in this procedure was enriched with important macroelements, N and P, and can be put into soil as a fertilizer supplement (Michalak et al., 2019).

Agricultural Wastewater Treatment

One unique remediation technique used to address pesticide pollution is biochar (Wang et al., 2020b). The agricultural industry's rapid growth has led to an increase in pesticides and harmful heavy metals being dumped onto farmlands, causing major contamination (Xiang et al., 2020). When untreated or inadequately treated wastewater is released into the environment, it can lead to harmful effects on organic compounds and enteric microbes that could contaminate sources of irrigation (Kaetzl et al., 2020). In agriculture, the two most widely used pesticides are atrazine and pentachlorophenol. Rice straw biochar and phosphoric acid-modified biochars effectively remove imidacloprid and atrazine from agricultural effluent. Soybean and corn straw biochar effectively remove atrazine, with adsorption capacity mostly determined by pore volume and pH (Xiang et al., 2020).

Stormwater Treatment

Biochar is used to treat urban stormwater runoff, which contains various contaminants and eventually enters surface water (e.g., streams, lakes), with the goal of removing indicator organisms and pathogens. Irrigation with polluted water can cause microbial contamination of vegetables. The use of biochar filters to remove microbes from water is generally used (Wang et al., 2020b). Urban stormwater runoff has become a major concern with the increasing rate of urbanization because of its influence on water quality. Stormwater runoff needs to be treated before being released because it can significantly degrade the quality of natural water sources. Biochar and its modified forms are effective media for stormwater treatment systems. According to a recent study, As^{5+} and other runoff pollutants including Pb^{2+}, Zn^{2+}, Cu^{2+}, and PO_4^{3-} can be successfully removed from contaminated urban water runoff using aluminum-impregnated biochar (Xiang et al., 2020). The removal of trace organic contaminants (TrOCs) is best achieved with hydrophobic biochar generated at a high temperature because of its high specific surface area, strong affinity, and abundance of hydrophobic moieties (He et al., 2022) (Table 4).

7.3 Biochar in Dye Degradation

Today, different dyes and additives are widely used in industrial applications such as textiles, paper, printing, paint, laundry, cosmetics, carpet, leather, food, and rubber. An estimated 7×10^5–1×10^6 tons of natural and synthetic dyes are generated annually worldwide, comprising over 10,000 distinct varieties. The main concerns with these dye-contaminated wastewaters are their chemical complexity, stability, and low biodegradability (Goswami et al., 2022). These pollutants have become a global threat in the last 10 years due to their effects on the environment and human health (Praveen et al., 2022). Synthetic dyes are polyaromatic unsaturated polymers containing 60–70% azo chemicals. Synthetic dye degradation products are extremely cytotoxic and genotoxic. In the past, several physicochemical treatment processes, including adsorption, sedimentation, oxidation, and flocculation, were modified for the degradation of dyes. However, these technologies have limitations in application, including high operational costs and the emission of hazardous

Table 4 Wastewater treatment using biochar

Type of wastewater treatment	Sources	Treatment	References
Industrial wastewater treatment	Mining, smelting, battery manufacture, chemical industries, leather manufacturing, dyes, and more	Biochar with chitosan for heavy metals Biochar of orange peel—organic pollutants biochar with LDH—pharmaceuticals	Wang et al. (2020a, b), Xiang et al. (2020) and Zubair et al. (2021)
Municipal wastewater treatment	Houses, commercial complexes, and public utilities	Biochar from rice husk and algal biochar used as biofilters to remove ammonium and other contaminants from wastewater	Kaetzl et al. (2019) and Xiang et al. (2020)
Agricultural wastewater treatment	Pesticides and harmful heavy metals	Rice straw biochar and phosphoric acid-modified biochars effectively remove imidacloprid and atrazine (major pesticide contaminants)	Kaetzl et al. (2020), Wang et al. (2020a, b) and Xiang et al. (2020)
Storm wastewater treatment	Urban stormwater runoff	Aluminum-impregnated biochar – removes Pb^{2+}, Zn^{2+}, Cu^{2+}, and PO_4^{3} Hydrophobic biochar—organic pollutants	Wang et al. (2020a, b) and Xiang et al. (2020)

intermediates (Ayyappan et al., 2018). Environmental degradation and global warming are currently important challenges that are causing widespread concern. There is a growing desire for eco-friendly and cost-effective ways to support a sustainable future.

Biochar's large specific surface area, pore volume, and hydrophobicity make it an effective biomaterial for carbon sequestration. Furthermore, it has received widespread recognition as a potentially highly efficient, low-cost, and environmentally friendly adsorbent for the removal of organic and inorganic contaminants from wastewater, particularly heavy metals and dyes (Goswami et al., 2022). Various techniques like impregnation, steam activation, and chemical or heat treatments have been used to enhance biochar properties. Post-modification falls into four categories: (i) increasing surface area and porosity, (ii) reducing positive surface charge, (iii) enhancing oxygen-containing functional groups, and (iv) magnetizing biochar for easier recovery. Each method can simultaneously improve multiple biochar characteristics (Goswami et al., 2022). The superficial properties of these materials are studied in relation to the operational usage of biochar because these qualities are critical to the materials interactions with target pollutants and their adsorptive performance (Praveen et al., 2022). Understanding surface structure interactions is essential to confirm biochar's unique ability to absorb various contaminants. Organic pollutants are absorbed through mechanisms like pore diffusion, hydrophobic action, hydrogen bonding, and ionic interactions. Biochar aids methyl blue degradation through adsorption, breakdown, and reformation, with its microporous structures primarily responsible for adsorption and decomposition. Dye degradation potentially increases when biochar precipitates onto bacterial surfaces (Ayyappan et al., 2018). pH significantly influences biochar's adsorption capacity. In a study on methylene blue removal from wastewater, dye removal efficiency increased from pH 2 to 7 but decreased from pH 7 to 11. The decline at higher pH levels was due to biochar's active sites combining with widely distributed hydroxyl groups, reducing efficiency (Srivatsav et al., 2020).

Efficiency of biochar as an adsorbent depends on factors like dye concentration, temperature, and pH. Studies showed that nickel-modified biochar effectively absorbed methylene blue, while biochar from organic peel waste was used to adsorb Congo red dye, with optimal adsorption at a pH of 2–3 due to H^+ ions. When pH was adjusted from 3 to 11, dye removal effectiveness only decreased by about 5% (Srivatsav et al., 2020). Biochar made from agricultural crop residuals and peanut shell effectively removes cationic dyes such as methyl violet, methylene blue, basic blue, crystal violet, and rhodamine through electrostatic adsorption (Praveen et al., 2022). In another study, sulfuric acid-modified corncob biochar (SA-MCB) was employed as a supplement in microbial fuel cells to degrade real dye wastewater (RDW) (Sonu et al., 2020). The application of biochar for textile dye adsorption from wastewater is facilitated by photocatalytic degradation of dyes using modified biochar, for example, the process of creating TiO_2/biochar photocatalysts and using them to break down sulfamethoxazole. The OH-free radical was the main active species responsible for sulfamethoxazole degradation (Lu et al., 2022). Another study focused on the breakdown of azo dye. Acid orange 7 (AO7), a type of model

Table 5 Dye degradation using biochar from different biomass

Biochar	Dye degradation	References
Nickel-modified biochar	Methylene blue	Srivatsav et al. (2020)
Organic peel waste biochar	Congo red dye	Srivatsav et al. (2020)
Food waste biochar	Organic dyes	Srivatsav et al. (2020)
Peanut shell biochar	Cationic dyes	Praveen et al. (2022)
Sulfuric acid-modified corncob biochar (SA-MCB)	Real dye wastewater (RDW)	Sonu et al. (2020)
TiO_2/biochar photocatalysts	Sulfamethoxazole	Lu et al. (2022)
Rice hull biochar	Acid orange 7 (AO7)	Li et al. (2019)
Fe_2O_3/TiO_2 functionalized biochar	Methylene blue (MB)	Chen et al. (2020)

compound, is typically used to examine the degradation of dyes. AO7 is degraded to an aqueous solution using rice hull biochar (Li et al., 2019) (Table 5).

7.4 Other Application of Biochar

7.4.1 Biochar as a Catalyst

Biochar is a versatile catalyst in chemical reactions, enhancing efficiency and product yield. It is used in biodiesel synthesis, syngas production, and catalytic cracking during coal pyrolysis. Fe/biochar is effective in breaking down toluene and is utilized in microbial fuel cells. Biochar from peanut hulls, bamboo, or pine chips serves as a cost-effective catalyst for biomass hydrolysis. Additionally, biochar catalyzes isomerization, hydroxylation, deoxygenation, and electrochemical reactions. It also generates sulfate and hydroxyl radicals, aiding in pollutant degradation and acting as an electron acceptor to boost catalyst effectiveness in photochemical processes (Lyu et al., 2020). According to studies, the MnO_x/rice straw-biochar combination showed a good NO_x removal effectiveness at low temperature SCR (Lee et al., 2017).

7.4.2 Biochar in Electrode Materials and Electrode Modifier

When investigated for asymmetric capacitive deionization, activated biochar electrodes showed increased charge efficiency and salt removal capacity due to its amine and sulfonate functional groups (Stephanie et al., 2021). The use of woody biomass-derived biochar as a material for the electrode resulted in quick charging-discharging and a high gravimetric capacitance. Activated biochar can give a high capacitance (Wang & Wang, 2019). Apart from this, biochar produced by pyrolyzing algal biomass can be used as an anode material to facilitate effective electron uptake (Wang et al., 2018).

Table 6 Other applications of biochar of different biomass

Feedstock	Process	Function	References
Wood	High-temperature gasification	Electrodes in microbial fuel cell	Huggins et al. (2014)
Wood	One pot pyrolysis and carbonization	Supercapacitor electrodes	Jiang et al. (2013)
Palm kernel shell	Thermal degradation	Catalyst in transesterification	Lee et al. (2017)
Peanut husk	Pyrolysis	Adsorbent to separate Cr (VI) from aqueous solution	Mahanty and Mondal (2021)
Alfalfa hays	Pyrolysis	Adsorption of tetracycline from water	Jang and Kan (2019)
Pinewood	Pyrolysis	Arsenic removal	Tripathi et al. (2017)
Rice husk and bagasse	Pyrolysis at 700 °C	Improve quality of building materials	Zeidabadi et al. (2018)

7.4.3 Biochar in Construction

Biochar in construction and building industries can reduce carbon dioxide emissions and achieve carbon neutrality and thus act as carbon sinks. Biochar's enhanced pore structure accelerates carbonation, reducing CO_2 emissions (Senadheera et al., 2023). Research findings indicated that the addition of biochar somewhat improved the cement hydration process; nevertheless, the comparatively big and brittle particles caused microcracks and reduced the strength of the sediment products (Wang et al., 2019). Engineered biochar can be used as a building material because of its flammability, chemical stability, and poor heat conductivity (Pandey et al., 2022).

7.4.4 Biochar in Medicine

Biochar is used in contaminant immobilization to clean up the components that are harmful to human health. In medical waste treatment and recovery of nutrients by recycling phosphate and nitrogen from human excreta, biochar can be utilized. Biochar can also be used as nanocarriers for drug delivery while treating diseases (Zhuo et al., 2023) (Table 6).

8 Conclusions

India is a leading agricultural producer and generates vast amounts of solid waste annually, rich in valuable compounds like complex carbohydrates, proteins, fibers, and polyphenols. This waste poses environmental risks but also offers potential for agricultural, food, and pharmaceutical applications. Biochar, a carbon-rich, porous material produced via pyrolysis of agricultural residues, is gaining attention for its roles in soil remediation, carbon sequestration, and sustainable fuels. The properties of biochar depend on its feedstock, pyrolysis temperature, and treatment. For example, wood-based biochar has the highest surface area, straw-based biochar excels in cation exchange capacity, and manure-based biochars are rich in nitrogen and phosphorus. Incorporating biochar into cement-based materials can sequester carbon and enhance physical properties, including thermal performance and strength development, even at low dosages. To improve biochar's sorption capacity, various modification methods are applied, considering factors like biochar type, preparation techniques, and environmental conditions. Biochar is gaining recognition for its role in sustainable fuels, soil remediation, and carbon sequestration.

References

Adejumo, I. O., & Adebiyi, O. A. (2021). Agricultural solid wastes: Causes, effects, and effective management. In *Strategies of sustainable solid waste management*. IntechOpen. https://doi.org/10.5772/intechopen.93601

Agricultural Waste Diversity and Sustainability Issues. (2021). Elsevier. https://doi.org/10.1016/C2020-0-02631-9

Ahmad, W., Nepal, J., Zou, Z., Munsif, F., Khan, A., Ahmad, I., Zaheer, S., Khan, M. S., Jadoon, S. A., & Tang, D. (2023). Biochar particle size coupled with biofertilizer enhances soil carbon-nitrogen microbial pools and CO2 sequestration in lentil. *Frontiers in Environmental Science, 11*. https://doi.org/10.3389/fenvs.2023.1114728

Ajien, A., Idris, J., Sofwan, N. M., Husen, R., & Seli, H. (2023). Coconut shell and husk biochar: A review of production and activation technology, economic, financial aspect and application. *Waste Management & Research: The Journal for a Sustainable Circular Economy, 41*(1), 37–51. https://doi.org/10.1177/0734242X221127167

Akinyemi, B. A., & Adesina, A. (2020). Recent advancements in the use of biochar for cementitious applications: A review. *Journal of Building Engineering, 32*, 101705. https://doi.org/10.1016/j.jobe.2020.101705

Alam, S. N., Khalid, Z., Guldhe, A., & Singh, B. (2022). Biodiesel and an overview of waste utilization at the various production stages. In B. Singh & A. Guldhe (Eds.), *Waste and biodiesel* (pp. 1–16). Elsevier. https://doi.org/10.1016/B978-0-12-823958-2.00005-7

Alsawy, T., Rashad, E., El-Qelish, M., & Mohammed, R. H. (2022). A comprehensive review on the chemical regeneration of biochar adsorbent for sustainable wastewater treatment. *Npj Clean Water, 5*(1), 29. https://doi.org/10.1038/s41545-022-00172-3

Amalina, F., Razak, A. S. A., Krishnan, S., Sulaiman, H., Zularisam, A. W., & Nasrullah, M. (2022). Biochar production techniques utilizing biomass waste-derived materials and environmental applications – A review. *Journal of Hazardous Materials Advances, 7*, 100134. https://doi.org/10.1016/j.hazadv.2022.100134

Arabi, Z., Eghtedaey, H., Gharehchmaghloo, B., & Faraji, A. (2018). Effects of biochar and biofertilizer on yield and qualitative properties of soybean and some chemical properties of soil. *Arabian Journal of Geosciences, 11*(21), 672. https://doi.org/10.1007/s12517-018-4041-1

Ayyappan, C. S., Bhalambaal, V. M., & Kumar, S. (2018). Effect of biochar on bio-electrochemical dye degradation and energy production. *Bioresource Technology, 251*, 165–170. https://doi.org/10.1016/j.biortech.2017.12.043

Bala, S., Garg, D., Sridhar, K., Inbaraj, B. S., Singh, R., Kamma, S., Tripathi, M., & Sharma, M. (2023). Transformation of agro-waste into value-added bioproducts and bioactive compounds: micro/nano formulations and application in the agri-food-pharma sector. *Bioengineering, 10*(2), 152. https://doi.org/10.3390/bioengineering10020152

Bertola, M., Mattarozzi, M., Sanangelantoni, A. M., Careri, M., & Visioli, G. (2019). PGPB colonizing three-year biochar-amended soil: Towards biochar-mediated biofertilization. *Journal of Soil Science and Plant Nutrition, 19*(4), 841–850. https://doi.org/10.1007/s42729-019-00083-2

Bhatnagar, A., Singhal, A., Tolvanen, H., Valtonen, K., Joronen, T., & Konttinen, J. (2022). Effect of pretreatment and biomass blending on bio-oil and biochar quality from two-step slow pyrolysis of rice straw. *Waste Management, 138*, 298–307. https://doi.org/10.1016/j.wasman.2021.12.013

Bhattacharya, T., Khan, A., Ghosh, T., Kim, J. T., & Rhim, J.-W. (2024). Advances and prospects for biochar utilization in food processing and packaging applications. *Sustainable Materials and Technologies, 39*, e00831. https://doi.org/10.1016/j.susmat.2024.e00831

Bong, C. P. C., Lim, L. Y., Lee, C. T., Ong, P. Y., Jaromír, J., Klemeš, C. L., & Gao, Y. (2020). Lignocellulosic biomass and food waste for biochar production and application: a review. *Chemical Engineering Transactions, 81*, 427–432.

Bušić, A., Kundas, S., Morzak, G., Belskaya, H., Marđetko, N., Ivančić Šantek, M., Komes, D., Novak, S., & Šantek, B. (2018). Recent trends in biodiesel and biogas production. *Food Technology and Biotechnology, 56*(2). https://doi.org/10.17113/ftb.56.02.18.5547

Cai, N., Zhang, H., Nie, J., Deng, Y., & Baeyens, J. (2020). Biochar from biomass slow pyrolysis. *IOP Conference Series: Earth and Environmental Science, 586*(1), 12001. https://doi.org/10.1088/1755-1315/586/1/012001

Campos, P., Miller, A. Z., Knicker, H., Costa-Pereira, M. F., Merino, A., la Rosa, J. M., & De. (2020). Chemical, physical and morphological properties of biochars produced from agricultural residues: Implications for their use as soil amendment. *Waste Management, 105*, 256–267. https://doi.org/10.1016/j.wasman.2020.02.013

Chen, X.-L., Li, F., Chen, H., Wang, H., & Li, G. (2020). Fe2O3/TiO2 functionalized biochar as a heterogeneous catalyst for dyes degradation in water under Fenton processes. *Journal of Environmental Chemical Engineering, 8*(4), 103905. https://doi.org/10.1016/j.jece.2020.103905

Conly, L., Hansen, & Cheong, D. Y. (2007). *Agricultural waste management in food processing*. Handbook of Farm, Dairy, and Food Machinery.

Cotta, M. A. (2012). Ethanol production from lignocellulosic biomass by recombinant Escherichia coli strain FBR5. *Bioengineered, 3*(4), 197–202. https://doi.org/10.4161/bioe.19874

Cui, J., Zhang, F., Li, H., Cui, J., Ren, Y., & Yu, X. (2020). Recent progress in biochar-based photocatalysts for wastewater treatment: Synthesis, mechanisms, and applications. *Applied Sciences, 10*(3), 1019. https://doi.org/10.3390/app10031019

Danapriatna, N., Ismarani, I., & Dede, M. (2023). Application of biochar and biological fertilizer to improve soil quality and *Oryza sativa* L. productivity. *Cogent Food & Agriculture, 9*(1). https://doi.org/10.1080/23311932.2023.2207416

Das, S. K., Ghosh, G. K., & Avasthe, R. (2021). Applications of biomass derived biochar in modern science and technology. *Environmental Technology & Innovation, 21*, 101306. https://doi.org/10.1016/j.eti.2020.101306

Datar, R., Huang, J., Maness, P., Mohagheghi, A., Czernik, S., & Chornet, E. (2007). Hydrogen production from the fermentation of corn stover biomass pretreated with a steam-explosion process. *International Journal of Hydrogen Energy, 32*(8), 932–939. https://doi.org/10.1016/j.ijhydene.2006.09.027

Deng, S., Chen, J., & Chang, J. (2021). Application of biochar as an innovative substrate in constructed wetlands/biofilters for wastewater treatment: Performance and ecological benefits. *Journal of Cleaner Production, 293*, 126156. https://doi.org/10.1016/j.jclepro.2021.126156

Dhyani, V., & Bhaskar, T. (2018). A comprehensive review on the pyrolysis of lignocellulosic biomass. *Renewable Energy, 129*, 695–716. https://doi.org/10.1016/j.renene.2017.04.035

Dhyani, V., & Bhaskar, T. (2019). Pyrolysis of biomass. In *Biofuels: alternative feedstocks and conversion processes for the production of liquid and gaseous biofuels* (pp. 217–244). Elsevier. https://doi.org/10.1016/B978-0-12-816856-1.00009-9

Domingues, R. R., Trugilho, P. F., Silva, C. A., de Melo, I. C. N. A., Melo, L. C. A., Magriotis, Z. M., & Sánchez-Monedero, M. A. (2017). Properties of biochar derived from wood and high-nutrient biomasses with the aim of agronomic and environmental benefits. *PLoS One, 12*(5), e0176884. https://doi.org/10.1371/journal.pone.0176884

Dróżdż, D., Malińska, K., Kacprzak, M., Mrowiec, M., Szczypiór, A., Postawa, P., & Stachowiak, T. (2020). Potential of fish pond sediments composts as organic fertilizers. *Waste and Biomass Valorization, 11*(10), 5151–5163. https://doi.org/10.1007/s12649-020-01074-6

Duque-Acevedo, M., Lancellotti, I., Andreola, F., Barbieri, L., Belmonte-Ureña, L. J., & Camacho-Ferre, F. (2022). Management of agricultural waste biomass as raw material for the construction sector: An analysis of sustainable and circular alternatives. *Environmental Sciences Europe, 34*(1), 70. https://doi.org/10.1186/s12302-022-00655-7

Enaime, G., Baçaoui, A., Yaacoubi, A., & Lübken, M. (2020). Biochar for wastewater treatment – Conversion technologies and applications. *Applied Sciences, 10*(10), 3492. https://doi.org/10.3390/app10103492

Ethaib, S., Omar, R., Kamal, S. M. M., Biak, D. R. A., & Zubaidi, S. L. (2020). Microwave-assisted pyrolysis of biomass waste: a mini review. *Processes, 8*(9), 1190. https://doi.org/10.3390/pr8091190

Fahmy, T. Y. A., Fahmy, Y., Mobarak, F., El-Sakhawy, M., & Abou-Zeid, R. E. (2020). Biomass pyrolysis: Past, present, and future. *Environment, Development and Sustainability, 22*(1), 17–32. https://doi.org/10.1007/s10668-018-0200-5

Gómez, N., Banks, S. W., Nowakowski, D. J., Rosas, J. G., Cara, J., Sánchez, M. E., & Bridgwater, A. V. (2018). Effect of temperature on product performance of a high ash biomass during fast pyrolysis and its bio-oil storage evaluation. *Fuel Processing Technology, 172*, 97–105. https://doi.org/10.1016/j.fuproc.2017.11.021

Gontard, N., Sonesson, U., Birkved, M., Majone, M., Bolzonella, D., Celli, A., Angellier-Coussy, H., Jang, G.-W., Verniquet, A., Broeze, J., Schaer, B., Batista, A. P., & Sebok, A. (2018). A research challenge vision regarding management of agricultural waste in a circular bio-based economy. *Critical Reviews in Environmental Science and Technology, 48*(6), 614–654. https://doi.org/10.1080/10643389.2018.1471957

González-Arias, J., Sánchez, M. E., Cara-Jiménez, J., Baena-Moreno, F. M., & Zhang, Z. (2022). Hydrothermal carbonization of biomass and waste: A review. *Environmental Chemistry Letters, 20*(1), 211–221. https://doi.org/10.1007/s10311-021-01311-x

Gopinath, A., Divyapriya, G., Srivastava, V., Laiju, A. R., Nidheesh, P. V., & Kumar, M. S. (2021). Conversion of sewage sludge into biochar: A potential resource in water and wastewater treatment. *Environmental Research, 194*, 110656. https://doi.org/10.1016/j.envres.2020.110656

Goswami, L., Kushwaha, A., Kafle, S. R., & Kim, B.-S. (2022). Surface modification of biochar for dye removal from wastewater. *Catalysts, 12*(8), 817. https://doi.org/10.3390/catal12080817

Gupta, G. K., Ram, M., Bala, R., Kapur, M., & Mondal, M. K. (2018). Pyrolysis of chemically treated corncob for biochar production and its application in Cr(VI) removal. *Environmental Progress & Sustainable Energy, 37*(5), 1606–1617. https://doi.org/10.1002/ep.12838

Gupta, C., Prakash, D., Gupta, S., & Nazareno, M. A. (2019). Role of vermicomposting in agricultural waste management. In *Sustainable Green Technologies for Environmental Management* (pp. 283–295). Springer. https://doi.org/10.1007/978-981-13-2772-8_15

Gupta, A. P., Upadhyay, P., Sen, T., & Dutta, J. (2023). Agricultural waste as a resource. In *Agriculture Waste Management and Bioresource* (pp. 1–20). Wiley. https://doi.org/10.1002/9781119808428.ch1

Haddad, K., Jeguirim, M., Jellali, S., Thevenin, N., Ruidavets, L., & Limousy, L. (2021). Biochar production from Cypress sawdust and olive mill wastewater: Agronomic approach. *Science of the Total Environment, 752*, 141713. https://doi.org/10.1016/j.scitotenv.2020.141713

Haider, F. U., Coulter, J. A., Cai, L., Hussain, S., Cheema, S. A., Wu, J., & Zhang, R. (2022). An overview on biochar production, its implications, and mechanisms of biochar-induced amelioration of soil and plant characteristics. *Pedosphere, 32*(1), 107–130. https://doi.org/10.1016/S1002-0160(20)60094-7

Hamdi, G. M. H., Nsaif Abbas, M., & Atalah Khaleefa Ali, S. (2024). Bioethanol production from agricultural waste: A review. *Journal of Engineering and Sustainable Development, 28*(02), 233–252. https://doi.org/10.31272/jeasd.28.2.7

Harinikumar, K. M., Kudahettige-Nilsson, R. L., Devadas, A., Holmgren, M., & Sellstedt, A. (2020). Bioethanol production from four abundant Indian agricultural wastes. *Biofuels, 11*(5), 607–613. https://doi.org/10.1080/17597269.2017.1387744

Harshwardhan, K., & Upadhyay, K. (2017). Effective utilization of agricultural waste: Review. *Journal of Fundamentals of Renewable Energy and Applications, 07*(05). https://doi.org/10.4172/2090-4541.1000237

He, C., Tang, C., Li, C., Yuan, J., Tran, K.-Q., Bach, Q.-V., Qiu, R., & Yang, Y. (2018). Wet torrefaction of biomass for high quality solid fuel production: A review. *Renewable and Sustainable Energy Reviews, 91*, 259–271. https://doi.org/10.1016/j.rser.2018.03.097

He, M., Xu, Z., Hou, D., Gao, B., Cao, X., Ok, Y. S., Rinklebe, J., Bolan, N. S., & Tsang, D. C. W. (2022). Waste-derived biochar for water pollution control and sustainable development. *Nature Reviews Earth & Environment, 3*(7), 444–460. https://doi.org/10.1038/s43017-022-00306-8

Hirst, E. A., Taylor, A., & Mokaya, R. (2018). A simple flash carbonization route for conversion of biomass to porous carbons with high CO_2 storage capacity. *Journal of Materials Chemistry A, 6*(26), 12393–12403. https://doi.org/10.1039/C8TA04409K

Hu, X., & Gholizadeh, M. (2019). Biomass pyrolysis: A review of the process development and challenges from initial researches up to the commercialisation stage. *Journal of Energy Chemistry, 39*, 109–143. https://doi.org/10.1016/j.jechem.2019.01.024

Huggins, T., Wang, H., Kearns, J., Jenkins, P., & Ren, Z. J. (2014). Biochar as a sustainable electrode material for electricity production in microbial fuel cells. *Bioresource Technology, 157*, 114–119. https://doi.org/10.1016/j.biortech.2014.01.058

Ibitoye, S. E., Mahamood, R. M., Jen, T.-C., Loha, C., & Akinlabi, E. T. (2023). An overview of biomass solid fuels: Biomass sources, processing methods, and morphological and microstructural properties. *Journal of Bioresources and Bioproducts, 8*(4), 333–360. https://doi.org/10.1016/j.jobab.2023.09.005

Ighalo, J. O., Iwuchukwu, F. U., Eyankware, O. E., Iwuozor, K. O., Olotu, K., Bright, O. C., & Igwegbe, C. A. (2022). Flash pyrolysis of biomass: A review of recent advances. *Clean Technologies and Environmental Policy, 24*(8), 2349–2363. https://doi.org/10.1007/s10098-022-02339-5

Ivanova, G., Rákhely, G., & Kovács, K. L. (2009). Thermophilic biohydrogen production from energy plants by Caldicellulosiruptor saccharolyticus and comparison with related studies. *International Journal of Hydrogen Energy, 34*(9), 3659–3670. https://doi.org/10.1016/j.ijhydene.2009.02.082

Jahromi, R., Rezaei, M., Samadi, S. H., & Jahromi, H. (2021). Biomass gasification in a downdraft fixed-bed gasifier: Optimization of operating conditions. *Chemical Engineering Science, 231*, 116249. https://doi.org/10.1016/j.ces.2020.116249

Jang, H. M., & Kan, E. (2019). Engineered biochar from agricultural waste for removal of tetracycline in water. *Bioresource Technology, 284*, 437–447. https://doi.org/10.1016/j.biortech.2019.03.131

Jatav, H. S., Rajput, V. D., Minkina, T., Singh, S. K., Chejara, S., Gorovtsov, A., Barakhov, A., Bauer, T., Sushkova, S., Mandzhieva, S., Burachevskaya, M., & Kalinitchenko, V. P. (2021). Sustainable approach and safe use of biochar and its possible consequences. *Sustainability, 13*(18), 10362. https://doi.org/10.3390/su131810362

Jiang, J., Zhang, L., Wang, X., Holm, N., Rajagopalan, K., Chen, F., & Ma, S. (2013). Highly ordered macroporous woody biochar with ultra-high carbon content as supercapacitor electrodes. *Electrochimica Acta, 113*, 481–489. https://doi.org/10.1016/j.electacta.2013.09.121

Joseph, S., Cowie, A. L., Van Zwieten, L., Bolan, N., Budai, A., Buss, W., Cayuela, M. L., Graber, E. R., Ippolito, J. A., Kuzyakov, Y., Luo, Y., Ok, Y. S., Palansooriya, K. N., Shepherd, J., Stephens, S., Weng, Z. H., & Lehmann, J. (2021). How biochar works, and when it doesn't: A review of mechanisms controlling soil and plant responses to biochar. *GCB Bioenergy, 13*(11), 1731–1764. https://doi.org/10.1111/gcbb.12885

Joshi, P. V. K. (2020). Fruit and vegetable processing waste management – An overview. *International Journal of Food and Fermentation Technology, 10*(2). https://doi.org/10.30954/2277-9396.02.2020.4

Kaab, A., Sharifi, M., Mobli, H., Nabavi-Pelesaraei, A., & Chau, K. (2019). Combined life cycle assessment and artificial intelligence for prediction of output energy and environmental impacts of sugarcane production. *Science of the Total Environment, 664*, 1005–1019. https://doi.org/10.1016/j.scitotenv.2019.02.004

Kachroo, H., Verma, V. K., Doddapaneni, T. R. K. C., Kaushal, P., & Jain, R. (2024). Organic/metallic component analysis of lignocellulosic biomass: A thermochemical-perspective-based study on rice and bamboo waste. *Bioresource Technology, 403*, 130835. https://doi.org/10.1016/j.biortech.2024.130835

Kaetzl, K., Lübken, M., Gehring, T., & Wichern, M. (2018). Efficient low-cost anaerobic treatment of wastewater using biochar and woodchip filters. *Water, 10*(7), 818. https://doi.org/10.3390/w10070818

Kaetzl, K., Lübken, M., Uzun, G., Gehring, T., Nettmann, E., Stenchly, K., & Wichern, M. (2019). On-farm wastewater treatment using biochar from local agroresidues reduces pathogens from irrigation water for safer food production in developing countries. *Science of the Total Environment, 682*, 601–610. https://doi.org/10.1016/j.scitotenv.2019.05.142

Kaetzl, K., Lübken, M., Nettmann, E., Krimmler, S., & Wichern, M. (2020). Slow sand filtration of raw wastewater using biochar as an alternative filtration media. *Scientific Reports, 10*(1), 1229. https://doi.org/10.1038/s41598-020-57981-0

Kalus, K., Koziel, J., & Opaliński, S. (2019). A review of biochar properties and their utilization in crop agriculture and livestock production. *Applied Sciences, 9*(17), 3494. https://doi.org/10.3390/app9173494

Kamali, M., Appels, L., Kwon, E. E., Aminabhavi, T. M., & Dewil, R. (2021). Biochar in water and wastewater treatment – A sustainability assessment. *Chemical Engineering Journal, 420*, 129946. https://doi.org/10.1016/j.cej.2021.129946

Kappler, G., & Alberto Mendes Moraes CMORAES, C. (n.d.). *Waste to energy and materials through pyrolysis: A review ABSTRACT*. https://periodicos.utfpr.edu.br/rts

Karam, D. S., Nagabovanalli, P., Rajoo, K. S., Ishak, C. F., Abdu, A., Rosli, Z., Muharam, F. M., & Zulperi, D. (2022). An overview on the preparation of rice husk biochar, factors affecting its properties, and its agriculture application. *Journal of the Saudi Society of Agricultural Sciences, 21*(3), 149–159. https://doi.org/10.1016/j.jssas.2021.07.005

Karanth, S., Feng, S., Patra, D., & Pradhan, A. K. (2023). Linking microbial contamination to food spoilage and food waste: the role of smart packaging, spoilage risk assessments, and date labeling. *Frontiers in Microbiology, 14*. https://doi.org/10.3389/fmicb.2023.1198124

Kavan Kumar, V., & Panwar, N. L. (2024). Pyrolysis technologies for biochar production in waste management: A review. *Clean Energy, 8*(4), 61–78. https://doi.org/10.1093/ce/zkae036

Kazawadi, D., Ntalikwa, J., & Kombe, G. (2021). A review of intermediate pyrolysis as a technology of biomass conversion for coproduction of biooil and adsorption biochar. *Journal of Renewable Energy, 2021*, 1–10. https://doi.org/10.1155/2021/5533780

Khawkomol, S., Neamchan, R., Thongsamer, T., Vinitnantharat, S., Panpradit, B., Sohsalam, P., Werner, D., & Mrozik, W. (2021). Potential of biochar derived from agricultural residues for sustainable management. *Sustainability, 13*(15), 8147. https://doi.org/10.3390/su13158147

Kumar, A., & Bhattacharya, T. (2021). Biochar: A sustainable solution. *Environment, Development and Sustainability, 23*(5), 6642–6680. https://doi.org/10.1007/s10668-020-00970-0

Kumari, K., Khalid, Z., Alam, S. N., Sweta, B. S., Guldhe, A., Shahi, D. K., & Bauddh, K. (2020). Biochar amendment in agricultural soil for mitigation of abiotic stress. In K. Bauddh, S. Kumar, R. Singh, & J. Korstad (Eds.), *Ecological and practical applications for sustainable agriculture* (pp. 305–344). Springer. https://doi.org/10.1007/978-981-15-3372-3_14

Kumari, S., Debbarma, R., Habibi, M., Haque, S., & Suprasana, P. (2025). Banana waste valorisation and the development of biodegradable biofilms. *Waste Management Bulletin, 3*, 100213. 10.1016/j.wmb.2025.100213.

Lahiri, A., Daniel, S., Kanthapazham, R., Vanaraj, R., Thambidurai, A., & Peter, L. S. (2023). A critical review on food waste management for the production of materials and biofuel. *Journal of Hazardous Materials Advances, 10*, 100266. https://doi.org/10.1016/j.hazadv.2023.100266

Lee, J., Kim, K.-H., & Kwon, E. E. (2017). Biochar as a catalyst. *Renewable and Sustainable Energy Reviews, 77*, 70–79. https://doi.org/10.1016/j.rser.2017.04.002

Li, F., Xie, Y., Wang, Y., Fan, X., Cai, Y., & Mei, Y. (2019). Improvement of dyes degradation using hydrofluoric acid modified biochar as persulfate activator. *Environmental Pollutants and Bioavailability, 31*(1), 32–37. https://doi.org/10.1080/26395940.2019.1578185

Li, W., Cheng, C., He, L., Liu, M., Cao, G., Yang, S., & Ren, N. (2021). Effects of feedstock and pyrolysis temperature of biochar on promoting hydrogen production of ethanol-type fermentation. *Science of the Total Environment, 790*, 148206. https://doi.org/10.1016/j.scitotenv.2021.148206

Li, Y., Gupta, R., Zhang, Q., & You, S. (2023). Review of biochar production via crop residue pyrolysis: Development and perspectives. In *Bioresource Technology* (Vol. 369). Elsevier Ltd.. https://doi.org/10.1016/j.biortech.2022.128423

Lu, Y., Cai, Y., Zhang, S., Zhuang, L., Hu, B., Wang, S., Chen, J., & Wang, X. (2022). Application of biochar-based photocatalysts for adsorption-(photo)degradation/reduction of environmental contaminants: mechanism, challenges and perspective. *Biochar, 4*(1), 45. https://doi.org/10.1007/s42773-022-00173-y

Lyu, H., Zhang, Q., & Shen, B. (2020). Application of biochar and its composites in catalysis. *Chemosphere, 240*, 124842. https://doi.org/10.1016/j.chemosphere.2019.124842

Mahanty, B., & Mondal, S. (2021). Synthesis of magnetic biochar using agricultural waste for the separation of Cr(VI) from aqueous solution. *Arabian Journal for Science and Engineering, 46*(11), 10803–10818. https://doi.org/10.1007/s13369-021-05572-0

Mallappa Munnoli, P., Teixeira da Silva, J. A., & Bhosle, S. (n.d.). *Dynamic soil, dynamic plant dynamics of the soil-earthworm-plant relationship: A review*.

Mamo, W., & Zewide, I. (2022). Citation: Wondimu Mamo and Israel Zewide (2022) Review on agricultural waste utilization. *British Journal of Earth Sciences Research, 10*(1), 25–34. https://doi.org/10.37745/bjesr.2013

Manyà, J. J., Azuara, M., & Manso, J. A. (2018). Biochar production through slow pyrolysis of different biomass materials: Seeking the best operating conditions. *Biomass and Bioenergy, 117*, 115–123. https://doi.org/10.1016/j.biombioe.2018.07.019

Manyuchi, M. M., Mbohwa, C., & Muzenda, E. (2018). Potential to use municipal waste bio char in wastewater treatment for nutrients recovery. *Physics and Chemistry of the Earth, Parts A/B/C, 107*, 92–95. https://doi.org/10.1016/j.pce.2018.07.002

Maraveas, C. (2020). Production of sustainable and biodegradable polymers from agricultural waste. *Polymers, 12*(5), 1127. https://doi.org/10.3390/polym12051127

Meng, F., Wang, D., & Zhang, M. (2021). Effects of different pretreatment methods on biochar properties from pyrolysis of corn stover. *Journal of the Energy Institute, 98*, 294–302. https://doi.org/10.1016/j.joei.2021.07.008

Michalak, I., Baśladyńska, S., Mokrzycki, J., & Rutkowski, P. (2019). Biochar from a freshwater macroalga as a potential biosorbent for wastewater treatment. *Water, 11*(7), 1390. https://doi.org/10.3390/w11071390

Misra, V., Mall, A. K., Srivastava, S., & Pathak, A. D. (2023). Expansion of agricultural residues to biofuel processing and production. In S. Singh, P. Singh, A. Sharma, & M. Choudhury (Eds.), *Agriculture waste management and bioresource*. https://doi.org/10.1002/9781119808428.ch13

Molenda, J., Swat, M., & Osuch-Słomka, E. (2018). Effect of thermal conditions of pyrolysis process on the quality of biochar obtained from vegetable waste. *Engineering and Protection of Environment, 21*(3), 289–302. https://doi.org/10.17512/ios.2018.3.7

Mousavi, S. A., Shokuhfar, A., Lak, S., Mojaddam, M., & Alavifazel, M. (2020). Integrated application of biochar and bio-fertilizer improves yield and yield components of Cowpea under water-deficient stress. *Italian Journal of Agronomy, 15*(2), 94–101. https://doi.org/10.4081/ija.2020.1581

Nadarajah, K., Asharp, T., & Jeganathan, Y. (2024). Biochar from waste biomass, its fundamentals, engineering aspects, and potential applications: An overview. *Water Science & Technology, 89*(5), 1211–1239. https://doi.org/10.2166/wst.2024.051

Niu, Y., Lv, Y., Lei, Y., Liu, S., Liang, Y., Wang, D., & Hui, S. (2019). Biomass torrefaction: Properties, applications, challenges, and economy. *Renewable and Sustainable Energy Reviews, 115*, 109395. https://doi.org/10.1016/j.rser.2019.109395

Nobaharan, K., Bagheri Novair, S., Asgari Lajayer, B., & van Hullebusch, E. (2021). Phosphorus removal from wastewater: The potential use of biochar and the key controlling factors. *Water, 13*(4), 517. https://doi.org/10.3390/w13040517

Núñez-Delgado, A. (Ed.). (2024). *Frontier studies in soil science*. Springer. https://doi.org/10.1007/978-3-031-50503-4

Obey, G., Adelaide, M., & Ramaraj, R. (2022). Biochar derived from non-customized matamba fruit shell as an adsorbent for wastewater treatment. *Journal of Bioresources and Bioproducts, 7*(2), 109–115. https://doi.org/10.1016/j.jobab.2021.12.001

Olatunji, K. O., Ahmed, N. A., & Ogunkunle, O. (2021). Optimization of biogas yield from lignocellulosic materials with different pretreatment methods: A review. *Biotechnology for Biofuels, 14*(1), 159. https://doi.org/10.1186/s13068-021-02012-x

Pandey, D., Daverey, A., & Arunachalam, K. (2020). Biochar: Production, properties and emerging role as a support for enzyme immobilization. *Journal of Cleaner Production, 255*, 120267. https://doi.org/10.1016/j.jclepro.2020.120267

Pandey, D., Chhimwal, M., & Srivastava, R. K. (2022). Engineered biochar as construction material. In *Engineered biochar fundamentals, preparation, characterization and applications* (pp. 303–318).

Panwar, N. L., Pawar, A., & Salvi, B. L. (2019). Comprehensive review on production and utilization of biochar. *SN Applied Sciences, 1*(2), 168. https://doi.org/10.1007/s42452-019-0172-6

Pereira, S. C., Maehara, L., Machado, C. M. M., & Farinas, C. S. (2015). 2G ethanol from the whole sugarcane lignocellulosic biomass. *Biotechnology for Biofuels, 8*(1), 44. https://doi.org/10.1186/s13068-015-0224-0

Perez-Mercado, L. F., Lalander, C., Berger, C., & Dalahmeh, S. S. (2018). Potential of biochar filters for onsite wastewater treatment: Effects of biochar type, physical properties and operating conditions. *Water, 10*(12), 1835. https://doi.org/10.3390/w10121835

Pradhan, S., Abdelaal, A. H., Mroue, K., Al-Ansari, T., Mackey, H. R., & McKay, G. (2020). Biochar from vegetable wastes: Agro-environmental characterization. *Biochar, 2*(4), 439–453. https://doi.org/10.1007/s42773-020-00069-9

Prakash Sharma, I., Kanta, B. C. G., Singh Gusain B. Y. G., Kanta, C., & Singh Gusain, Y. (2018). *Crop residues utilization: Wheat, paddy, cotton, sugarcane and groundnut*. www.botanyjournals.com

Prasanna Kumar, D. J., Mishra, R. K., Chinnam, S., Binnal, P., & Dwivedi, N. (2024). A comprehensive study on anaerobic digestion of organic solid waste: A review on configurations, operating parameters, techno-economic analysis and current trends. In *Biotechnology notes* (Vol. 5, pp. 33–49). KeAi Communications Co.. https://doi.org/10.1016/j.biotno.2024.02.001

Praveen, S., Jegan, J., Pushpa, T. B., Gokulan, R., & Bulgariu, L. (2022). Biochar for removal of dyes in contaminated water: An overview. *Biochar, 4*(1), 10. https://doi.org/10.1007/s42773-022-00131-8

Puglia, D., Pezzolla, D., Gigliotti, G., Torre, L., Bartucca, M. L., & Del Buono, D. (2021). The opportunity of valorizing agricultural waste, through its conversion into biostimulants, biofertilizers, and biopolymers. *Sustainability, 13*(5), 2710. https://doi.org/10.3390/su13052710

Rayne, N., & Aula, L. (2020). Livestock manure and the impacts on soil health: A review. *Soil Systems, 4*(4), 64. https://doi.org/10.3390/soilsystems4040064

Riseh, R. S., Vazvani, M. G., Hassanisaadi, M., & Thakur, V. K. (2024). Agricultural wastes: A practical and potential source for the isolation and preparation of cellulose and application in agriculture and different industries. *Industrial Crops and Products, 208*, 117904. https://doi.org/10.1016/j.indcrop.2023.117904

Robinson, J., Binner, E., Vallejo, D. B., Perez, N. D., Al Mughairi, K., Ryan, J., Shepherd, B., Adam, M., Budarin, V., Fan, J., Gronnow, M., & Peneranda-Foix, F. (2022). Unravelling the mechanisms of microwave pyrolysis of biomass. *Chemical Engineering Journal, 430*, 132975. https://doi.org/10.1016/j.cej.2021.132975

Román, S., Libra, J., Berge, N., Sabio, E., Ro, K., Li, L., Ledesma, B., Álvarez, A., & Bae, S. (2018). Hydrothermal carbonization: Modeling, final properties design and applications: a review. *Energies, 11*(1), 216. https://doi.org/10.3390/en11010216

Rudoy, D., Odabashyan, M., Olshevskaya, A., Rybak, A., Kovalchuk, D., Kulikova, N., Prutskov, A., & Sarkisian, D. (2024). Methods for ensuring the processing and disposal of agricultural waste. *BIO Web of Conferences, 113*, 5011. https://doi.org/10.1051/bioconf/202411305011

Safarian, S., Unnþórsson, R., & Richter, C. (2019). A review of biomass gasification modelling. *Renewable and Sustainable Energy Reviews, 110*, 378–391. https://doi.org/10.1016/j.rser.2019.05.003

Saha, B. C., & Cotta, M. A. (2008). Lime pretreatment, enzymatic saccharification and fermentation of rice hulls to ethanol. *Biomass and Bioenergy, 32*(10), 971–977. https://doi.org/10.1016/j.biombioe.2008.01.014

Saini, J. K., Saini, R., & Tewari, L. (2015). Lignocellulosic agriculture wastes as biomass feedstocks for second-generation bioethanol production: Concepts and recent developments. *3 Biotech, 5*(4), 337–353. https://doi.org/10.1007/s13205-014-0246-5

Sakib, K. M. N., Ahmed, S. M. N., Mubdee, A., & Kirtania, K. (2021). Biochar production from waste biomass using modular pyrolyzer for soil amendment. *Chemical Engineering Research Bulletin*, 14–19. https://doi.org/10.3329/cerb.v22i1.54293

Searchinger, T., Heimlich, R., Houghton, R. A., Dong, F., Elobeid, A., Fabiosa, J., Tokgoz, S., Hayes, D., & Yu, T.-H. (2008). Use of U.S. croplands for biofuels increases greenhouse gases through emissions from land-use change. *Science, 319*(5867), 1238–1240. https://doi.org/10.1126/science.1151861

Senadheera, S. S., Gupta, S., Kua, H. W., Hou, D., Kim, S., Tsang, D. C. W., & Ok, Y. S. (2023). Application of biochar in concrete – A review. *Cement and Concrete Composites, 143*, 105204. https://doi.org/10.1016/j.cemconcomp.2023.105204

Shikha, F. S., Rahman, M. M., Sultana, N., Mottalib, M. A., & Yasmin, M. (2023). Effects of biochar and biofertilizer on groundnut production: A perspective for environmental sustainability in Bangladesh. *Carbon Research, 2*(1), 10. https://doi.org/10.1007/s44246-023-00043-7

Shyamsundar, P., Springer, N. P., Tallis, H., Polasky, S., Jat, M. L., Sidhu, H. S., Krishnapriya, P. P., Skiba, N., Ginn, W., Ahuja, V., Cummins, J., Datta, I., Dholakia, H. H., Dixon, J., Gerard, B., Gupta, R., Hellmann, J., Jadhav, A., Jat, H. S., et al. (2019). Fields on fire: Alternatives to crop residue burning in India. *Science, 365*(6453), 536–538. https://doi.org/10.1126/science.aaw4085

Singh, Y., & Sidhu, H. S. (2014). Management of cereal crop residues for sustainable rice-wheat production system in the Indo-gangetic Plains of India. *Proceedings of the Indian National Science Academy, 80*(1), 95. https://doi.org/10.16943/ptinsa/2014/v80i1/55089

Soares, J. F., Confortin, T. C., Todero, I., Mayer, F. D., & Mazutti, M. A. (2020). Dark fermentative biohydrogen production from lignocellulosic biomass: Technological challenges and future prospects. *Renewable and Sustainable Energy Reviews, 117*, 109484. https://doi.org/10.1016/j.rser.2019.109484

Sonu, K., Sogani, M., Syed, Z., Dongre, A., & Sharma, G. (2020). Effect of corncob derived biochar on microbial electroremediation of dye wastewater and bioenergy generation. *ChemistrySelect, 5*(31), 9793–9798. https://doi.org/10.1002/slct.202002652

SriBala, G., Carstensen, H., Van Geem, K. M., & Marin, G. B. (2019). Measuring biomass fast pyrolysis kinetics: State of the art. *WIREs Energy and Environment, 8*(2). https://doi.org/10.1002/wene.326

Srivatsav, P., Bhargav, B. S., Shanmugasundaram, V., Arun, J., Gopinath, K. P., & Bhatnagar, A. (2020). Biochar as an eco-friendly and economical adsorbent for the removal of colorants (dyes) from aqueous environment: A review. *Water, 12*(12), 3561. https://doi.org/10.3390/w12123561

Stephanie, H., Mlsna, T. E., & Wipf, D. O. (2021). Functionalized biochar electrodes for asymmetrical capacitive deionization. *Desalination, 516*, 115240. https://doi.org/10.1016/j.desal.2021.115240

Sun, H., Zhang, Y., Yang, Y., Chen, Y., Jeyakumar, P., Shao, Q., Zhou, Y., Ma, M., Zhu, R., Qian, Q., Fan, Y., Xiang, S., Zhai, N., Li, Y., Zhao, Q., & Wang, H. (2021). Effect of biofertilizer and wheat straw biochar application on nitrous oxide emission and ammonia volatilization from paddy soil. *Environmental Pollution, 275*, 116640. https://doi.org/10.1016/j.envpol.2021.116640

Tasim, B., Masood, T., Shah, Z. A., Arif, M., Ullah, A., Miraj, G., & uhammad Samiullah. (2019). Quality evaluation of biochar prepared from different agricultural residues. *Sarhad Journal of Agriculture, 35*(1). https://doi.org/10.17582/journal.sja/2019/35.1.134.143

Tezer, Ö., Karabağ, N., Öngen, A., Çolpan, C. Ö., & Ayol, A. (2022). Biomass gasification for sustainable energy production: A review. *International Journal of Hydrogen Energy, 47*(34), 15419–15433. https://doi.org/10.1016/j.ijhydene.2022.02.158

Tomczyk, A., Sokołowska, Z., & Boguta, P. (2020). Biochar physicochemical properties: Pyrolysis temperature and feedstock kind effects. *Reviews in Environmental Science and Bio/Technology, 19*(1), 191–215. https://doi.org/10.1007/s11157-020-09523-3

Tripathi, M., Tripathi, M., Mubarak, N. M., Mubarak, N. M., Sahu, J. N., Sahu, J. N., Sahu, J. N., Sahu, J. N., & Ganesan, P. (2017). Overview on synthesis of magnetic bio char from discarded agricultural biomass. In *Handbook of composites from renewable materials* (pp. 435–460). Wiley. https://doi.org/10.1002/9781119441632.ch16

Uddin, M. N., Techato, K., Taweekun, J., Rahman, M. M., Rasul, M. G., Mahlia, T. M. I., & Ashrafur, S. M. (2018). An overview of recent developments in biomass pyrolysis technologies. *Energies, 11*(11), 3115. https://doi.org/10.3390/en11113115

Ugwuoke, C. U., Monwuba, N., Onu, F. M., Shimave, A. G., Okonkwo, E. N., & Opurum, C. C. (2018). Impact of agricultural waste on sustainable environment and health of rural women. *Civil and Environmental Research, 10*(9).

Upadhyay, K. (n.d.). *Effective utilization of agricultural waste – Review paper*. www.ijert.org

Usman, M. N., Suleiman, M. A., & Mi, B. (2021). Anaerobic digestion of agricultural wastes: A potential remedy for energy shortfalls in Nigeria. *Journal of Waste Management and Disposal, 4*, 104. www.scholarena.com

Venderbosch, R. H. (2019). Fast pyrolysis. In *Thermochemical processing of biomass* (pp. 175–206). Wiley. https://doi.org/10.1002/9781119417637.ch6

Vikram, S., Rosha, P., & Kumar, S. (2021). Recent modeling approaches to biomass pyrolysis: A review. *Energy & Fuels, 35*(9), 7406–7433. https://doi.org/10.1021/acs.energyfuels.1c00251

Wang, J., & Wang, S. (2019). Preparation, modification and environmental application of biochar: A review. *Journal of Cleaner Production, 227*, 1002–1022. https://doi.org/10.1016/j.jclepro.2019.04.282

Wang, Y.-S., Li, D.-B., Zhang, F., Tong, Z.-H., & Yu, H.-Q. (2018). Algal biomass derived biochar anode for efficient extracellular electron uptake from Shewanella oneidensis MR-1. *Frontiers of Environmental Science & Engineering, 12*(4), 11. https://doi.org/10.1007/s11783-018-1072-5

Wang, L., Chen, L., Tsang, D. C. W., Kua, H. W., Yang, J., Ok, Y. S., Ding, S., Hou, D., & Poon, C. S. (2019). The roles of biochar as green admixture for sediment-based construction products. *Cement and Concrete Composites, 104*, 103348. https://doi.org/10.1016/j.cemconcomp.2019.103348

Wang, S., Zhang, H., Huang, H., Xiao, R., Li, R., & Zhang, Z. (2020a). Influence of temperature and residence time on characteristics of biochars derived from agricultural residues: A comprehensive evaluation. *Process Safety and Environmental Protection, 139*, 218–229. https://doi.org/10.1016/j.psep.2020.03.028

Wang, X., Guo, Z., Hu, Z., & Zhang, J. (2020b). Recent advances in biochar application for water and wastewater treatment: A review. *PeerJ, 8*, e9164. https://doi.org/10.7717/peerj.9164

Wang, L., Olsen, M. N. P., Moni, C., Dieguez-Alonso, A., de la Rosa, J. M., Stenrød, M., Liu, X., & Mao, L. (2022). Comparison of properties of biochar produced from different types of lignocellulosic biomass by slow pyrolysis at 600°C. *Applications in Energy and Combustion Science, 12*, 100090. https://doi.org/10.1016/j.jaecs.2022.100090

Wang, N., He, Y., Zhao, K., Lin, X., He, X., Chen, A., Wu, G., Zhang, J., Yan, B., Luo, L., & Xu, D. (2024). Greenhouse gas emission characteristics and influencing factors of agricultural waste composting process: A review. *Journal of Environmental Management, 354*, 120337. https://doi.org/10.1016/j.jenvman.2024.120337

Waqas, M., Hashim, S., Humphries, U. W., Ahmad, S., Noor, R., Shoaib, M., Naseem, A., Hlaing, P. T., & Lin, H. A. (2023). Composting processes for agricultural waste management: A comprehensive review. In *Processes* (Vol. 11, Issue 3). Multidisciplinary Digital Publishing Institute (MDPI). https://doi.org/10.3390/pr11030731

Weber, K., & Quicker, P. (2018). Properties of biochar. *Fuel, 217*, 240–261. https://doi.org/10.1016/j.fuel.2017.12.054

Weidner, E., Karbassiyazdi, E., Altaee, A., Jesionowski, T., & Ciesielczyk, F. (2022). Hybrid metal oxide/biochar materials for wastewater treatment technology: A review. *ACS Omega, 7*(31), 27062–27078. https://doi.org/10.1021/acsomega.2c02909

Wijitkosum, S. (2022). Biochar derived from agricultural wastes and wood residues for sustainable agricultural and environmental applications. *International Soil and Water Conservation Research, 10*(2), 335–341. https://doi.org/10.1016/j.iswcr.2021.09.006

Xiang, W., Zhang, X., Chen, J., Zou, W., He, F., Hu, X., Tsang, D. C. W., Ok, Y. S., & Gao, B. (2020). Biochar technology in wastewater treatment: A critical review. *Chemosphere, 252*, 126539. https://doi.org/10.1016/j.chemosphere.2020.126539

Yazid, M., Pusfasari, W., & Wildayana, E. (2020). Social, economic and ecological benefits and farmers' perception of agricultural waste processing in Banyuasin Regency. *IOP Conference Series: Earth and Environmental Science, 473*(1), 12020. https://doi.org/10.1088/1755-1315/473/1/012020

Zadeh, Z. E., Abdulkhani, A., Aboelazayem, O., & Saha, B. (2020). Recent Insights into lignocellulosic biomass pyrolysis: A critical review on pretreatment, characterization, and products upgrading. *Processes, 8*(7), 799. https://doi.org/10.3390/pr8070799

Zeidabadi, Z. A., Bakhtiari, S., Abbaslou, H., & Ghanizadeh, A. R. (2018). Synthesis, characterization and evaluation of biochar from agricultural waste biomass for use in building materials. *Construction and Building Materials, 181*, 301–308. https://doi.org/10.1016/j.conbuildmat.2018.05.271

Zhang, C., Liu, L., Zhao, M., Rong, H., & Xu, Y. (2018). The environmental characteristics and applications of biochar. *Environmental Science and Pollution Research, 25*(22), 21525–21534. https://doi.org/10.1007/s11356-018-2521-1

Zheng, Y., Zhang, Q., Zhang, Z., Jing, Y., Hu, J., He, C., & Lu, C. (2022). A review on biological recycling in agricultural waste-based biohydrogen production: Recent developments. *Bioresource Technology, 347*, 126595. https://doi.org/10.1016/j.biortech.2021.126595

Zhou, Y., Zhang, L., & Cheng, Z. (2015). Removal of organic pollutants from aqueous solution using agricultural wastes: A review. *Journal of Molecular Liquids, 212*, 739–762. https://doi.org/10.1016/j.molliq.2015.10.023

Zhuo, Q., Liang, Y., Hu, Y., Shi, M., Zhao, C., & Zhang, S. (2023). Applications of biochar in medical and related environmental fields: current status and future perspectives. *Carbon Research, 2*(1), 32. https://doi.org/10.1007/s44246-023-00066-0

Zubair, M., Ihsanullah, I., Aziz, H. A., Ahmad, M. A., & Al-Harthi, M. A. (2021). Sustainable wastewater treatment by biochar/layered double hydroxide composites: Progress, challenges, and outlook. *Bioresource Technology, 319*, 124128. https://doi.org/10.1016/j.biortech.2020.124128

Biochar from Animal Waste and Its Applications

Sharmistha Majumder, Swetanjana Ghosh, Urvashi Lama, Archita Dey, Debadrita Das, Antara Das, Barun Kumar Thakur, Monojit Mondal, and Tarit Roychowdhury

Abstract

Animal waste-derived biochar, a natural product, can be used in a wide range of agricultural and biological strata for sustainable environmental management. It represents a promising method to enhance environmental fitness by potential ways of waste reduction, greenhouse gas mitigation, land reclamation, arresting of toxic chemicals from soil-plant system, etc. This article has sumptuously discussed animal waste biochar sources, significant preparative processes, biochar characteristics, parameters regulating biochar quality, application and impacts on soil properties, soil-plant nutrients, plant growth and yield, waste water treatment, waste management, etc., through a massive literature review. Biochar derived from various animal wastes suggests distinct advantages, high nutrient content (N, P, K), and varied chemical compositions that cater to the particular requirements of soils. Applying an ideal pyrolysis temperature for precise feedstock can improve the effectiveness of biochar while also decreasing the toxic substance of its raw ingredients. Being discussed the chemistry, application, and pros and cons of animal-origin biochar at a single platform will help future researchers for finding the optimum features for best output.

S. Majumder · S. Ghosh · U. Lama · A. Dey · D. Das · M. Mondal · T. Roychowdhury (✉)
School of Environmental Studies, Jadavpur University, Kolkata, India

A. Das
School of Environmental Studies, Jadavpur University, Kolkata, India

Department of Economics, FLAME University, Pune, Maharashtra, India

B. K. Thakur
Department of Economics, FLAME University, Pune, Maharashtra, India

S. Singh, S. Srivastava (eds.), *Biochar: A Precious Resource from Biological Waste*,
https://doi.org/10.1007/978-981-95-0425-1_7

Keywords

Animal waste-derived biochar · Biochar feedstock · Pyrolysis temperature · Soil-plant system · Environmental management

1 Introduction

Biochar, a carbon-dense substance produced by pyrolyzing organic materials, has gained considerable interest for its contributions to sustainable agriculture and environmental management. Produced from a range of feedstocks, such as animal waste, biochar is valued for its ability to boost soil health, increase agricultural productivity, and tackle environmental issues related to waste management and greenhouse gas emissions (Mandal et al., 2016; Côrtes et al., 2019). Animal waste, including manure from poultry, cattle, pigs, sheep, and goats, is an excellent feedstock for biochar production because of its rich organic content and nutrient density. During pyrolysis, the organic material is heated in a low-oxygen environment, converting the waste into biochar, a stable carbon compound that preserves many of the nutrients from the original material. This transformation not only helps in managing the waste but also converts it into a valuable soil amendment (Man et al., 2021; Ro, 2016; Lehmann et al., 2015).

Biochar enhances soil structure, boosts nutrient and water retention, and encourages beneficial microbial activity. For example, biochar derived from poultry litter is noted for its high nutrient content, which enhances soil fertility and supports plant growth. Similarly, biochar from cow manure, characterized by its moderate nutrient levels and rich organic matter, improves soil structure and serves as a slow-release nutrient source (Zhao et al., 2013; Qin et al., 2019). Also, biochar derived from animal waste is vital for effective waste management. Animal manure can be difficult to handle due to its large volume and its potential to cause environmental issues like water pollution and greenhouse gas emissions. Transforming this waste into biochar not only facilitates its management but also generates a product with environmental advantages. This conversion minimizes the environmental impact of animal waste and turns it into a valuable resource with diverse applications (Wang et al., 2021; Kalus et al., 2019; Irmak Yilmaz & Ergun, 2019). Biochar also plays a significant role in mitigating greenhouse gas emissions. Its stable carbon structure allows it to sequester carbon for extended periods, thereby lowering atmospheric CO_2 levels. Additionally, applying biochar to soil has been found to reduce emissions of nitrous oxide and methane—two potent greenhouse gases. By integrating biochar into agricultural soils, it helps decrease the greenhouse gas emissions typically associated with farming practices, thereby supporting climate change mitigation efforts (Roberts et al., 2023; Xiang et al., 2021).

The growing acknowledgment of these benefits highlights biochar's potential as a crucial tool in advancing sustainable agricultural and environmental practices. Application and proficiency of biochar are influenced by factors such as source material, pyrolysis conditions, and post-processing treatments (Hou et al., 2022; Schimmelpfennig & Glaser, 2012).

This chapter deals with details of biochar including animal waste sources, biochar pretreatment and production methods, pyrolysis techniques, parameters influencing biochar quality, the strengths and weaknesses of different biochar synthesis process, etc. Simultaneously, application and impacts of animal waste-derived biochar, impacts on soil properties, soil-plant nutrients, plant growth and yield, decreasing heavy metals in plants, wastewater treatment, and other related topics are also discussed.

2 Feedstock for Animal Waste-Derived Biochar and Their Characteristics

Biochar is a carbon-rich product derived from the thermal decomposition of organic materials under limited oxygen conditions, a process known as pyrolysis. Animal waste, such as manure from poultry, cattle, pigs, sheep, and horses, is a valuable feedstock for biochar production due to its high nutrient content and availability. Utilizing animal waste for biochar production addresses waste management challenges and enhances soil fertility by recycling nutrients (Parthasarathy et al., 2022; Kalus et al., 2019).

2.1 Sources of Animal Waste

Animal waste originates from various agricultural and husbandry activities, including livestock farming (cattle, swine, poultry), dairy operations, and slaughterhouses. Manure, a predominant form of animal waste, is rich in organic matter and nutrients like nitrogen (N), phosphorus (P), and potassium (K). Poultry litter, consisting of manure mixed with bedding materials such as sawdust or straw, is another abundant source. Additionally, carcasses from livestock or poultry processing facilities can also be utilized for biochar production, contributing to waste management and environmental sustainability. Few types of animal waste as feedstock along with their characteristics and beneficial properties are discussed in Table 1.

Each type of animal waste offers unique chemical compositions and nutrient profiles, influencing the properties and potential applications of the biochar produced. Feedstocks such as poultry litter, cow manure, pig manure, sheep and goat manure, and horse manure offer diverse nutrient profiles and physical properties that can influence soil enhancement, waste management, and greenhouse gas mitigation. Biochar derived from these feedstocks can significantly improve soil fertility, structure, and water retention while promoting beneficial microbial activity. Additionally, it provides a sustainable solution for sequestering carbon to mitigate climate change. The production of biochar from animal waste through pyrolysis presents a sustainable solution for waste management and resource recovery in agriculture. Understanding the sources, pyrolysis process, influencing parameters, and

Table 1 Types and characteristics of animal waste as feedstock and beneficial properties

Name	Key composition	Elemental composition (in mass %)	Benefits	Ref
Poultry litter	High levels of N, P, and K, along with organic matter and trace elements	C = 73.18, O = 11.24, N = 4.3, Na = 1.72, Mg = 1.27, P = 2.91, K = 6.06, Ca = 2.04	Can produce biochar with significant agronomic benefits, such as improving soil fertility and structure	Pontiroli et al. (2019) Dalólio et al. (2017)
Cow manure	Rich in organic matter and nutrients, with moderate N content and also beneficial microorganisms that contribute to soil health	C = 40, H = 4.96, O = 33.22, N = 2.30, P = 5963 $mgkg^{-1}$, K = 15,415 $mgkg^{-1}$, Cr = 17.6 $mgkg^{-1}$	The high moisture content makes it suitable for slow pyrolysis, which can maximize biochar yield	Guo et al. (2021)
Pig manure	Has a balanced nutrient profile with moderate levels of N and P	C = 40.37, H = 6.33 N = 2.31, S = 0.36, O = 50.63	Pretreatment such as drying or composting is often required to reduce the moisture content before pyrolysis	Kalus et al. (2019) Wu et al. (2020)
Goat manure	High in organic matter but has lower N content compared to poultry litter	C = 40.09, O = 41.16 H = 5.85, N = 1.95 Ca = 3.52, P = 1.86 K = 1.64, Mg = 1.29	In general, has low moisture content, making it suitable for various pyrolysis methods	Touray et al. (2014)
Horse manure	High in fiber with moderate nutrient content, particularly nitrogen	C = 45.21, H = 4.91 O = 44.83, N = 0.50 Ca = 0.42, P = 0.37, Mg = 0.18	Due to high moisture content, horse manure requires drying before pyrolysis.	Tsai et al. (2015)

types of animal waste used provides insights into optimizing biochar production for various agricultural and environmental applications (Yadav & Jagadevan, 2019; Daful et al., 2022).

Feedstock pretreatment methods are essential for optimizing biochar production by improving the properties of the feedstock and enhancing the efficiency of the pyrolysis process (Orlando & Borja, 2020). Table 2 discusses some key pretreatment methods used in biochar production.

3 Methods of Biochar Production from Animal Waste

Animal waste, such as manure and carcasses, can be converted into biochar through various methods, each influencing the properties and efficacy of the resulting biochar product (Fig. 1). The primary techniques employed include pyrolysis, hydrothermal carbonization (HTC), and gasification, each with distinct advantages and challenges (Kloss et al., 2012).

Pyrolysis, a thermal decomposition process without oxygen, is widely used for biochar production from animal waste. The process involves heating the waste material to high temperatures (300–800 °C) to decompose organic compounds into biochar, bio-oil, and syngas. Fast pyrolysis operates at higher temperatures for

Table 2 Pretreatment methods of feedstock

Methods	Process	Benefits	References
Drying	Sun drying, mechanical drying	Reduces moisture content, increases pyrolysis efficiency	Demirbas (2004)
Composting	Aerobic decomposition of organic matter	Stabilizes nutrients, reduces pathogens, lowers moisture content	Sadeghpour et al. (2016)
Grinding	Mechanical grinding to reduce particle size	Increases surface area, improves heat transfer during pyrolysis	Chen and Kuo (2010)
Blending	Mixing different types of animal waste to balance nutrient content	Creates a more uniform feedstock, optimizes biochar properties	Jindo et al. (2012)

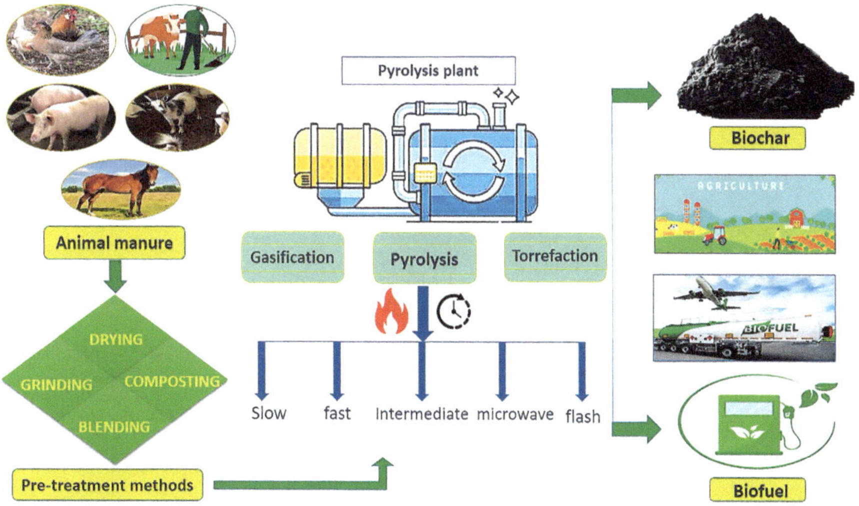

Fig. 1 Biochar preparation from different types of animal manure using various pretreatment methods

shorter durations, yielding higher bio-oil quantities, whereas slow pyrolysis produces higher-quality biochar with more stable carbon structures and reduced bio-oil (Xiang et al., 2022; Parthasarathy et al., 2022; Hossain et al., 2021; Fahmy et al., 2020).

Hydrothermal carbonization, alternatively, employs water under high temperatures (180–250 °C) and pressure to convert organic materials into biochar. This method is advantageous for animal waste because it can process wet and heterogeneous feedstocks efficiently, reducing energy requirements compared to pyrolysis. Hydrothermal carbonization produces hydrochar, a form of biochar with higher moisture content and potential nutrient retention capabilities, suitable for specific soil amendments (Jang et al., 2022; Gabhane et al., 2020).

Gasification converts animal waste into syngas (a mixture of hydrogen, carbon monoxide, and methane) through partial oxidation at high temperatures (700–1200 °C) and limited oxygen supply. The resulting biochar contains a higher proportion of fixed carbon, making it more stable and less prone to decomposition. Gasification offers flexibility in feedstock types and produces biochar with different chemical compositions suitable for diverse agricultural applications (Tezer et al., 2022; Mishra & Upadhyay, 2021).

Each method of biochar production from animal waste presents unique advantages and considerations, influencing the quality, yield, and environmental impact of the biochar produced. Proper selection of the method depends on feedstock characteristics, desired biochar properties, and operational considerations, ensuring optimal utilization of animal waste resources for sustainable agricultural practices (Kumar & Vyas, 2024).

3.1 Pyrolysis Process

Pyrolysis is a thermochemical conversion process that breaks down organic materials under controlled temperature conditions in the absence of oxygen or with limited oxygen supply (Table 3). The process can be categorized into different types based on operating conditions and residence times, influencing the yield and properties of the resulting biochar (Rex et al., 2023; Tomczyk et al., 2020).

Table 3 Comparison of pyrolysis methods (Amalina et al., 2022)

Parameter	Slow pyrolysis	Fast pyrolysis	Gasification
Temperature range	300–500 °C	500–700 °C	700–1000 °C
Heating rate	Low	High	High
Residence time	Hours	Seconds	Minutes
Biochar yield	High	Medium	Low
Energy production	Low	Medium	High
Suitable feedstocks	Poultry litter, cow manure	Dry manure, lower moisture waste	Dry manure

3.2 Description of Pyrolysis

Pyrolysis is a thermal decomposition process that converts organic materials, such as animal waste, into three primary products: biochar, bio-oil, and syngas (Table 4). This process occurs under controlled temperatures (300–1000 °C) in the absence of oxygen or with limited oxygen supply. Each product has unique properties and potential applications, making pyrolysis a versatile technology for waste management and resource recovery (Parthasarathy et al., 2022; Hassen-Trabelsi et al., 2014).

3.3 Optimizing the Pyrolysis Process

The process of pyrolysis can be tailored to optimize the production of biochar with specific properties, such as surface area, porosity, and nutrient retention capacity, depending on the intended application in agriculture or environmental remediation.

- Temperature: High temperatures favor the production of syngas and bio-oil, while lower temperatures increase biochar yield.
- Residence Time: Longer residence times enhance the carbon content and stability of biochar.

Table 4 Three main products of pyrolysis of animal waste, their properties, and applications (Lee et al., 2020; Imam & Capareda, 2012)

	Formation	Properties	Applications
Biochar	Carbon-rich solid residue retains much of the feedstock's original carbon content and structural components.	High pyrolysis temperatures generally increase the and porosity of biochar, enhancing its effectiveness as a soil amendment	Widely used in agriculture to improve soil fertility, enhance water retention, and sequester carbon, environmental remediation, where it can immobilize heavy metals and reduce soil contamination
Bio-oil	Liquid product consisting of complex mixtures of organic compounds, produced during the intermediate and fast pyrolysis of animal waste	Contains various chemical compounds that can be used as renewable fuel or chemical feedstock; its composition can be influenced by the type of feedstock and the pyrolysis conditions	Can be utilized directly as a renewable fuel or further processed to extract valuable chemicals for industrial applications
Syngas	Gaseous mixture of hydrogen, carbon monoxide, methane, and other gasses produced during the pyrolysis process and can vary in composition based on the pyrolysis conditions	Versatile energy carrier that can be combusted directly for heat and power generation or further processed into fuels or chemicals	Used as a renewable energy source and can be converted into synthetic natural gas or liquid fuels through processes like Fischer-Tropsch synthesis

- Heating Rate: Fast heating rates favor bio-oil production, while slow heating rates increase biochar yield.

By adjusting these parameters, the pyrolysis process can be optimized to produce biochar with desired characteristics, enhancing its effectiveness for specific agricultural or environmental applications (Rex et al., 2023; Tomczyk et al., 2020).

3.4 Parameters Influencing Biochar Quality

Several parameters influence the quality and characteristics of biochar produced from animal waste through pyrolysis:

- Temperature: High temperatures generally result in higher biochar yield and increased carbon stability.
- Heating Rate: The speed at which the material is heated affects the distribution of products and biochar properties.
- Residence Time: The duration the material remains at pyrolysis temperature influences biochar properties, with longer residence times typically leading to higher carbon content and stability.
- Feedstock Characteristics: The composition and moisture content of animal waste influence biochar properties, including nutrient content and stability (Yadav & Jagadevan, 2019; Daful et al., 2022).

A feedstock with high water-holding capacity demands extra energy to handle the elevated heat of vaporization as the biomass is heated to pyrolysis temperatures. Additionally, the gases and liquids produced during pyrolysis of water-rich biomass are diluted with steam, which decreases their calorific value. Hydrothermal carbonization methods can effectively process wet biomass with a water content of at least 70%. Among the various pyrolysis techniques, slow and fast pyrolysis are the most common, with slow pyrolysis proving to be the most effective for generating high-yield biochar (Tomczyk et al., 2020). Details of the techniques are mentioned in Table 5.

3.5 Pyrolysis Reactors

Pyrolysis reactors are specialized systems designed to convert organic materials into biochar, bio-oil, and syngas through pyrolysis, a thermal decomposition process carried out in the absence of oxygen. The design and operation of these reactors are crucial for optimizing the yield and quality of the products, as well as for ensuring efficient and controlled processing of feedstocks (Mong et al., 2022; Lewandowski et al., 2019). Fixed-bed reactors are one of the simplest and most traditional types of pyrolysis reactors. In these systems, the feedstock is placed in a stationary bed, and heat is applied to drive the pyrolysis reactions. Its advantages include simple design, relatively low cost, and ease of operation. It also has some

Table 5 Pyrolysis processes temperature differences and the by-product formation fractions

Technique	Temperature (°C)	Biochar (%)	Bio-oil (%)	Syngas (%)	Ref.
Slow pyrolysis	300–550	35	30	35	Khan et al. (2023) Premchand et al. (2023)
Intermediate pyrolysis	450–550	25	50	25	Chandraratne and Daful (2022)
Fast pyrolysis	450–600	12	75	13	Amalina et al. (2022) Rasaq et al. (2021)
Flash pyrolysis	750–1000	–	–	–	Ighalo et al. (2022)
Gasification	> 800	10	5	85	Nahar Myyas et al. (2023)
Hydrothermal carbonization	< 200	35	30	35	Brown et al. (2022)
Microwave-assisted pyrolysis	400–800	–	–	–	Zaker (2021)
Torrefaction	450–550	75	20	5	Abdulyekeen et al. (2022)

disadvantages of limited capacity for handling large volumes of feedstock and less efficient in terms of heat transfer and product uniformity (Liu & Tsai, 2016; Hassen-Trabelsi et al., 2014).

3.5.1 Fluidized-Bed Reactors

Fluidized-bed reactors utilize a stream of hot gas to keep the feedstock particles suspended and in continuous motion. This enhances heat transfer and reaction efficiency. Its advantages include improved heat transfer, uniform temperature distribution, and better control over reaction conditions. It also has some disadvantages of more complex design and higher operational costs (Guo et al., 2020; Yıldız et al., 2019).

3.5.2 Rotary Kiln Reactors

Rotary kiln reactors consist of a cylindrical drum that rotates while the feedstock is heated. The rotation ensures continuous mixing and uniform heating of the material. Its advantages include effective for processing large quantities of feedstock and achieving uniform product quality. It also has some disadvantages of requiring significant energy input and has higher maintenance needs (Moser et al., 2023; Panwar et al., 2019).

3.5.3 Auger Reactors

Auger reactors use a screw conveyor to transport the feedstock through a heated chamber where pyrolysis occurs. The continuous movement of the feedstock ensures consistent processing. Advantages of this includes continuous feed and product removal and suitable for processing various types of biomass, and disadvantages are potential issues with clogging and uneven heating. Its applications are ideal for processing smaller, more uniform feedstocks and for continuous operations (Rego et al., 2022; Mao et al., 2019).

3.5.4 Microwave Pyrolysis Reactors

Microwave pyrolysis reactors use microwave radiation to heat the feedstock directly, providing rapid and uniform heating. Its advantages include fast heating rates, high energy efficiency, and potentially lower production costs. Some disadvantages are limited to specific types of feedstocks and relatively high equipment costs (Zhang et al., 2022b; Mong et al., 2020).

Each type of pyrolysis reactor offers distinct advantages and challenges, making the choice of reactor dependent on factors such as feedstock characteristics, scale of operation, and desired biochar properties.

3.6 The Strengths and Weaknesses of Different Biochar Synthesis Processes

Pyrolysis is a versatile thermal decomposition process with several advantages, such as high biochar yield, use of any organic feedstock, and environmental friendliness, but it also has disadvantages (Chandraratne & Daful, 2022; Ndirangu et al., 2019). The hydrothermal carbonization (HTC) process involves converting organic waste into biochar by subjecting it to high temperature and pressure in a water-rich environment, effectively mimicking the natural coal formation process but in a much shorter timeframe (Pan, 2023; Yu et al., 2018). The gasification process for making biochar involves converting organic materials into biochar and syngas through partial combustion at high temperatures in a controlled oxygen environment (Yaashikaa et al., 2020; Gu et al., 2019). The advantages and disadvantages are discussed in Table 6.

4 Application and Impact of Animal Waste-Derived Biochar in Soil-Plant System/Health

Biochar is mostly utilized to improve the quality of soils. It is a potent conditioner that improves the chemical and physical characteristics of the soil. It serves as a nutrient reservoir, enhancing fertilizer efficiency, reducing nutrient loss, and ultimately increasing crop yields (Akhtar et al., 2014; Awad et al., 2021; Palansooriya et al., 2019). According to Robb et al. (2020), recent economic evaluations confirm the benefits of mixing biochar into fertilizers made for certain crops' growth resulting in increased crop yield and decreased production costs. In some cases, biochar can even replace traditional fertilizers entirely (Hossain et al., 2021). Biochar derived from the pyrolysis of biomass enhances soil fertility and increases water retention, thus leading to the availability of nutrients required by plants (El-Naggar et al., 2019). This double advantage promotes high agricultural productivity while potentially minimizing additional needs for fertilizers. Moreover, biochar

Table 6 Advantages and disadvantages of different biochar synthesis processes

		Advantages	Disadvantages
9.3.6.1	Pyrolysis	(i) Simple (ii) Low residence time (iii) High biochar yield (iv) Any organic feedstock can be utilized (v) Large surface area, high conductivity, and the porous structure (vi) Environmentally friendly	(i) Necessitates dry feedstock that have a moisture content of <30% (ii) It needs an inert environment (iii) Costly
9.3.6.2	Hydrothermal carbonization	(i) Eliminates energy-intensive drying methods (ii) Suitable for biomasses with high moisture content (i) The char formed has a high concentration of oxygen functional groups	(i) Low toxic substances and less stability
9.3.6.3	Gasification	(i) High syngas yield (ii) Short residence time (iii) Inexpensive	(i) Low biochar yield

stimulates microbial activity, improves nutrient cycling, as well as decreases nitrogen leaching and volatilization rates. For the improvement of crop biomass productivity, biochar plays a critical role in various aspects of growth and development. It facilitates seed germination, enhances plant growth and flowering, and boosts the resistance of plants to diseases and environmental stresses (Ibrahim et al., 2022; Lehmann et al., 2011). Specifically, biochar derived from animal waste has notable effects on both soil and plant systems, as detailed below. This underscores its multifaceted role in sustainable agriculture, where it not only improves soil health but also enhances overall crop productivity and resilience. Different study observations regarding the impact of animal waste-derived biochar on soil properties and plant growth are shown in Table 7.

4.1 Impact on Physical Properties of Soil

Soil processes have a significant impact on influencing soil physical properties, such as temperature, porosity, density, stability, texture, structure, aggregate stability, and drainage. Besides, nutrient availability, erosion resistance, and biological activity are all influenced by these characteristics (Sun & Lu, 2014). For instance, soil structure and aggregate stability affect root penetration and erosion control, while soil texture controls water retention and aeration. Moreover, porosity and density of soil determine the flow of air and water, which is essential for plant growth (Sharma & Kumar, 2023). All of these characteristics impact soil health, production, and sustainability from an ecological and agricultural perspective. Applying porous biochar enhances the quality of the soil, and its porosity is

Table 7 Different study observations about the application outcomes of animal waste-derived biochar on improving soil properties

Sl. no.	Feedstock type	Pyrolysis temperature and time	Dose of biochar	Observed changes in soil conditions	References
1.	Cow manure	Pyrolyzed for 2 h at 500 °C with 10 °C min^{-1} heating rate	3.0 and 6.0 wt%	Cr accumulation ranging from 34.3% to 699.9% Lowered Cd from the red, acidic soil	Kiran et al (2017)
2.	Poultry manure	Pyrolyzed with an increasing rate of 15 °C min^{-1} heating up to 300 °C and for 4 h	2 wt%	In calcareous soil, the availability of soil nutrients, OC, EC, and CEC increased.	Zolfi-Bavariani et al. (2016)
3.	Black soldier fly (*Hermetia illucens*)	Pyrolyzed at temperatures 500, 600, and 700 °C for 30 min	–	Acted as a carrier for fertilizers and stimulating the reproduction of beneficial bacteria (except biochar produced at 600 °C)	Bulak et al. (2023)
4.	Cow dung	Pyrolyzed at different temperatures (550–750 °C for 1 h with 10 °C min^{-1} heating rate	2 wt%	Enhanced soil pH, organic carbon content, soil moisture, and accessible P	Chen et al. (2018)
5.	Swine manure	Pyrolyzed at 500 °C for 2 hours at a rate of 15 °C min^{-1}	1 wt% and 5 wt%	Raised soil pH, lowered NH4+ concentration, and inhibited nitrate reductase activity to control the nitrogen cycle in paddy soil and reduce N_2O emission	Yang et al. (2020)
6.	Swine manure	Pyrolyzed at 350 °C	2 wt%	The activity of sucrase, phosphatase, catalase, and urease in tea garden soil rose by 63.3, 23.2, 50.3, and 27.9%, respectively. The C and N contents of the microbial biomass also rose by 36.4 and 34.3%, respectively.	Jiang et al. (2021)
7.	Swine manure digestate	Pyrolyzed with heating rate of 35 °C min^{-1} between 40 and 900 °C	25 t·ha^{-1}	Improved water-holding capacity and C utilization which leads to increase soil fertility and the sustainability	Ayaz et al. (2022)

(continued)

Table 7 (continued)

Sl. no.	Feedstock type	Pyrolysis temperature and time	Dose of biochar	Observed changes in soil conditions	References
8.	Cow bone	Pyrolyzed at 500 °C	10 wt%	Increase the variety of microbial biomass and bacterial proliferation of a particular phylum that induces bioremediation of soil damaged by smelters, i.e., pseudomonas (up to 793%), Saccharibacteria (583%), Parcubacteria (138%), Actinobacteria (65%), and Firmicutes (48%)	Azeem et al. (2023)
9.	Whole pig carcass	Pyrolyzed at 650 °C for 2 h	2 wt%	Boost the nutrients in the soil and the activities of sucrase, catalase, and urease in low OC soil co-contaminated with metal-phthalic acid ester	Chen et al. (2020)
10.	Animal carcass	Pyrolyzed at 500 °C with an increasing rate of 5 °C min-1 for 2 h	1%, 2%, 5%, and 10 wt%	Rose soil pH, organic matter, P and CEC. Phyto-availability of Cd decreases by 26 to 98% with increase in biochar dose.	Yun et al. (2023)

distinguished by its large specific surface area and low bulk density (Downie et al., 2009). Since biochar is porous, it increases soil surface area and promotes microbial activity and nutrient absorption (Zhang et al.,2018). Its compact size improves soil weight and moisture content and reduces soil erosion. This improves root penetration and nutrient uptake while mitigating erosion risk (Liu et al., 2017). To maintain soil integrity and minimize runoff, biochar enhances aggregate formation and stabilizes soil structure. Additionally, the higher moisture content of soils treated with biochar helps retain water and is especially useful in soils that are vulnerable to drought stress (Purakayastha et al., 2019; Tiwari, 2023). Several research has demonstrated the effectiveness of biochar formed from animal waste in improving many aspects of soil qualities. Biochar made from poultry manure was found greatly to enhance soil quality (Huang et al., 2021). A reduction in the specific gravity of bare soil from 2.67 to 2.54–2.62 and 2.45–2.58, respectively, was observed upon the addition of 5 wt% and 10 wt% poultry manure biochar. This decline therefore occurs due to the replacement of soil particles with less dense biochar. Also, biochar-treated soils showed greater capacity to retain water than untreated soils, with leaching rates ranging from 1.89% to 3.14% in soils with increased soil saturation (Huang et al., 2021). Similar to this study, pig manure biochar also has shown a noticeable increment in the peak water content while mixed in 5 wt% and 8 wt% biochar content. However, the water retention capacity in soil treated with pig manure biochar was lower compared to peach shell biochar due to the higher roughness and shape of pig manure biochar particles (An et al., 2021). Additionally, the impact of cow dung biochar on water retention and wheat cultivation in sandy loam soil was examined by Rehman et al. (2020). According to their research, adding 1 wt% biochar to the soil made it 1.5% more capable of retaining moisture, which prolonged crop development by 0.5 to 1 day. Moreover, incorporating 12 wt% of biochar into the soil extended water availability for crops by an additional 10 to 12 days, resulting in enhanced crop yields.

Likewise, several further studies have demonstrated the effectiveness of biochar produced from animal waste in enhancing the physical characteristics of soil. For example, as compared to non-amended soils, biochar-amended soil had better water usage efficiency (Uzoma et al., 2011), saturated hydraulic conductivity (Gavili et al., 2018), and moisture retention (Rahman et al., 2014). These findings highlight the potential of cow manure biochar as a beneficial soil amendment for improving water retention and agricultural productivity in sandy loamy soils. These beneficial physical changes in soil supplemented with biochar are caused by the fine-particle biochar being mixed with the coarse soil, which reduces pore size and promotes moisture retention by capillary action (Bordoloi et al., 2018). Additionally, the ability of the soil-biochar composite to retain water is improved by the hydrophilic surface linkages, which are mostly hydroxyl (OH) groups, and the interior pores of the biochar (Gray et al., 2014).

4.2 Impact on Chemical Properties of Soil

Plant growth may be significantly impacted by the composition of the soil. Plant development can be influenced by a variety of factors, including salinity, sodicity, nutrient levels, acidity, and alkalinity. A pH of less than 5.5 is regarded as acidic soil, which can cause metals like manganese (Mn) and aluminum (Al) to become more soluble and perhaps damage plants. On the other hand, vital elements like calcium (Ca), magnesium (Mg), and phosphorus (P), which are vital for plant health, become less available and more difficult for plants to properly absorb and use (Dai et al., 2017). Additionally, Dai et al. (2021) showed that applying biochar raises the pH of soil by using mineral components such as Ca, K, Mg, Na, and Si, which form carbonates or oxides during pyrolysis. These substances improve the chemical composition of vital elements in acidic soils by neutralizing H+ ions and monomeric Al species by decreasing acidity (Dai et al., 2017). In this context, animal-based biochar affects soil pH more than plant-based biochar (Chen et al., 2020). This effect reduces soil acidity, as supported by Shetty and Prakash (2020), thereby enhancing soil fertility and promoting healthier plants. Butnan et al. (2015) demonstrated that in addition to increasing soil pH, the addition of biochar to soil also leads to elevated electrical conductivity (ECe) and cation exchange capacity (CEC).

The source of feedstock and the particular functional groups generated during biochar production procedure through pyrolysis are two variables that affect the CEC, and these variables have a major effect on the surface characteristics of biochar (Suliman et al., 2016). In comparison to chicken manure biochar, which had a greater CEC of 81.4 cmol kg^{-1} when pyrolyzed at 500 °C, biochar generated from pig manure showed a lower CEC of 32.7 cmol kg^{-1} (Cely et al., 2015). This diversity demonstrates how biochars with varying CEC values may be produced from various feedstock. Similarly, another study also found chicken manure biochar more proficient in increasing CEC by 7% in silt-loam soil and 11% in sandy soil demonstrating the superior effectiveness of chicken manure biochar in soil applications (Clark et al., 2019). In saline and sodic soils which are mainly characterized by higher ECe, exchangeable sodium percentage (ESP), and pH levels, biochar application offers a solution to mitigate salt effects and enhance soil fertility. Biochar application in salt-affected soils serves multiple roles: Acidic biochars reduce soil pH, high CEC biochars enhance cation uptake by roots, and biochar increases surface charges, replacing Na with K, Ca, and Mg to lower ESP levels (Chaganti et al., 2015; Kim et al., 2016; Lashari et al., 2015; Luo et al., 2017; Yu et al., 2019). Farmyard manure biochar can lessen Na^+ toxicity in saline soils by increasing K^+ uptake and reducing Na^+ absorption (Bin Yousaf et al., 2022). Besides, the biochar made from shrimp waste had a high proportion of organic matter and calcium (30%), which also helps in reducing the uptake of harmful salts in the salty sandy soil (Abo-Elyousr et al., 2022).

Furthermore, the application of biochar to soil has shown significant increases in soil organic matter (SOM) and organic carbon levels (Rathnayake et al., 2023). Sukartono et al. (2011) reported that the conversion of cow manure to biochar had a positive effect on soil organic carbon content, which was greater than the application of cow manure at a rate of 15 Mg ha^{-1}. Besides, Zolfi-Bavariani et al. (2016) found a significant increase in the organic carbon content, ECe, and CEC in the soils treated with biochar of poultry manure compared to those treated with raw poultry manure (applied at 2 wt%). The studies showed a positive influence of biochar conversion on the improvement of soil carbon status and the improvement of other soil properties that are crucial for agricultural production. The study carried out by Gasco et al. (2016) examined the influence of pig manure biochar and showed a significant diminish of carbon (C) mineralization in comparison to treatments with raw pig manure. It was explained that biochar lacks labile carbon which would have been easily mineralized in the soil. On the other hand, Mavi et al. (2020) observed the characteristics of biochar obtained from chicken manure under salinity stress and found it to be higher in C mineralization than biochar derived from the plant. This complex relationship illustrates the different types of biochars exhibit different responses to environmental pressures, thus reflecting their unique compositions and sources. Furthermore, Clark et al. (2019) also revealed that the biochar of chicken manures is beneficial to soil C content, increasing by 16% in sandy soils and 30% in silt-loam soils. The subsequent studies by Abbasi and Anwar (2015), Abo-Elyousr et al. (2022), Ayaz et al. (2021), and Zhang et al. (2018) confirmed these results, showing soil organic carbon (SOC) content in soil treated with animal manure biochar higher compared to raw manure-treated soil. The boost in SOC as a result of biochar signified that it could be advantageous in enhancing the storage of carbon and soil fertility of various types. The impact of animal waste-derived biochar application on different soil conditions has been described cumulatively in Fig. 2.

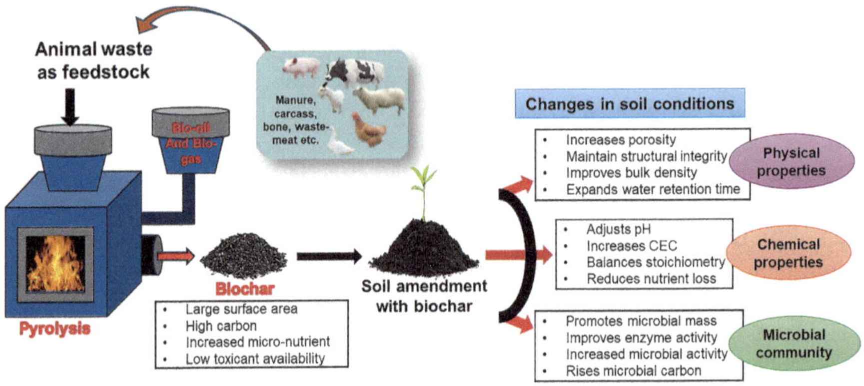

Fig. 2 Impact of animal waste-derived biochar application on soil conditions

4.3 Impact on Soil Microbiota

Biochar has different effects on soil properties, depending on plant species, soil types, and the type of biochar feedstock (Uzoma et al., 2011). By changing nutrient availability, biochar alters the soil microbial community. Biochar can better aerate soil aeration and hold in more nutrients, as well as provide a niche for microorganisms, due to its porous nature (Lu et al., 2017; Zhu et al., 2017). Microbes benefit from this physical support as it favors the growth of microbes and their activities (Ding et al., 2016). It nurtures bacteria that are beneficial to soil microbial proliferation and functioning (Lu et al., 2020; Otsuka et al., 2008). They found that, in comparison to untreated soils, the use of biochar increases the diversity of soil bacterial communities by 25%.

Moreover, Kolb et al. (2009) found that the surface characteristics (charges and functional groups) of biochar promote the absorption of soil organic matter leading to the colonization of soil microbial communities. While biochar application increases nutrient availability, microbial biomass can simultaneously reduce C and N mineralization (Rutigliano et al., 2014; Ayaz et al., 2021). This trade-off highlights the importance of balanced management strategies to maximize agricultural productivity while minimizing the risk of adverse effects on soil microorganisms. Mavi et al. (2020) also highlight the effect of biochar on soil microbial performance under salinity stress. In their study, microbial activity was found to decrease by 28–45% in unamended soils under high salinity. Specifically, the biochar formed from poultry manure has demonstrated an enhanced resistance to salt stress. Similar to this, when mixed with earthworm supplements, cow dung biochar generated under oxygen-limited conditions showed notable improvements in phosphatase activity and the abundance of P-solubilizing bacteria (Nahidan & Ghasemzadeh, 2022). This demonstrates how beneficial interactions between biochar and soil organisms might improve nutrient cycling and plant-available P in agricultural soils.

4.4 Impact on Soil-Plant Nutrient Enrichment

Biochar production is a well-known method for increasing nutrient availability in soil by enhancing its nutrient content and enhancing soil nutrient retention through its unique physical, chemical, and biological properties (Ippolito et al., 2015). Nitrogen (N) is a vital component of proteins, enzymes, and chlorophyll and is necessary for the development of plants. On the other hand, there are conditions where the N in the soil can be very volatile and likely to be lost; for example, due to ammonia volatilization in alkaline soil conditions, nitrate leaching, particularly in sandy or poorly structured soils, N runoff into the waterbodies (Naz, 2023; Tegeder & Masclaux-Daubresse, 2018). Thus, sustaining a steady supply of N is necessary for a sizable crop production. While biomass usually has a low N content, moisture removal during the pyrolysis process of biochar formation causes it to become enriched with several sources of nitrogen (Liu et al., 2018). Numerous studies showed that the source of the biochar and the thermal treatment techniques employed can have a substantial impact on the amount of N. For example, depending on the precise pyrolysis techniques, biochars made from chicken litter have been reported to have N concentrations varying from 3.5% to 5.85% (Chang et al., 2015; Clark et al., 2019). The overall N content rose by 26% in sandy soil when fractions of 53 to 250 μm and 250 μm to 2 mm of poultry dung biochar were treated, in comparison to control samples. On the other hand, in silt-loam soil, notable increases in N content were noted mainly by using the bigger fractions (>2 mm and 250 μm to 2 mm) of biochar, leading to a 25% increase in total N storage (Clark et al., 2019). Phosphorus (P) is also essential for plant growth, energy transfer, photosynthesis, and maintaining genetic traits (Ajmera et al., 2019).

Biochar with high ash content is commonly high in phosphorus. Animal waste biochar was found with higher N and P content than plant biochar (Hou et al., 2022). Heat anaerobic pyrolysis of poultry feces fosters greater synthesis of P-based compounds at high temperatures, highlighting the importance of feedstock selection and treatment (Hou et al., 2022; Li et al., 2020). In a study by Hosseini et al. (2019), sheep manure biochar produced at 500 °C for 2 hours under restricted O and N flow conditions produced a total P content of 4.78 g kg^{-1}. In contrast, pig carcass-derived biochar, pyrolyzed at 650 °C for 2 hours, showed exceptionally high total phosphorus concentration (>80 g kg^{-1}) (Yang et al., 2021). Conversely, dairy manure conversion at 450 °C for 4 hours forms less-soluble whitlockite [$(Ca, Mg)_3(PO_4)_2$], reducing P release. P release further decreases by up to 76% with extended pyrolysis time exceeding 240 hours (Liang et al., 2014). The size of biochar is also important for nutrient enrichment in plants. Utilizing fine biochar derived from dairy cattle carcasses (<1 mm) improved P absorption by corn crops in soil with low P levels. Postharvest soil analysis showed elevated pH levels and increased availability of P, as well as higher exchangeable Ca and Mg contents. Conversely, larger-sized biochar particles (2–4 mm and > 4 mm) had minimal impact on P uptake, likely due to their slower P release rates (Ma & Matsunaka, 2013).

Potassium (K) influences several physiological processes and is essential for plant growth and development (Wang & Wu, 2017). Research has indicated that biochar produced from chicken dung and litter had substantially higher K levels than biochar made from plant biomass (Agegnehu et al., 2017; Das et al., 2021; Hossain et al., 2020). Potassium concentrations of poultry litter biochar were around 3.88% at 400 °C and 5.88% at 600 °C (Subedi et al., 2016). Similarly, Tsai et al. (2012) reported that potassium concentrations in swine manure biochar, pyrolyzed at temperatures between 400 and 800 °C, ranged from 2.7% to 3.1%.

Moreover, micronutrients like Ca and Mg are primarily involved in redox reactions, photosynthesis, and other metabolic activities; for this reason, the presence of these minerals in soil and plants is thought to be crucial. A study conducted in 2019 by Le et al. discovered that biochar derived from animal carcasses, specifically swine and cattle, had a greater Ca content (30.02% and 27.23%, respectively), compared to biochar derived from manure, which included swine (5.29%), poultry (12.12%), and cattle (5.05%). But for Mg, the tendency was the opposite: The levels were lower in biochar obtained from whole animals (swine, 0.27% and cattle, 0.22%) than in biochar derived from manure (poultry, 0.93%; swine, 0.89%; and cattle, 1.24%). Similarly, shrimp waste-derived biochar has been found to significantly enhance soil quality and increase nutrient availability, particularly Ca and Mg, according to recent research (Abo-Elyousr et al., 2022). Chen et al. (2020) found that incorporating bone char into soil enhanced C mineralization, making it a valuable organic fertilizer. The char's hydroxyapatite properties and ample Ca and P content contribute significantly to soil enrichment and crop growth potential. Biochar also serves as a reservoir for essential trace elements crucial for plant health. These mostly include iron (Fe), copper (Cu), boron (B), zinc (Zn), and manganese (Mn) (Vatansever et al., 2017). These trace elements play vital roles in various plant functions, such as enhancing nutrient absorption and supporting mechanisms that boost resistance against diseases and stress.

4.5 Impact on Plant Growth and Yield

Several research investigations have indicated that incorporating biochar derived from animal waste into soils, compared to non-amended soils, enhances nutrient availability by improving a favorable soil environment (Uzoma et al., 2011; Xiao et al., 2018; Zhang et al., 2018. The impact of biochar amendments on soil-crop systems is influenced by the conditions under which pyrolysis occurs (Biederman & Harpole, 2013). Biochar produced under high-temperature pyrolysis tend to be alkaline in nature and have lower levels of biologically active volatile compounds that might hinder their ability to effectively accelerate crop productivity (Lou et al., 2016; Gundale & DeLuca, 2006). Biochar derived from animal waste has been recognized for its role in enhancing plant growth, primarily through direct nutrient contributions and indirect improvements in nutrient use efficiency. The unique structure of biochar allows it to act as a slow-release nutrient source, with its porous network effectively delaying nutrient release and facilitating their gradual availability for plant uptake (Xiao et al., 2018; Lehmann & Joseph, 2015; Agegnehu et al., 2017; Zhu et al., 2017). For instance, Zhang et al. (2018) investigated yak manure-derived biochar and highlighted its high surface area, alkaline pH, porous structure, and nutrient richness (N, P, Ca, Mg, K), suggesting its potential as a soil conditioner. Similarly, Laila et al. (2021) observed significant growth improvements in ridge gourd due to essential nutrients (N, P, K, Ca, Si) in waste bone biochar from chicken, beef, and mutton bones. Their study indicated that biochar application significantly enhanced the morphological and biochemical parameters of ridge gourd compared to controls or chemical fertilizers. Moreover, Azeem et al. (2021a) demonstrated that cow bone biochar at specific application rates (2.5 wt%) improved soil properties and plant attributes, such as chlorophyll content, albeit higher application rates (10 wt%) negatively impacted soil enzymatic activities and plant growth. Interestingly, plants grown in biochar-amended soils exhibited accelerated flowering, possibly linked to organic volatiles on biochar surfaces triggering hormonal changes in plants (Laila et al., 2021). Joardar et al. (2020) reported enhanced plant growth with poultry litter biochar (3 wt%) compared to raw litter and control treatments. This is further supported by several other research where biochar was found to be more effective compared to its feedstock, i.e., poultry manure (Bhattarai et al., 2015; Dróżdż et al., 2023; Ewulo et al., 2008; Usman, 2015). Krounbi et al. (2021) observed increased biomass in tomato, marigold, and radish with dairy manure biochar enriched with nitrogen (N). Moreover, Abbasi and Anwar (2015) used poultry manure biochar in soil and noticed about a 26% increase in maize shoot nitrogen levels. In arid sandy soils, poultry manure biochar application, especially in combination with N fertilizer, improved soil biochemical properties, organic matter content, and crop productivity for wheat and forage sorghum (Mohamed & Hammam, 2019). In terms of application rates, farmyard manure biochar at 6 wt% proved most effective in enhancing physiological attributes in salt-affected soils, with a clear trend observed across species (Bin Yousaf et al., 2022). Furthermore, feedlot manure biochar was shown to increase the plant (subtropical grassland) efficiency in N utilization (Slavich et al., 2013) while cow manure biochar increases water and nutrient uptake by maize plant (Uzoma et al., 2011). However, not all studies report positive outcomes. Some researchers have noted mixed or negligible effects on plant growth post-biochar application (Are et al., 2017; Novak et al., 2018; Schmidt et al., 2021), emphasizing the variability in biochar's effectiveness depending on soil conditions and biochar properties. Impact of animal waste-derived biochar application on plant growth and yield is summarized in Fig. 3.

4.6 Impact on Soil-Plant Heavy Metal Uptake

Heavy metals are adsorbed from the soil by porous biochar, which lowers their availability and consequently reduces absorption into the growing crops (Ibrahim et al., 2022). When compared to untreated cow manure, the application of biochar (3 wt% and 6 wt%) made from cow dung demonstrated a greater capacity to remove cadmium (Cd) from acidic red soil with 34.3 to 69.9% Cd elimination efficiency (Kiran et al., 2017). Yang et al. (2021) observed that phosphorus-rich biochar

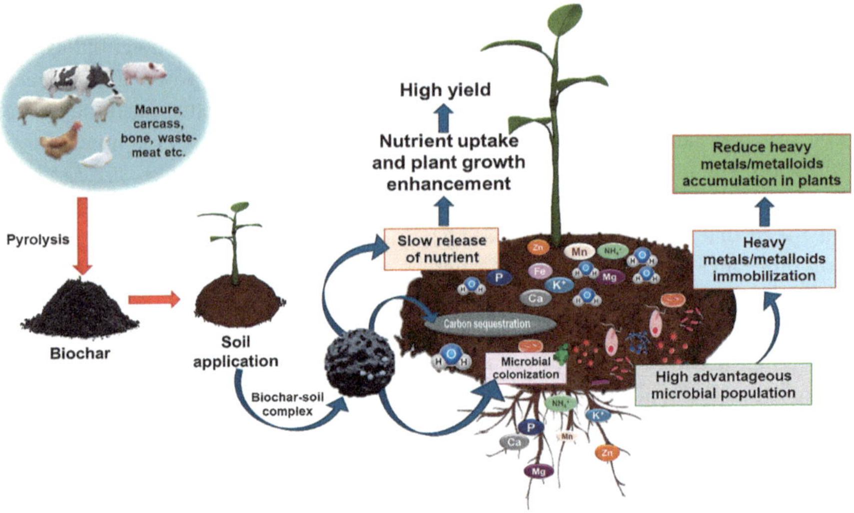

Fig. 3 Impact of animal waste-derived biochar application on plant growth and yield

significantly reduced dissolved Cd concentrations by 8–19% under reducing conditions and less than 3% under oxidizing conditions compared to the control. In mildly reducing conditions (Eh < 0 mV), it effectively immobilized lead (Pb), achieving reductions of 55–82%. However, the initial amount of heavy metal present in the feedstock used to make biochar might affect the metal content in the final biochar, which could have a detrimental effect on plant development and soil pollution. Kale, lettuce, and mustard greens accumulated more chromium (Cr) when cultivated in soil treated with biochar made from rabbit manure. Moreover, the availability of Pb and Cd in the tissues of the studied plants decreased when compost and biochar were combined with soil, suggesting a reduction of metal absorption. On the other hand, spinach and lettuce absorb 20% more copper (Cu) and iron (Cr) as a result of the same alteration (Medyńska-Juraszek et al., 2022). Various studies have examined the effects of using biochar made from animal waste on plant growth, yield, and the accumulation of heavy metals and metalloids as described in Table 8.

5 Other Applications of Animal Waste-Derived Biochar

The sustainable processes of animal waste-derived biochar applications in different environmental strata have been shown figuratively in Fig. 4.

5.1 Biodiesel Production

Biodiesel is produced through the transesterification process from used cooking oils, animal fats, and both edible and nonedible vegetable oils with the use of

Table 8 Different study observations about the application outcomes of animal waste-derived biochar on improving plant growth, yield, and heavy metal/metalloid accumulation

Sl. no.	Feedstock type	Pyrolysis temperature and time	Dose of biochar	Observed changes in plant growth and yield	References
1.	Cow manure	Pyrolysis at 380 °C and 420 °C for 2 h	1 wt%	Increase the amount of water available to the crop for an extra 10–12 days leading to more crop yield	Rehman et al. (2020)
2.	Cow manure	Pyrolysis at 380 °C and 420 °C for 2 h	12 wt%	Increase the soil's ability to retain water by 1.5% and provide extra 0.5–1 days of support for the crop development	Rehman et al. (2020)
3.	Poultry litter	Pyrolysis at 300 °C for 10 min	4 t ha^{-1}	Decrease of nutrient loss, enhance long-term nutrient availability and soil fertility, and development of Gima kalmi with high biomass output	Sikder and Joardar (2019)
4.	Cow manure	Pyrolyzed for 2 h at 500 °C with an increase in temperature 10 °C min^{-1}	3.0 and 6.0% (w/w).	Decrease the amount of Cd that was available in the soil by 34.3–69.9% and the amount that was bioaccumulated in plants by 51.2 and 67.4%, respectively. The addition of biochar greatly enhanced plant biomass production and improve the extractability and accumulation of trace metals (Zn, Mn, Fe, and Cu) by plants.	Kiran et al. (2017)
5.	Cow bones	Pyrolysis at 550 °C for 2 h	10% (w/w)	Lower the enzymatic activity of the soil, which in turn lowered plant development	Azeem et al. (2021a)
6.	Cow bones	Pyrolysis at 500 °C for 2 h	2.5%	Reduce the amount of Zn (55 and 40%) and Cu (57 and 67%) in the maize roots and shoots, respectively, by changing the acid-soluble fraction of both metals to the residual fraction. At the lower dosages of biochar, an increase in the activity of alkaline phosphatase (71%) and β-glucosidase (51%) was noticed and helps in plant growth by improving antioxidants.	Azeem et al. (2021a)
7.	Chicken, beef, and mutton bones	Pyrolysis at 550 °C for 25 min with an increase rate of 30 °C min^{-1}	15%	Produce a notable impact on the accumulation of chlorophyll content, which may have been brought about by the bioavailability of nutrients (N, Mg, and P) that were released slowly leading to change in photosynthetic activity	Hussain et al. (2021)

(continued)

Table 8 (continued)

Sl. no.	Feedstock type	Pyrolysis temperature and time	Dose of biochar	Observed changes in plant growth and yield	References
8.	Sheep bone	Pyrolysis at 800 °C for 2 h	10%	Decrease the amounts of Zn and Cu in maize roots by 57 and 60%, and in shoots by 42 and 61%, respectively.	Azeem et al. (2021b)
9.	Whole pig carcasses	Pyrolysis at 650 °C for 2 h	2%	Increase phyto-availabilities of nutrients and enhance the pak choi growth in low organic carbon soils co-contaminated with metal and phthalic acid ester	Chen et al. (2020)
10.	Pig carcass	Pyrolysis at 650 °C for 2 h	–	Alleviate the toxicity of as in mildly reducing soil environment, but in paddy soils that are strongly reducing and oxidizing, it may raise the risk	Yang et al. (2022)

alcohol and a catalyst (Balajii & Niju, 2019). Traditional methods using strong acids or alkalis often result in acidic or basic waste products (Yang et al., 2023). To address these environmental problems and find a long-term replacement for costly nonrenewable catalysts, researchers have proposed using biochar derived from animal waste like waste pig meat and bonemeal (Wang et al., 2017), swine manure (Jung et al., 2020), and chicken manure (Jung et al., 2017) as a renewable heterogeneous catalyst. According to Jung et al. (2017), biochar made by pyrolyzing chicken excrement at 550 °C included a sizable amount of Ca. In the transesterification of used cooking oil, it was discovered that the presence of calcium carbonate ($CaCO_3$) in the biochar made from chicken dung increased the catalytic activity (Jung et al., 2018). Thus, the efficiency of biodiesel production processes may be improved by using biochar not only as a catalyst but also as a source of advantageous minerals. Moreover, using biochar made from swine manure as a catalyst to convert waste cooking oil, 99% biodiesel was found to be produced (Jung et al., 2020). Wang et al. (2017) also used biochar made from pig meat and bonemeal to reach a high biodiesel output of 98%. In addition to aiding the circular economy by lowering the quantity of biomass that would otherwise be lost, biochar-based catalysts assist to boost the efficiency of biodiesel production by recycling organic waste into products with extra value.

5.2 Hydrogen Gas Production

One thermochemical method for producing hydrogen (H)-rich gas from biomass sources is catalytic pyrolysis (Singh et al., 2023). To maximize H generation, Xin et al. (2017) used a two-step pyrolysis-gasification strategy with cattle dung, finding the ideal temperatures at 500 °C for pyrolysis-carbonization and 850 °C for

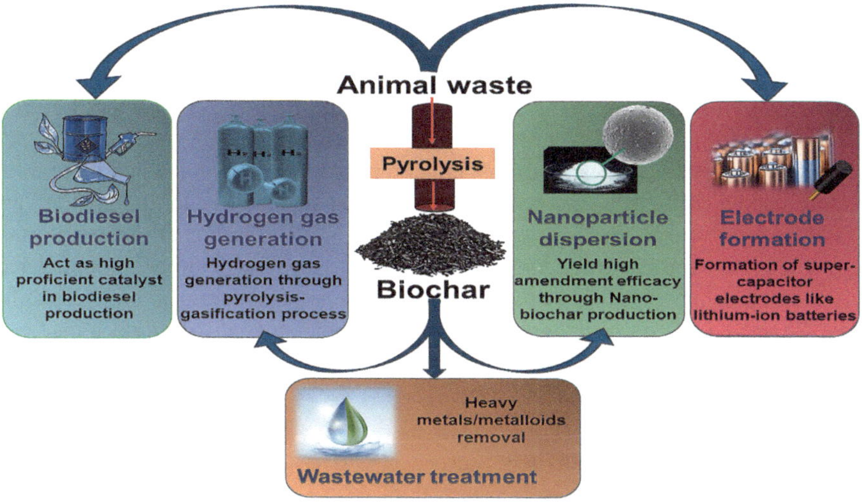

Fig. 4 Sustainable environmental application of animal waste-derived biochar

gasification. Zeng et al. (2022) used biochar catalysts generated from cow dung that was impregnated with heavy metals (Cu, Pb, and Zn) to catalytically pyrolyze agricultural biomass, such as wheat straw. They found a Cu-loaded biochar catalyst to produce the highest gaseous product yield (~51.6 wt%). In another study, Deng et al. (2021) successfully synthesized an in situ N-, P-, and Ca-codoped biochar from pig bone through thermal treatment at 800 °C. This catalyst exhibited high atomic contents and synergistic effects of N, P, and Ca, with characteristics including a large electrochemically active surface area, low charge-transfer resistance, high conductivity, and large specific area. These electrocatalytic properties endowed the catalyst with excellent activity for the hydrogen evolution reaction (HER) and stability in acidic H_2SO4 solution. Furthermore, animal waste-derived biochar catalysts, rich in hydroxyl, carbonyl, and other carbon-oxygen functional groups, act as proton donors (Su et al., 2020), thereby enhancing catalytic reforming reactions (Lu et al., 2013). The combination of surface functional groups and metal active sites promotes the thermal decomposition of lignocellulose-derived pyrolytic volatiles into non-condensable gases such as hydrogen (Zeng et al., 2022).

5.3 Nanoparticle Dispersion

Nano-biochar exhibits superior physical and chemical characteristics compared to macro-biochar, making it highly effective as a catalyst for a wide array of reactions involving functionalized carbon materials (Xu et al., 2024). Animal waste-based biochar shows promise in dispersing nanoparticles effectively due to its strong catalytic capabilities. In a different approach, Moradi et al. (2019) utilized chicken manure-derived biochar as a catalyst support for dispersing palladium nanoparticles. The chicken manure-derived biochar underwent modification with (3-chloropropyl) trimethoxysilane and 2-(thiophene-2-yl)-1H-benzo[d]imidazole, followed by impregnation with palladium (II) acetate nanoparticles. This modified catalyst was employed in Suzuki-Miyaura and Heck-Mizoroki cross-coupling reactions, achieving high turnover numbers (>110) attributed to the presence of Pd species. However, shrimp shell nano-biochar currently faces challenges stemming from impurities and structural inconsistencies, limiting its stability in applications for catalysis and energy storage. To address these issues, researchers are exploring modifications such as introducing functional groups through heteroatom doping and developing composites to enhance catalytic efficiency and energy storage capabilities. Wu et al. (2023) and Zhang et al. (2022a, b) independently developed methods to enhance shrimp shell nano-biochar.

5.4 Electrode Formation

According to Yang et al. (2023), biochar made from animal waste works well for producing supercapacitor electrodes. In this instance, cattle manure, insect excreta, dairy manure, poultry litter, and swine manure can be used as a promising feedstock for biochar preparation (Kim et al., 2022; Wang et al., 2022; Pontiroli et al., 2019, b; Qin et al., 2021). Ni-laden dairy manure biochar production followed by microwave treatment showed remarkable stability in specific capacitance, with less than 2% loss even after 1000 charge-discharge cycles (Wang et al., 2017). Thus, supercapacitor electrode generation using biochar produced from animal manure might be an efficient waste management solution (Yang et al., 2023). Zhang et al. (2022a, b) examined the electrocatalytic potential of biochar made from dairy dung and heated to 500°, 600°, and 700 °C. They found that the biochar electrodes exhibited differing levels of electrocatalytic activity in oxygen reduction processes. The electrode composed of cattle dung had a hierarchical porous carbon composition and a bicontinuous structure that allowed for the rapid transit of electrons and ions during charging and discharging. Chicken litter was also used as a raw material in a more recent study (Park et al., 2023) to create electrodes for lithium-ion batteries.

5.5 Wastewater Treatment

Biochar made from animal waste biomass is widely used in treating wastewater to remove harmful pollutants and heavy metals/metalloids (Gupta et al., 2020). In one study, raw guinea fowl manure and cattle manure were subjected to pyrolysis at 500 °C for 3 h to produce biochar for Cr removal (Idrees et al., 2018). The biochar obtained demonstrated high effectiveness in the removal of Cu (II), achieving maximum capacities of 43.60 and 44.50 mg g^{-1} for each type, respectively. Analysis with FTIR spectroscopy revealed that both biochar exhibited irregular structural characteristics and contained significant oxygen-rich groups, aiding in the formation of stable complexes such as CuO and $CuSO_4$. It is suggested that electrostatic interaction (chemisorption) was one of the main mechanisms for Cu removal (Idrees et al., 2018). However, whole frozen livestock carcasses, including naturally deceased cattle and swine, exhibited higher Ca-P compared to biochar from manure and were more efficient in precipitating Pb (II). The immobilization of Cd (II)/Cu (II) on biochar derived from whole animals is facilitated by ion exchange with Ca (II) in hydroxyapatite (HAP), whereas Hg (II) sorption is attributed to the reduction of Hg (II) to Hg_2Cl_2 (Lei et al., 2019). For arsenic removal from aqueous solutions, biochar derived from animal sources (e.g., crayfish shells) with O- and N-containing functional groups plays a crucial role (Sun et al., 2022).

6 Constraints of Animal Waste-Derived Biochar Application and Recommendations

Despite having positive effects of animal-origin biochar on environmental health and waste management, it is important to recognize the limitations associated with its application. Animal waste biochars may contain hazardous metals that pose ecological risks (Lin et al., 2017). Although the risk can be reduced by higher pyrolysis temperatures (Wang et al., 2019), it is important to avoid excessively high temperatures to prevent N losses and decreases in P availability (Xu et al., 2016). However, existing research mainly compares manure biochar with untreated manure or with soils supplemented with mineral fertilizer, making it difficult to draw definitive conclusions about nutritional perspectives, and consequently, it is currently difficult to make clear recommendations for waste pyrolysis as a plant nutritional source. It is necessary to conduct more studies and experiments to analyze the observed results based on the amount of waste required to produce a certain amount of biochar. In simpler terms, different manure pyrolysis techniques should be compared using the same initial amount of specific manure content. This information will help to evaluate the ecological and economic benefits between conventional manure systems and those using manure biochar, thus helping in the decision-making process for farmers (Rathnayake et al., 2023). Additionally, there is a lack of research on the various methods of preparing biochar from the same source and using the same adsorbents for adsorption. To further understand the long-term consequences of biochar applications to polluted soils, long-term field trials are needed. To increase the efficiency of biochar, many techniques are being used to improve its surface functional characteristics. On the other hand, the scientific data on the industrial manufacturing of modified biochar are also inadequate. As a result, assessing the effectiveness of current biochar modification techniques on a larger scale is essential for more active and eco-friendly biochar production.

7 Conclusion

This article emphasizes several advantages of using biochar, including its ability to enhance soil's physical, chemical, and biological characteristics. Biochar makes a substantial contribution to sustainable farming techniques by strengthening soil structure, boosting water retention capacity, and improving nutrient availability. Its capacity to lessen the negative environmental effects of conventional agricultural methods is further demonstrated by its involvement in improving fertilizer usage efficiency and lowering nutrient leaching. The diverse impacts on soil microbiota further highlight the complex interactions of biochar within soil ecosystems, promoting microbial diversity and activity crucial for nutrient cycling. Moreover, its application in other functions which include biodiesel and hydrogen gas production, electrode generation, and heavy metal elimination is also promising. Besides, most toxic heavy metals and minerals are converted into less available forms during manure pyrolysis at optimum temperature. Hence, they exhibit decreased bioavailable toxic elements and a more gradual release of nutrients than conventionally treated manure. Despite these advantages, there are also drawbacks, such as the efficiency of biochar varies according to the feedstock and pyrolysis conditions, as well as potential concerns over heavy metal content is also present. To overcome these obstacles, more investigation is needed to produce biochar in optimum conditions and evaluate its long-term effects on ecosystem health and soil fertility.

References

Abbasi, M. K., & Anwar, A. A. (2015). Ameliorating effects of biochar derived from poultry manure and white clover residues on soil nutrient status and plant growth promotion-greenhouse experiments. *PLoS One, 10*(6), e0131592.

Abdulyekeen, K. A., Daud, W. M. A. W., Patah, M. F. A., & Abnisa, F. (2022). Torrefaction of organic municipal solid waste to high calorific value solid fuel using batch reactor with helical screw induced rotation. *Bioresource Technology, 363*, 127974.

Ajmera, I., Hodgman, T. C., & Lu, C. (2019). An integrative systems perspective on plant phosphate research. Genes, 10(2), 139

Abo-Elyousr, K. A., Mousa, M. A., Ibrahim, O. H., Alshareef, N. O., & Eissa, M. A. (2022). Calcium-rich biochar stimulates salt resistance in pearl millet (Pennisetum glaucum L.) plants by improving soil quality and enhancing the antioxidant defense. *Plants, 11*(10), 1301.

Agegnehu, G., Srivastava, A. K., & Bird, M. I. (2017). The role of biochar and biochar-compost in improving soil quality and crop performance: A review. *Applied Soil Ecology, 119*, 156–170.

Akhtar, S. S., Li, G., Andersen, M. N., & Liu, F. (2014). Biochar enhances yield and quality of tomato under reduced irrigation. *Agricultural Water Management, 138*, 37–44.

Amalina, F., Abd Razak, A. S., Krishnan, S., Sulaiman, H., Zularisam, A. W., & Nasrullah, M. (2022). Biochar production techniques utilizing biomass waste-derived materials and environmental applications–A review. *Journal of Hazardous Materials Advances, 7*, 100134.

An, Y., Lu, J., Niu, R., Li, M., Zhao, X., Huang, X., Huang, H., Garg, A., & Zhussupbekov, A. (2021). Exploring effects of novel chemical modification of biochar on soil water retention and crack suppression: Towards commercialization of production of biochar for soil remediation. *Biomass Conversion and Biorefinery, 13*, 13897–13910.

Are, K. S., Adelana, A. O., Fademi, I. O., & Aina, O. A. (2017). Improving physical properties of degraded soil: Potential of poultry manure and biochar. *Agriculture and Natural Resources, 51*(6), 454–462.
Awad, M., Liu, Z., Skalicky, M., Dessoky, E. S., Brestic, M., Mbarki, S., Rastogi, A., & El Sabagh, A. (2021). Fractionation of heavy metals in multi-contaminated soil treated with biochar using the sequential extraction procedure. *Biomolecules, 11*(3), 448.
Ayaz, M., Feizienė, D., Tilvikienė, V., Akhtar, K., Stulpinaitė, U., & Iqbal, R. (2021). Biochar role in the sustainability of agriculture and environment. *Sustainability, 13*(3), 1330.
Ayaz, M., Feizienė, D., Feiza, V., Tilvikienė, V., Baltrėnaitė-Gedienė, E., & Khan, A. (2022). The impact of swine manure biochar on the physical properties and microbial activity of loamy soils. *Plants, 11*(13), 1729.
Azeem, M., Ali, A., Jeyasundar, P. G. S. A., Li, Y., Abdelrahman, H., Latif, A., Li, R., Basta, N., Li, G., Shaheen, S. M., Rinklebe, J., & Zhang, Z. (2021a,b). Bone-derived biochar improved soil quality and reduced Cd and Zn phytoavailability in a multi-metal contaminated mining soil. *Environmental Pollution, 277*, 116800.
Azeem, M., Ali, A., Jeyasundar, P. G. A., Bashir, S., Hussain, Q., Wahid, F., Ali, E. F., Abdelrahman, H., Li, R., Antoniadis, V., Rinklebe, J., Shaheen, S. M., Li, G., & Zhang, Z. (2021b). Effects of sheep bone biochar on soil quality, maize growth, and fractionation and phytoavailability of Cd and Zn in a mining-contaminated soil. *Chemosphere, 282*, 131016.
Azeem, M., Jeyasundar, P. G. S. A., Ali, A., Riaz, L., Khan, K. S., Hussain, Q., Kareem, H. A., Abbas, F., Latif, A., Majrashi, A., Ali, E. F., Li, R., Shaheen, S. M., Li, G., Zhang, Z., & Zhu, Y. (2023). Cow bone-derived biochar enhances microbial biomass and alters bacterial community composition and diversity in a smelter contaminated soil. *Environmental Research, 216*, 114278.
Balajii, M., & Niju, S. (2019). Biochar-derived heterogeneous catalysts for biodiesel production. *Environmental Chemistry Letters, 17*(4), 1447–1469.
Bhattarai, B., Neupane, J., & Dhakal, S. P. (2015). Effect of biochar from different origin on physio-chemical properties of soil and yield of garden pea (Pisum sativum L.) at Paklihawa, Rupandehi, Nepal. *World Journal of Agricultural Research, 3*(4), 129–138.
Biederman, L. A., & Harpole, W. S. (2013). Biochar and its effects on plant productivity and nutrient cycling: A meta-analysis. *GCB Bioenergy, 5*(2), 202–214.
Bin Yousaf, M. T., Nawaz, M. F., Yasin, G., Cheng, H., Ahmed, I., Gul, S., Rizwan, M., Rehim, A., Xuebin, Q., & Ur Rahman, S. (2022). Determining the appropriate level of farmyard manure biochar application in saline soils for three selected farm tree species. *PLoS One, 17*(4), e0265005.
Bordoloi, S., Garg, A., Sreedeep, S., Lin, P., & Mei, G. (2018). Investigation of cracking and water availability of soil-biochar composite synthesized from invasive weed water hyacinth. *Bioresource Technology, 263*, 665–677.
Brown, A. E., Hammerton, J. M., Camargo-Valero, M. A., & Ross, A. B. (2022). Integration of hydrothermal carbonisation and anaerobic digestion for the energy valorisation of grass. *Energies, 15*(10), 3495.
Bulak, P., Proc-Pietrycha, K., Kaczor, M., Złotko, K., Polakowski, C., Wiącek, D., Waniak-Nowicka, H., Zięba, E., Waśko, A., Oleszczuk, P., & Bieganowski, A. (2023). A novel type of biochar from chitinous Hermetia illucens waste with a built-in stimulating effect on plants and soil arthropods. *Scientific Reports, 13*(1), 8306.
Butnan, S., Deenik, J. L., Toomsan, B., Antal, M. J., & Vityakon, P. (2015). Biochar characteristics and application rates affecting corn growth and properties of soils contrasting in texture and mineralogy. *Geoderma, 237*, 105–116.
Cely, P., Gascó, G., Paz-Ferreiro, J., & Méndez, A. (2015). Agronomic properties of biochars from different manure wastes. *Journal of Analytical and Applied Pyrolysis, 111*, 173–182.
Chaganti, V. N., Crohn, D. M., & Šimůnek, J. (2015). Leaching and reclamation of a biochar and compost amended saline–sodic soil with moderate SAR reclaimed water. *Agricultural Water Management, 158*, 255–265.

Chandraratne, M. R., & Daful, A. G. (2022). Advances in bioenergy production using fast pyrolysis and hydrothermal processing. In *Biomass, biorefineries and bioeconomy*. IntechOpen.

Chang, Y. M., Tsai, W. T., & Li, M. H. (2015). Chemical characterization of char derived from slow pyrolysis of microalgal residue. *Journal of Analytical and Applied Pyrolysis, 111*, 88–93.

Chen, W. H., & Kuo, P. C. (2010). A study on torrefaction of various biomass materials and its impact on lignocellulosic structure simulated by a thermogravimetry. *Energy, 35*(6), 2580–2586.

Chen, Q., Qin, J., Sun, P., Cheng, Z., & Shen, G. (2018). Cow dung-derived engineered biochar for reclaiming phosphate from aqueous solution and its validation as slow-release fertilizer in soil-crop system. *Journal of Cleaner Production, 172*, 2009–2018.

Chen, H., Yang, X., Wang, H., Sarkar, B., Shaheen, S. M., Gielen, G., Bolan, N., Guo, J., Che, L., Sun, H., & Rinklebe, J. (2020). Animal carcass-and wood-derived biochars improved nutrient bioavailability, enzyme activity, and plant growth in metal-phthalic acid ester co-contaminated soils: A trial for reclamation and improvement of degraded soils. *Journal of Environmental Management, 261*, 110246.

Clark, M., Hastings, M. G., & Ryals, R. (2019). Soil carbon and nitrogen dynamics in two agricultural soils amended with manure-derived biochar. *Journal of Environmental Quality, 48*(3), 727–734.

Côrtes, L. N., Druzian, S. P., Streit, A. F. M., Godinho, M., Perondi, D., Collazzo, G. C., Oliveira, M. L. S., Cadaval Jr, T. R. S., & Dotto, G. L. (2019). Biochars from animal wastes as alternative materials to treat colored effluents containing basic red 9. *Journal of Environmental Chemical Engineering, 7*(6), 103446.

Daful, A. G., Chandraratne, M. R., & Loridon, M. (2022). Recent perspectives in biochar production, characterization and applications. In *Recent perspectives in pyrolysis research.*

Dai, Z., Zhang, X., Tang, C., Muhammad, N., Wu, J., Brookes, P. C., & Xu, J. (2017). Potential role of biochars in decreasing soil acidification-a critical review. *Science of the Total Environment, 581*, 601–611.

Dai, Z., Xiong, X., Zhu, H., Xu, H., Leng, P., Li, J., et al. (2021). Association of biochar properties with changes in soil bacterial, fungal and fauna communities and nutrient cycling processes. *Biochar, 3*, 239–254.

Ding, Y., Liu, Y., Liu, S., Li, Z., Tan, X., Huang, X., Zeng, G., Zhou, L. and Zheng, B. (2016). Biochar to improve soil fertility. A review. *Agronomy for Sustainable Development, 36*(2), 36.

Dalólio, F. S., da Silva, J. N., de Oliveira, A. C. C., Tinôco, I. D. F. F., Barbosa, R. C., de Oliveira Resende, M., Albino, L. F. T., & Coelho, S. T. (2017). Poultry litter as biomass energy: A review and future perspectives. *Renewable and Sustainable Energy Reviews, 76*, 941–949.

Das, S. K., Ghosh, G. K., & Avasthe, R. (2021). Applications of biomass derived biochar in modern science and technology. *Environmental Technology & Innovation, 21*, 101306.

Demirbas, A. (2004). Effect of initial moisture content on the yields of oily products from pyrolysis of biomass. *Journal of Analytical and Applied Pyrolysis, 71*(2), 803–815.

Deng, L., Zhang, Y., Wang, Y., Yuan, H., Chen, Y., & Wu, Y. (2021). In situ N-, P-and Ca-codoped biochar derived from animal bones to boost the electrocatalytic hydrogen evolution reaction. *Resources, Conservation and Recycling, 170*, 105568.

Downie, A., Munroe, P., & Crosky, A. (2009). Characteristics of biochar—Physical and structural properties. In *Biochar for environmental management: Science and technology* (pp. 13–29).

Dróżdż, D., Malińska, K., Wystalska, K., Meers, E., & Robles-Aguilar, A. (2023). The influence of poultry manure-derived biochar and compost on soil properties and plant biomass growth. *Materials, 16*(18), 6314.

El-Naggar, A., Lee, S. S., Rinklebe, J., Farooq, M., Song, H., Sarmah, A. K., Zimmerman, A. R., Ahmad, M., Shaheen, S. M., & Ok, Y. S. (2019). Biochar application to low fertility soils: A review of current status, and future prospects. *Geoderma, 337*, 536–554.

Ewulo, B. S., Ojeniyi, S. O., & Akanni, D. A. (2008). Effect of poultry manure on selected soil physical and chemical properties, growth, yield and nutrient status of tomato. *African Journal of Agricultural Research, 3*(9), 612–616.

Fahmy, T. Y., Fahmy, Y., Mobarak, F., El-Sakhawy, M., & Abou-Zeid, R. E. (2020). Biomass pyrolysis: Past, present, and future. *Environment, Development and Sustainability, 22*, 17–32.

Gabhane, J. W., Bhange, V. P., Patil, P. D., Bankar, S. T., & Kumar, S. (2020). Recent trends in biochar production methods and its application as a soil health conditioner: A review. *SN Applied Sciences, 2*, 1–21.

Gasco, G., Paz-Ferreiro, J., Cely, P., Plaza, C., & Mendez, A. (2016). Influence of pig manure and its biochar on soil CO2 emissions and soil enzymes. *Ecological Engineering, 95*, 19–24.

Gavili, E., Moosavi, A. A., & Moradi Choghamarani, F. (2018). Cattle manure biochar potential for ameliorating soil physical characteristics and spinach response under drought. *Archives of Agronomy and Soil Science, 64*(12), 1714–1727.

Gray, M., Johnson, M. G., Dragila, M. I., & Kleber, M. (2014). Water uptake in biochars: The roles of porosity and hydrophobicity. *Biomass and Bioenergy, 61*, 196–205.

Gu, H., Tang, Y., Yao, J., & Chen, F. (2019). Study on biomass gasification under various operating conditions. *Journal of the Energy Institute, 92*(5), 1329–1336.

Gundale, M. J., & DeLuca, T. H. (2006). Temperature and source material influence ecological attributes of ponderosa pine and Douglas-fir charcoal. *Forest Ecology and Management, 231*(1–3), 86–93.

Guo, M., Li, H., Baldwin, B., & Morrison, J. (2020). Thermochemical processing of animal manure for bioenergy and biochar. *Animal manure: Production, Characteristics, Environmental Concerns, and Management, 67*, 255–274.

Guo, J., Zheng, L., Li, Z., Zhou, X., Cheng, S., Zhang, L., & Zhang, Q. (2021). Effects of various pyrolysis conditions and feedstock compositions on the physicochemical characteristics of cow manure-derived biochar. *Journal of Cleaner Production, 311*, 127458.

Gupta, S., Sireesha, S., Sreedhar, I., Patel, C. M., & Anitha, K. L. (2020). Latest trends in heavy metal removal from wastewater by biochar based sorbents. *Journal of Water Process Engineering, 38*, 101561.

Hassen-Trabelsi, A. B., Kraiem, T., Naoui, S., & Belayouni, H. (2014). Pyrolysis of waste animal fats in a fixed-bed reactor: Production and characterization of bio-oil and bio-char. *Waste Management, 34*(1), 210–218.

Hossain, M. Z., Bahar, M. M., Sarkar, B., Donne, S. W., Ok, Y. S., Palansooriya, K. N., Kirkham, M. B., Chowdhury, S., & Bolan, N. (2020). Biochar and its importance on nutrient dynamics in soil and plant. *Biochar, 2*, 379–420.

Hossain, M. Z., Bahar, M. M., Sarkar, B., Donne, S. W., Wade, P., & Bolan, N. (2021). Assessment of the fertilizer potential of biochars produced from slow pyrolysis of biosolid and animal manures. *Journal of Analytical and Applied Pyrolysis, 155*, 105043.

Hosseini, S. H., Liang, X., Niyungeko, C., Miaomiao, H., Li, F., Khan, S., & Eltohamy, K. M. (2019). Effect of sheep manure-derived biochar on colloidal phosphorus release in soils from various land uses. *Environmental Science and Pollution Research, 26*, 36367–36379.

Hou, J., Pugazhendhi, A., Sindhu, R., Vinayak, V., Thanh, N. C., Brindhadevi, K., Chi, N. T. L., & Yuan, D. (2022). An assessment of biochar as a potential amendment to enhance plant nutrient uptake. *Environmental Research, 214*, 113909.

Huang, H., Reddy, N. G., Huang, X., Chen, P., Wang, P., Zhang, Y., Huang, Y., Lin, P., & Garg, A. (2021). Effects of pyrolysis temperature, feedstock type and compaction on water retention of biochar amended soil. *Scientific Reports, 11*(1), 7419.

Hussain, A., Nazir, A., & Shafiq, M. (2021). Potential application of biochar composite derived from rice straw and animal bones to improve plant growth. *Sustainability, 13*(19), 11104.

Ibrahim, E. A., El-Sherbini, M. A., & Selim, E. M. M. (2022). Effects of biochar on soil properties, heavy metal availability and uptake, and growth of summer squash grown in metal-contaminated soil. *Scientia Horticulturae, 301*, 111097.

Idrees, M., Batool, S., Kalsoom, T., Yasmeen, S., Kalsoom, A., Raina, S., Zhuang, Q., & Kong, J. (2018). Animal manure-derived biochars produced via fast pyrolysis for the removal of divalent copper from aqueous media. *Journal of Environmental Management, 213*, 109–118.

Ighalo, J. O., Iwuchukwu, F. U., Eyankware, O. E., Iwuozor, K. O., Olotu, K., Bright, O. C., & Igwegbe, C. A. (2022). Flash pyrolysis of biomass: A review of recent advances. *Clean Technologies and Environmental Policy, 24*(8), 2349–2363.

Imam, T., & Capareda, S. (2012). Characterization of bio-oil, syn-gas and bio-char from switchgrass pyrolysis at various temperatures. *Journal of Analytical and Applied Pyrolysis, 93*, 170–177.
Ippolito, J. A., Spokas, K. A., Novak, J. M., Lentz, R. D., & Cantrell, K. B. (2015). Biochar elemental composition and factors influencing nutrient retention. In *Biochar for environmental management* (pp. 139–163). Routledge.
Irmak Yilmaz, F., & Ergun, Y. A. (2019). Impact of biochar and animal manure on some biological and chemical properties of soil. *Applied Ecology and Environmental Research, 17*, 8865–8876.
Jang, E. S., Ryu, D. Y., & Kim, D. (2022). Hydrothermal carbonization improves the quality of biochar derived from livestock manure by removing inorganic matter. *Chemosphere, 305*, 135391.
Jiang, Y., Wang, X., Zhao, Y., Zhang, C., Jin, Z., Shan, S., & Ping, L. (2021). Effects of biochar application on enzyme activities in tea garden soil. *Frontiers in Bioengineering and Biotechnology, 9*, 728530.
Jindo, K., Suto, K., Matsumoto, K., García, C., Sonoki, T., & Sanchez-Monedero, M. A. (2012). Chemical and biochemical characterisation of biochar-blended composts prepared from poultry manure. *Bioresource Technology, 110*, 396–404.
Joardar, J. C., Mondal, B., & Sikder, S. (2020). Comparative study of poultry litter and poultry litter biochar application in the soil for plant growth. *SN Applied Sciences, 2*, 1–9.
Jung, J. M., Lee, S. R., Lee, J., Lee, T., Tsang, D. C., & Kwon, E. E. (2017). Biodiesel synthesis using chicken manure biochar and waste cooking oil. *Bioresource Technology, 244*, 810–815.
Jung, J. M., Oh, J. I., Baek, K., Lee, J., & Kwon, E. E. (2018). Biodiesel production from waste cooking oil using biochar derived from chicken manure as a porous media and catalyst. *Energy Conversion and Management, 165*, 628–633.
Jung, S., Kim, M., Jung, J. M., & Kwon, E. E. (2020). Valorization of swine manure biochar as a catalyst for transesterifying waste cooking oil into biodiesel. *Environmental Pollution, 266*, 115377.
Kalus, K., Koziel, J. A., & Opaliński, S. (2019). A review of biochar properties and their utilization in crop agriculture and livestock production. *Applied Sciences, 9*(17), 3494.
Kolb, S. E., Fermanich, K. J., & Dornbush, M. E. (2009). Effect of charcoal quantity on microbial biomass and activity in temperate soils. *Soil Science Society of America Journal, 73*(4), 1173-1181.
Khan, A., Ali, I., Naqvi, S. R., AlMohamadi, H., Shahbaz, M., Ali, A. M., & Shahzad, K. (2023). Assessment of thermokinetic behaviour of tannery sludge in slow pyrolysis process through artificial neural network. *Chemosphere, 337*, 139226.
Kim, H. S., Kim, K. R., Yang, J. E., Ok, Y. S., Owens, G., Nehls, T., Wessolek, G., & Kim, K. H. (2016). Effect of biochar on reclaimed tidal land soil properties and maize (Zea mays L.) response. *Chemosphere, 142*, 153–159.
Kim, J., Park, C., Park, H., Han, J., Lee, J., & Kim, S. K. (2022). Upcycling of cattle manure for simultaneous energy recovery and supercapacitor electrode production. *Energy, 258*, 124877.
Kiran, Y. K., Barkat, A., Cui, X. Q., Ying, F. E. N. G., Pan, F. S., Lin, T. A. N. G., & Yang, X. E. (2017). Cow manure and cow manure-derived biochar application as a soil amendment for reducing cadmium availability and accumulation by Brassica chinensis L. in acidic red soil. *Journal of Integrative Agriculture, 16*(3), 725–734.
Kloss, S., Zehetner, F., Dellantonio, A., Hamid, R., Ottner, F., Liedtke, V., Schwanninger, M., Gerzabek, M. H., & Soja, G. (2012). Characterization of slow pyrolysis biochars: Effects of feedstocks and pyrolysis temperature on biochar properties. *Journal of Environmental Quality, 41*(4), 990–1000.
Krounbi, L., Enders, A., Gaunt, J., Ball, M., & Lehmann, J. (2021). Plant uptake of nitrogen adsorbed to biochars made from dairy manure. *Scientific Reports, 11*(1), 15001.
Kumar, J., & Vyas, S. (2024). Comprehensive review of biomass utilization and gasification for sustainable energy production. *Environment, Development and Sustainability, 27*, 1–40.
Laila, U., Hussain, A., Nazir, A., & Shafiq, M. (2021). Potential application of biochar composite derived from rice straw and animal bones to improve plant growth. *Sustainability, 13*(19), 11104.

Lashari, M. S., Ye, Y., Ji, H., Li, L., Kibue, G. W., Lu, H., et al. (2015). Biochar–manure compost in conjunction with pyroligneous solution alleviated salt stress and improved leaf bioactivity of maize in a saline soil from Central China: A 2-year field experiment. *Journal of the Science of Food and Agriculture, 95*(6), 1321–1327.

Lee, X. J., Ong, H. C., Gan, Y. Y., Chen, W. H., & Mahlia, T. M. I. (2020). State of art review on conventional and advanced pyrolysis of macroalgae and microalgae for biochar, bio-oil and bio-syngas production. *Energy Conversion and Management, 210*, 112707.

Lehmann, J., & Joseph, S. (2015). Biochar for environmental management: An introduction. In *Biochar for environmental management* (pp. 1–13). Routledge.

Lehmann, J., Rillig, M. C., Thies, J., Masiello, C. A., Hockaday, W. C., & Crowley, D. (2011). Biochar effects on soil biota–a review. *Soil Biology and Biochemistry, 43*(9), 1812–1836.

Lehmann, J., Kuzyakov, Y., Pan, G., & Ok, Y. S. (2015). Biochars and the plant-soil interface. *Plant and Soil, 395*, 1–5.

Lei, S., Shi, Y., Qiu, Y., Che, L., & Xue, C. (2019). Performance and mechanisms of emerging animal-derived biochars for immobilization of heavy metals. *Science of the Total Environment, 646*, 1281–1289.

Lewandowski, W. M., Januszewicz, K., & Kosakowski, W. (2019). Efficiency and proportions of waste Tyre pyrolysis products depending on the reactor type—A review. *Journal of Analytical and Applied Pyrolysis, 140*, 25–53.

Li, H., Li, Y., Xu, Y., & Lu, X. (2020). Biochar phosphorus fertilizer effects on soil phosphorus availability. *Chemosphere, 244*, 125471.

Liang, Y., Cao, X., Zhao, L., Xu, X., & Harris, W. (2014). Phosphorus release from dairy manure, the manure-derived biochar, and their amended soil: Effects of phosphorus nature and soil property. *Journal of Environmental Quality, 43*(4), 1504–1509.

Lin, Q., Xu, X., Wang, L., Chen, Q., Fang, J., Shen, X., Lou, L., & Tian, G. (2017). The speciation, leachability and bioaccessibility of Cu and Zn in animal manure-derived biochar: Effect of feedstock and pyrolysis temperature. *Frontiers of Environmental Science & Engineering, 11*, 1–12.

Liu, S. C., & Tsai, W. T. (2016). Thermochemical characteristics of dairy manure and its derived biochars from a fixed-bed pyrolysis. *International Journal of Green Energy, 13*(10), 963–968.

Liu, Q., Liu, B., Zhang, Y., Lin, Z., Zhu, T., Sun, R., Wang, X., Ma, J., Bei, Q., Liu, G., & Lin, X. (2017). Can biochar alleviate soil compaction stress on wheat growth and mitigate soil N2O emissions? *Soil Biology and Biochemistry, 104*, 8–17.

Liu, L., Tan, Z., Gong, H., & Huang, Q. (2018). Migration and transformation mechanisms of nutrient elements (N, P, K) within biochar in straw–biochar–soil–plant systems: A review. *ACS Sustainable Chemistry & Engineering, 7*(1), 22–32.

Lou, K., Rajapaksha, A. U., Ok, Y. S., & Chang, S. X. (2016). Pyrolysis temperature and steam activation effects on sorption of phosphate on pine sawdust biochars in aqueous solutions. *Chemical Speciation & Bioavailability, 28*(1–4), 42–50.

Lu, Q., Zhang, Z. B., Yang, X. C., Dong, C. Q., & Zhu, X. F. (2013). Catalytic fast pyrolysis of biomass impregnated with K3PO4 to produce phenolic compounds: Analytical Py-GC/MS study. *Journal of Analytical and Applied Pyrolysis, 104*, 139–145.

Lu, K., Yang, X., Gielen, G., Bolan, N., Ok, Y. S., Niazi, N. K., Xu, S., Yuan, G., Chen, X., Zhang, X., Liu, D., Song, Z., Liu, X., & Wang, H. (2017). Effect of bamboo and rice straw biochars on the mobility and redistribution of heavy metals (Cd, Cu, Pb and Zn) in contaminated soil. *Journal of Environmental Management, 186*, 285–292.

Lu, H., Yan, M., Wong, M. H., Mo, W. Y., Wang, Y., Chen, X. W., & Wang, J. J. (2020). Effects of biochar on soil microbial community and functional genes of a landfill cover three years after ecological restoration. *Science of the Total Environment, 717*, 137133.

Luo, X., Liu, G., Xia, Y., Chen, L., Jiang, Z., Zheng, H., & Wang, Z. (2017). Use of biochar-compost to improve properties and productivity of the degraded coastal soil in the Yellow River Delta, China. *Journal of Soils and Sediments, 17*, 780–789.

Ma, Y. L., & Matsunaka, T. (2013). Biochar derived from dairy cattle carcasses as an alternative source of phosphorus and amendment for soil acidity. *Soil Science and Plant Nutrition, 59*(4), 628–641.

Man, K. Y., Chow, K. L., Man, Y. B., Mo, W. Y., & Wong, M. H. (2021). Use of biochar as feed supplements for animal farming. *Critical Reviews in Environmental Science and Technology, 51*(2), 187–217.

Mandal, S., Kunhikrishnan, A., Bolan, N. S., Wijesekara, H., & Naidu, R. (2016). Application of biochar produced from biowaste materials for environmental protection and sustainable agriculture production. In *Environmental materials and waste* (pp. 73–89). Academic Press.

Mao, X., Kang, Q., Liu, Y., Siyal, A. A., Ao, W., Ran, C., Fu, J., Deng, Z., Song, Y., & Dai, J. (2019). Microwave-assisted pyrolysis of furfural residue in a continuously operated auger reactor: Biochar characterization and analysis. *Energy, 168*, 573–584.

Mavi, M. S., Bhullar, R. S., & Choudhary, O. P. (2020). Differential ability of pyrolysed biomass derived from diverse feedstocks in alleviating salinity stress. *Biomass Conversion and Biorefinery*, 1–12.

Medyńska-Juraszek, A., Marcinkowska, K., Gruszka, D., & Kluczek, K. (2022). The effects of rabbit-manure-derived biochar co-application with compost on the availability and heavy metal uptake by green leafy vegetables. *Agronomy, 12*(10), 2552.

Mishra, S., & Upadhyay, R. K. (2021). Review on biomass gasification: Gasifiers, gasifying mediums, and operational parameters. *Materials Science for Energy Technologies, 4*, 329–340.

Mohamed, W. S., & Hammam, A. A. (2019). Poultry manure-derived biochar as a soil amendment and fertilizer for sandy soils under arid conditions. *Egyptian Journal of Soil Science, 59*(1), 1–14.

Mong, G. R., Chong, C. T., Ng, J. H., Chong, W. W. F., Lam, S. S., Ong, H. C., & Ani, F. N. (2020). Microwave pyrolysis for valorisation of horse manure biowaste. *Energy Conversion and Management, 220*, 113074.

Mong, G. R., Chong, C. T., Chong, W. W. F., Ng, J. H., Ong, H. C., Ashokkumar, V., Tran, M. V., Karmakar, S., Goh, B. H. H., & Yasin, M. F. M. (2022). Progress and challenges in sustainable pyrolysis technology: Reactors, feedstocks and products. *Fuel, 324*, 124777.

Moradi, P., Hajjami, M., & Valizadeh-Kakhki, F. (2019). Biochar as heterogeneous support for immobilization of Pd as efficient and reusable biocatalyst in C–C coupling reactions. *Applied Organometallic Chemistry, 33*(11), e5205.

Moser, K., Wopienka, E., Pfeifer, C., Schwarz, M., Sedlmayer, I., & Haslinger, W. (2023). Screw reactors and rotary kilns in biochar production–A comparative review. *Journal of Analytical and Applied Pyrolysis, 174*, 106112.

Nahar Myyas, R. E., Tostado-Véliz, M., Gómez-González, M., & Jurado, F. (2023). Review of bioenergy potential in Jordan. *Energies, 16*(3), 1393.

Nahidan, S., & Ghasemzadeh, M. (2022). Biochemical phosphorus transformations in a calcareous soil as affected by earthworm, cow manure and its biochar additions. *Applied Soil Ecology, 170*, 104310.

Naz, F. (2023). Plant nutrition, transport, mechanism and sensing in plants. In *Sustainable Plant Nutrition* (pp. 209–228). Academic Press.

Ndirangu, S. M., Liu, Y., Xu, K., & Song, S. (2019). Risk evaluation of pyrolyzed biochar from multiple wastes. *Journal of Chemistry, 2019*(1), 4506314.

Novak, J. M., Johnson, M. G., & Spokas, K. A. (2018). Concentration and release of phosphorus and potassium from lignocellulosic-and manure-based biochars for fertilizer reuse. *Frontiers in Sustainable Food Systems, 2*, 54.

Orlando, M. Q., & Borja, V. M. (2020). Pretreatment of animal manure biomass to improve biogas production: A review. *Energies, 13*(14), 3573.

Otsuka, S., Sudiana, I., Komori, A., Isobe, K., Deguchi, S., Nishiyama, M., Shimizu, H., & Senoo, K. (2008). Community structure of soil bacteria in a tropical rainforest several years after fire. *Microbes and Environments, 23*(1), 49–56.

Palansooriya, K. N., Ok, Y. S., Awad, Y. M., Lee, S. S., Sung, J. K., Koutsospyros, A., & Moon, D. H. (2019). Impacts of biochar application on upland agriculture: A review. *Journal of Environmental Management, 234*, 52–64.

Pan, R. (2023). The heavy metals transformation during the pyrolysis and hydrothermal carbonation of municipal sewage sludge. In *E3S web of conferences* (Vol. 393, p. 03001). EDP Sciences.

Panwar, N. L., Pawar, A., & Salvi, B. L. (2019). Comprehensive review on production and utilization of biochar. *SN Applied Sciences, 1*, 1–19.

Park, H., Joo, J., Kim, J., Lee, J., & Kim, S. K. (2023). A sustainable strategy for organic waste upcycling: Concurrent production of energy and Li-ion battery anode from chicken litter. *Energy, 278*, 127797.

Parthasarathy, P., Al-Ansari, T., Mackey, H. R., Narayanan, K. S., & McKay, G. (2022). A review on prominent animal and municipal wastes as potential feedstocks for solar pyrolysis for biochar production. *Fuel, 316*, 123378.

Pontiroli, D., Scaravonati, S., Magnani, G., Fornasini, L., Bersani, D., Bertoni, G., Milanese, C., Girella, A., Ridi, F., Verucchi, R., Mantovani, L., Malcevschi, A., & Riccò, M. (2019). Super-activated biochar from poultry litter for high-performance supercapacitors. *Microporous and Mesoporous Materials, 285*, 161–169.

Premchand, P., Demichelis, F., Chiaramonti, D., Bensaid, S., & Fino, D. (2023). Biochar production from slow pyrolysis of biomass under CO2 atmosphere: A review on the effect of CO2 medium on biochar production, characterisation, and environmental applications. *Journal of Environmental Chemical Engineering, 11*(3), 110009.

Purakayastha, T. J., Bera, T., Bhaduri, D., Sarkar, B., Mandal, S., Wade, P., Kumari, S., Biswas, S., Menon, M., Pathak, H., & Tsang, D. C. (2019). A review on biochar modulated soil condition improvements and nutrient dynamics concerning crop yields: Pathways to climate change mitigation and global food security. *Chemosphere, 227*, 345–365.

Qin, J., Qian, S., Chen, Q., Chen, L., Yan, L., & Shen, G. (2019). Cow manure-derived biochar: Its catalytic properties and influential factors. *Journal of Hazardous Materials, 371*, 381–388.

Qin, L., Wu, Y., Hou, Z., Zhang, S., & Jiang, E. (2021). Synthesis of heteroatom and metallic compound self-co-doped porous carbon derived from swine manure for supercapacitor electrodes and lead ion adsorbents. *Journal of Industrial and Engineering Chemistry, 102*, 195–205.

Rahman, L., Whitelaw-Weckert, M. A., & Orchard, B. (2014). Impact of organic soil amendments, including poultry-litter biochar, on nematodes in a Riverina, New South Wales, vineyard. *Soil Research, 52*(6), 604–619.

Rasaq, W. A., Golonka, M., Scholz, M., & Białowiec, A. (2021). Opportunities and challenges of high-pressure fast pyrolysis of biomass: A review. *Energies, 14*(17), 5426.

Rathnayake, D., Schmidt, H. P., Leifeld, J., Mayer, J., Epper, C. A., Bucheli, T. D., & Hagemann, N. (2023). Biochar from animal manure: A critical assessment on technical feasibility, economic viability, and ecological impact. *GCB Bioenergy, 15*(9), 1078–1104.

Rego, F., Xiang, H., Yang, Y., Ordovás, J. L., Chong, K., Wang, J., & Bridgwater, A. (2022). Investigation of the role of feedstock properties and process conditions on the slow pyrolysis of biomass in a continuous auger reactor. *Journal of Analytical and Applied Pyrolysis, 161*, 105378.

Rehman, A., Nawaz, S., Alghamdi, H. A., Alrumman, S., Yan, W., & Nawaz, M. Z. (2020). Effects of manure-based biochar on uptake of nutrients and water holding capacity of different types of soils. *Case Studies in Chemical and Environmental Engineering, 2*, 100036.

Rex, P., Mohammed Ismail, K. R., Meenakshisundaram, N., Barmavatu, P., & Sai Bharadwaj, A. V. S. L. (2023). Agricultural biomass waste to biochar: A review on biochar applications using machine learning approach and circular economy. *ChemEngineering, 7*(3), 50.

Ro, K. S. (2016). Kinetics and energetics of producing animal manure-based biochar. *Bioenergy Research, 9*, 447–453.

Robb, S., Joseph, S., Abdul Aziz, A., Dargusch, P., & Tisdell, C. (2020). Biochar's cost constraints are overcome in small-scale farming on tropical soils in lower-income countries. *Land Degradation & Development, 31*(13), 1713–1726.

Roberts, C., Greene, J., & Nemet, G. F. (2023). Key enablers for carbon dioxide removal through the application of biochar to agricultural soils: Evidence from three historical analogues. *Technological Forecasting and Social Change, 195*, 122704.
Rutigliano, F. A., Romano, M., Marzaioli, R., Baglivo, I., Baronti, S., Miglietta, F., & Castaldi, S. (2014). Effect of biochar addition on soil microbial community in a wheat crop. *European Journal of Soil Biology, 60*, 9–15.
Sadeghpour, A., Ketterings, Q. M., Vermeylen, F., Godwin, G. S., & Czymmek, K. J. (2016). Soil properties under nitrogen-vs. phosphorus-based manure and compost management of corn. *Soil Science Society of America Journal, 80*(5), 1272–1282.
Schimmelpfennig, S., & Glaser, B. (2012). One step forward toward characterization: Some important material properties to distinguish biochars. *Journal of Environmental Quality, 41*(4), 1001–1013.
Schmidt, H. P., Kammann, C., Hagemann, N., Leifeld, J., Bucheli, T. D., Sánchez Monedero, M. A., & Cayuela, M. L. (2021). Biochar in agriculture–A systematic review of 26 global meta-analyses. *GCB Bioenergy, 13*(11), 1708–1730.
Sharma, P. K., & Kumar, S. (2023). Soil structure and plant growth. In *Soil physical environment and plant growth: Evaluation and management* (pp. 125–154). Springer International Publishing.
Shetty, R., & Prakash, N. B. (2020). Effect of different biochars on acid soil and growth parameters of rice plants under aluminium toxicity. *Scientific Reports, 10*(1), 12249.
Sikder, S., & Joardar, J. C. (2019). Biochar production from poultry litter as management approach and effects on plant growth. *International Journal of Recycling of Organic Waste in Agriculture, 8*, 47–58.
Singh, P., Kumar, S., Srivastava, N., Srivastava, R., & Chattopadhyay, J. (2023). Thermochemical hydrogen production. In *Solar-driven green hydrogen generation and storage* (pp. 333–345). Elsevier.
Slavich, P. G., Sinclair, K., Morris, S. G., Kimber, S. W. L., Downie, A., & Van Zwieten, L. (2013). Contrasting effects of manure and green waste biochars on the properties of an acidic ferralsol and productivity of a subtropical pasture. *Plant and Soil, 366*, 213–227.
Su, Y., Liu, L., Xu, D., Du, H., Xie, Y., Xiong, Y., & Zhang, S. (2020). Syngas production at low temperature via the combination of hydrothermal pretreatment and activated carbon catalyst along with value-added utilization of tar and bio-char. *Energy Conversion and Management, 205*, 112382.
Subedi, R., Taupe, N., Ikoyi, I., Bertora, C., Zavattaro, L., Schmalenberger, A., Leahy, J. J., & Grignani, C. (2016). Chemically and biologically-mediated fertilizing value of manure-derived biochar. *Science of the Total Environment, 550*, 924–933.
Sukartono, U. W., Kusuma, Z., & Nugroho, W. H. (2011). Soil fertility status, nutrient uptake, and maize (Zea mays L.) yield following biochar and cattle manure application on sandy soils of Lombok, Indonesia. *Journal of Tropical Agriculture, 49*(1–2), 47–52.
Suliman, W., Harsh, J. B., Abu-Lail, N. I., Fortuna, A. M., Dallmeyer, I., & Garcia-Perez, M. (2016). Influence of feedstock source and pyrolysis temperature on biochar bulk and surface properties. *Biomass and Bioenergy, 84*, 37–48.
Sun, F., & Lu, S. (2014). Biochars improve aggregate stability, water retention, and pore-space properties of clayey soil. *Journal of Plant Nutrition and Soil Science, 177*(1), 26–33.
Sun, T., Pei, P., Sun, Y., Xu, Y., & Jia, H. (2022). Performance and mechanism of as (III/V) removal from aqueous solution by novel positively charged animal-derived biochar. *Separation and Purification Technology, 290*, 120836.
Tegeder, M., & Masclaux-Daubresse, C. (2018). Source and sink mechanisms of nitrogen transport and use. *New Phytologist, 217*(1), 35–53.
Tezer, Ö., Karabağ, N., Öngen, A., Çolpan, C. Ö., & Ayol, A. (2022). Biomass gasification for sustainable energy production: A review. *International Journal of Hydrogen Energy, 47*(34), 15419–15433.
Tiwari, T. (2023). Biochar for soil, ecology, and crop production. *Rajat Yadav, 63*, 31.

Tomczyk, A., Sokołowska, Z., & Boguta, P. (2020). Biochar physicochemical properties: Pyrolysis temperature and feedstock kind effects. *Reviews in Environmental Science and Bio/Technology, 19*(1), 191–215.

Touray, N., Tsai, W. T., Chen, H. R., & Liu, S. C. (2014). Thermochemical and pore properties of goat-manure-derived biochars prepared from different pyrolysis temperatures. *Journal of Analytical and Applied Pyrolysis, 109*, 116–122.

Tsai, W. T., Liu, S. C., Chen, H. R., Chang, Y. M., & Tsai, Y. L. (2012). Textural and chemical properties of swine-manure-derived biochar pertinent to its potential use as a soil amendment. *Chemosphere, 89*(2), 198–203.

Tsai, W. T., Huang, C. N., Chen, H. R., & Cheng, H. Y. (2015). Pyrolytic conversion of horse manure into biochar and its thermochemical and physical properties. *Waste and Biomass Valorization, 6*, 975–981.

Usman, M. (2015). Cow dung, goat and poultry manure and their effects on the average yields and growth parameters of tomato crop. *Journal of Biology, Agriculture and Healthcare, 5*(5), 7–10.

Uzoma, K. C., Inoue, M., Andry, H., Fujimaki, H., Zahoor, A., & Nishihara, E. (2011). Effect of cow manure biochar on maize productivity under sandy soil condition. *Soil Use and Management, 27*(2), 205–212.

Vatansever, R., Ozyigit, I. I., & Filiz, E. (2017). Essential and beneficial trace elements in plants, and their transport in roots: A review. *Applied Biochemistry and Biotechnology, 181*, 464–482.

Wang, Y., & Wu, W. H. (2017). Regulation of potassium transport and signaling in plants. *Current Opinion in Plant Biology, 39*, 123–128.

Wang, S., Yuan, H., Wang, Y., & Shan, R. (2017). Transesterification of vegetable oil on low cost and efficient meat and bone meal biochar catalysts. *Energy Conversion and Management, 150*, 214-221.

Wang, A., Zou, D., Zhang, L., Zeng, X., Wang, H., Li, L., Liu, F., Ren, B., & Xiao, Z. (2019). Environmental risk assessment in livestock manure derived biochars. *RSC Advances, 9*(69), 40536–40545.

Wang, A., Zou, D., Zeng, X., Chen, B., Zheng, X., Li, L., et al. (2021). Speciation and environmental risk of heavy metals in biochars produced by pyrolysis of chicken manure and water-washed swine manure. *Scientific Reports, 11*(1), 11994.

Wang, S., Shi, Y., Xiang, H., Liu, R., Su, L., Zhang, L., & Ji, R. (2022). Functional utilization of biochar derived from Tenebrio molitor feces for CO 2 capture and supercapacitor applications. *RSC Advances, 12*(35), 22760–22769.

Wu, Q., Wang, H., Zheng, X., Liu, F., Wang, A., Zou, D., Yuan, J., & Xiao, Z. (2020). Thermochemical liquefaction of pig manure: Factors influencing on oil. *Fuel, 264*, 116884.

Wu, M., Teng, X., Liang, X., Zhang, Y., Huang, Z., & Yin, Y. (2023). Supporting nanoscale zero-valent iron onto shrimp shell-derived N-doped biochar to boost its reactivity and electron utilization for selenite sequestration. *Chemosphere, 319*, 137979.

Xiang, L., Liu, S., Ye, S., Yang, H., Song, B., Qin, F., Shen, M., Tan, C., Zeng, G., & Tan, X. (2021). Potential hazards of biochar: The negative environmental impacts of biochar applications. *Journal of Hazardous Materials, 420*, 126611.

Xiang, W., Zhang, X., Cao, C., Quan, G., Wang, M., Zimmerman, A. R., & Gao, B. (2022). Microwave-assisted pyrolysis derived biochar for volatile organic compounds treatment: Characteristics and adsorption performance. *Bioresource Technology, 355*, 127274.

Xiao, X., Chen, B., Chen, Z., Zhu, L., & Schnoor, J. L. (2018). Insight into multiple and multi-level structures of biochars and their potential environmental applications: A critical review. *Environmental Science & Technology, 52*(9), 5027–5047.

Xin, Y., Cao, H., Yuan, Q., & Wang, D. (2017). Two-step gasification of cattle manure for hydrogen-rich gas production: Effect of biochar preparation temperature and gasification temperature. *Waste Management, 68*, 618–625.

Xu, G., Zhang, Y., Shao, H., & Sun, J. (2016). Pyrolysis temperature affects phosphorus transformation in biochar: Chemical fractionation and 31P NMR analysis. *Science of the Total Environment, 569*, 65–72.

Xu, J., Xie, Y., Yao, Q., Lv, L., & Chu, H. (2024). Advances in sustainable nano-biochar: Precursors, synthesis methods and applications. *Nanoscale, 16*, 15009.

Yaashikaa, P. R., Kumar, P. S., Saravanan, A., Varjani, S., & Ramamurthy, R. (2020). Bioconversion of municipal solid waste into bio-based products: A review on valorisation and sustainable approach for circular bioeconomy. *Science of the Total Environment, 748*, 141312.

Yadav, K., & Jagadevan, S. (2019). Influence of process parameters on synthesis of biochar by pyrolysis of biomass: An alternative source of energy. In *Recent advances in pyrolysis*.

Yang, Z., Yu, Y., Hu, R., Xu, X., Xian, J., Yang, Y., Liu, L., & Cheng, Z. (2020). Effect of rice straw and swine manure biochar on N2O emission from paddy soil. *Scientific Reports, 10*(1), 10843.

Yang, X., Pan, H., Shaheen, S. M., Wang, H., & Rinklebe, J. (2021). Immobilization of cadmium and lead using phosphorus-rich animal-derived and iron-modified plant-derived biochars under dynamic redox conditions in a paddy soil. *Environment International, 156*, 106628.

Yang, X., Hinzmann, M., Pan, H., Wang, J., Bolan, N., Tsang, D. C., Ok, Y. S., Wang, S. L., Shaheen, S. M., Wang, H., & Rinklebe, J. (2022). Pig carcass-derived biochar caused contradictory effects on arsenic mobilization in a contaminated paddy soil under fluctuating controlled redox conditions. *Journal of Hazardous Materials, 421*, 126647.

Yang, W., Park, Y. K., & Lee, J. (2023). Biochar catalysts from animal manure: Production and application. *Energy & Environment*. https://doi.org/10.1177/0958305X231219787

Yıldız, Z., Kaya, N., Topcu, Y., & Uzun, H. (2019). Pyrolysis and optimization of chicken manure wastes in fluidized bed reactor: CO2 capture in activated bio-chars. *Process Safety and Environmental Protection, 130*, 297–305.

Yu, Y., Lei, Z., Yang, X., Yang, X., Huang, W., Shimizu, K., & Zhang, Z. (2018). Hydrothermal carbonization of anaerobic granular sludge: Effect of process temperature on nutrients availability and energy gain from produced hydrochar. *Applied Energy, 229*, 88–95.

Yu, H., Zou, W., Chen, J., Chen, H., Yu, Z., Huang, J., Tang, H., Wei, X., & Gao, B. (2019). Biochar amendment improves crop production in problem soils: A review. *Journal of Environmental Management, 232*, 8–39.

Yun, J. J., Lee, D. W., Shim, J. H., Jeon, S. H., Lee, Y. H., Kwon, S. I., Park, J. H., Kang, S. W., Cho, J. S., & Kim, S. H. (2023). Evaluation of crop growth and soil chemical properties in cadmium contaminated soil using biochar derived from rendered animal carcass residues. *Korean Journal of Soil Science and Fertilizer, 56*(4), 533–543.

Zaker, A. (2021). *Applied pyrolysis for conversion of sewage sludge into value-added products* (Doctoral dissertation). Concordia University.

Zeng, C., Jiang, Y., Xu, R., Han, L., & Zhang, X. (2022). Phenols-enriched biofuel and H2-rich gas from catalytic fast pyrolysis/gasification of agricultural biomass over a novel heavy metals-containing livestock manure biochar catalyst. *Journal of Analytical and Applied Pyrolysis, 167*, 105680.

Zhang, J., Huang, B., Chen, L., Li, Y., Li, W., & Luo, Z. (2018). Characteristics of biochar produced from yak manure at different pyrolysis temperatures and its effects on the yield and growth of highland barley. *Chemical Speciation & Bioavailability, 30*(1), 57–67.

Zhang, Y., Fan, S., Liu, T., Fu, W., & Li, B. (2022a). A review of biochar prepared by microwave-assisted pyrolysis of organic wastes. *Sustainable Energy Technologies and Assessments, 50*, 101873.

Zhang, Y., Zhang, J., Song, K., & Lv, W. (2022b). Potential of biochar derived from three biomass wastes as an electrode catalyzing oxygen reduction reaction. *Environmental Pollutants and Bioavailability, 34*(1), 42–50.

Zhao, L., Cao, X., Wang, Q., Yang, F., & Xu, S. (2013). Mineral constituents profile of biochar derived from diversified waste biomasses: Implications for agricultural applications. *Journal of Environmental Quality, 42*(2), 545–552.

Zhu, X., Chen, B., Zhu, L., & Xing, B. (2017). Effects and mechanisms of biochar-microbe interactions in soil improvement and pollution remediation: A review. *Environmental Pollution, 227*, 98–115.

Zolfi-Bavariani, M., Ronaghi, A., Ghasemi-Fasaei, R., & Yasrebi, J. (2016). Influence of poultry manure–derived biochars on nutrients bioavailability and chemical properties of a calcareous soil. *Archives of Agronomy and Soil Science, 62*(11), 1578–1591.

Part II

Biochar Application for the Management of Soil Health and Crop Productivity

Drought Stress Mitigation with the Use of Biochar

Mohammad Faizan, Yamshi Arif, Priyadarshani Rajput, Saif Ahmad, Shamsul Hayat, Tatiana Minkina, Pravej Alam, and S. Maqbool Ahmed

Abstract

Around the world, drought stress is a major abiotic element that is drastically lowering crop yield. The intensity of drought has grown due to the rapid evolution of climate change, which poses a serious danger to the world's food security. To ensure sustained agricultural production, this issue must be resolved. In order to improve agricultural performance under stress, a variety of tactics are being used. One popular method for enhancing soil health and increasing crop yields under stress is biochar. The application of biochar improves soil organic matter, boosts nutrient and water retention, and strengthens aggregate stability and soil structure. The activity of helpful fungi and bacteria is also encouraged, which together increase the plant's resistance to abiotic stressors. Biochar greatly increases tolerance to stress by improving membrane stability, water absorption, nutritional balance, and antioxidant activity, which lowers the generation of reactive oxygen species (ROS). Furthermore, the addition of biochar to soil improves gene expression, photosynthesis, chlorophyll synthesis, and the activity of stress-

M. Faizan (✉) · S. M. Ahmed
Department of Botany, School of Sciences, Maulana Azad National Urdu University, Hyderabad, India
e-mail: mohammadfaizan@manuu.edu.in

Y. Arif · S. Ahmad · S. Hayat
Department of Botany, Plant Physiology Section, Faculty of Life Sciences, Aligarh Muslim University, Aligarh, India

P. Rajput · T. Minkina
Academy of Biology and Biotechnology, Southern Federal University, Rostov-on-Don, Russia

P. Alam
Department of Biology, College of Science and Humanities, Prince Sattam Bin Abdulaziz University, Alkharj, Saudi Arabia

S. Singh, S. Srivastava (eds.), *Biochar: A Precious Resource from Biological Waste*,
https://doi.org/10.1007/978-981-95-0425-1_8

responsive proteins. Additionally, it supports tolerance to osmotic and ionic stressors by regulating osmolytes and preserving hormonal homeostasis. To sum up, biochar exhibits a lot of promise as a soil supplement that can improve plants' resistance to drought stress. We have looked at a number of ways that biochar helps to increase drought resilience in this chapter. These include improving water retention, stabilizing cell membranes, encouraging homeostasis and nutrient uptake, increasing antioxidant activity to lessen oxidative damage, and facilitating improved photosynthetic efficiency. Furthermore, biochar controls osmolytes, affects gene expression, and preserves hormonal equilibrium—all of which are essential for enhancing a plant's resistance to drought. Biochar is a useful tool for encouraging sustainable farming practices in areas that are prone to drought because of its many advantages.

Keywords

Antioxidants · Germination · Osmolytes · Photosynthesis · Reactive oxygen species

1 Introduction

Crop production and global food security are seriously threatened by abiotic pressures (Faizan et al., 2024; Liu et al., 2023; Bhat et al., 2022). Drought, temperature, heavy metals, and salinity are among the stresses that are becoming more severe, which has a negative effect on agricultural productivity (Hassan et al., 2021; Xu et al., 2023; Faisal et al., 2024; Inam et al., 2024). One of the most severe abiotic factors that significantly lowers crop production globally is drought (Singh & Thakur, 2018). One important abiotic factor that drastically lowers crop productivity and poses a big challenge to agricultural output is drought stress (Latif et al., 2022). Plant functions are affected at the morphological, physiological, and anatomical levels by drought stress (Zhao et al., 2022). It causes a considerable reduction in photosynthesis and, eventually, a drop in agricultural output by hastening leaf senescence and lowering chlorophyll production (Ma et al., 2021). Proline, total carotenoid, ascorbic acid, polyphenols, flavonoids, and enzymatic antioxidant activity were all regulated by the increased electrolyte leakage, reactive oxygen species (ROS), and malondialdehyde (MDA) content caused by drought stress (Sarker & Oba, 2018). The formation of ROS, which harms proteins, lipids, DNA, and enzymatic processes, increased under drought stress (Cui et al., 2017). Grain productivity is decreased as a result of the negative effects of drought stress on physiological functions and agronomic characteristics (Zhou et al., 2007). However, the degree of yield loss varies according to the plant species and the length and severity of the drought (Agarwal et al., 2016). Furthermore, stomatal closure brought on by drought stress lowers stomatal conductance, reducing water loss while adversely affecting transpiration and photosynthetic rates (Mahmoud & Swaefy, 2020; Faizan et al., 2023a).

Biochar is a carbon-rich material that improves soil carbon sequestration and overall soil quality by acting as an efficient soil conditioner (Rajput et al., 2023, 2024). Biochar is made from organic materials by pyrolyzing them at temperatures between 250 and 700 °C. It has unique qualities that make it an economical, environmentally responsible, and sustainable way to enhance soil health (Kravchenko et al., 2024). It is now known that biochar is a useful technique for increasing agricultural productivity and abiotic stress resilience (Haiying et al., 2022; Sultan et al., 2024). It is a useful tool for reducing the detrimental effects of difficult environmental circumstances on crops because its use in agriculture not only increases plant growth but also fortifies stress tolerance systems (Moragues-Saitua et al., 2023). Biochar is a useful amendment for recovering salt-affected soils and enhancing soil health and production due to its high cation exchange capacity and alkaline qualities (Lashari et al., 2013). According to Singh et al. (2019), biochar is a useful strategy for increasing crop productivity and water use efficiency (WUE). Its use has been demonstrated to increase antioxidant activity, carbon assimilation, and nutrient uptake, all of which considerably accelerate plant development in drought-stressed environments (Wang et al., 2018). Furthermore, biochar supports plant growth and resistance in drought situations by improving stomatal conductance, WUE, and chlorophyll production (Haider et al., 2020). Biochar is essential in reducing the negative impacts of drought stress on crops by enhancing these physiological functions.

Applying biochar has been shown to dramatically increase soil structure, water-holding capacity, and nutrient availability, fostering an environment that is favorable for plant growth, particularly in areas that are prone to drought (Jatav et al., 2021). These improvements in the physicochemical and biological characteristics of the soil aid plants in maintaining improved root development, nutrient uptake, and water usage efficiency during times of water scarcity. Further supporting plant resilience by encouraging symbiotic connections that support nutrient cycling and stress tolerance, biochar also increases microbial activity in the soil Gorovtsov et al., 2020).

This chapter examined the various ways that biochar can help plants that are suffering from drought stress by examining a number of mechanisms, including as improved osmotic regulation, increased antioxidant activities, and the modification of important growth-promoting pathways. We emphasized how biochar improves drought resistance in plants. Furthermore, this chapter provides a thorough explanation of how biochar functions as a potent weapon in sustainable agriculture to battle drought stress by highlighting its ability to lower oxidative damage, control hormonal balance, and boost efficient photosynthesis under stress.

2 Effect of Drought Stress on Soil Particles

Plant growth, soil properties, and overall productivity are all adversely affected by drought and salt stress (Ohashi et al., 2015; Rath & Rousk, 2015; Faizan et al., 2023b; Alam et al., 2022). The physical characteristics of the topsoil layer are significantly impacted by drought. According to Zhang et al. (2019), it causes a significant decrease in soil aggregate stability, which may erode the soil's structural integrity. Repeated cycles of wetting and drying brought on by drought have a major impact on soil respiration, fungal dynamics, microbial activity, and organic matter degradation (Geng et al., 2015; Mariotte et al., 2015). Furthermore, organic carbon (OC) and nitrogen concentrations in soil aggregates are considerably reduced under drought circumstances (Zhang et al., 2019). According to research by Ohashi et al. (2015), CO_2 emissions from tropical seasonal rainforest soil were not significantly impacted by brief drought stress. Total nitrogen, mineral nitrogen, bray phosphorus, and exchangeable potassium were among the nutrients that were significantly lower in soils exposed to drought than in soils that were regularly irrigated (Zaheer et al., 2021). The decomposition of organic matter is slowed down by the decreased microbial activity caused by the lower soil moisture during drought circumstances (Zaheer et al., 2021). When organic matter accumulates during a drought, hydrogen ions are released, which can lower the pH of the soil. Moreover, soil drying may result in a higher concentration of cations, which could further lower pH (Zaheer et al., 2021). According to Song et al. (2012), a decrease in soil pH under water stress may indicate that moisture is crucial for strengthening the connection between soil phosphatase activity and nutrient availability. Drought has a negative overall impact and can eventually worsen the health and characteristics of the soil (Zhang et al., 2019).

3 Production and Physicochemical Properties of Biochar Amendments

Low-oxygen thermochemical conversion of organic molecules produces biochar. Charcoal has historically been produced by burning large quantities of wood in earthen mounds that were built aboveground (Mohan et al., 2006). The kind of feedstock used in the manufacturing of biochar has a significant impact on its characteristics. Terre Preta do Indio soils are rich in nutrients because pre-Columbian Indigenous people frequently added charcoal to the soil to increase fertility (Glaser et al., 2002). These days, using biochar is thought to be a promising technique with many benefits for sustainable farming. An outline of the sustainable biochar concept was given by Woolf et al. (2010). Hydrothermal carbonization, gasification, vertical silo reactors, rotating kilns, and both fast and slow pyrolysis are common processes used to produce biochar. According to Jindo et al. (2014) and Qayyum et al. (2015), biochar is typically defined by large concentrations of minerals (such as sodium, potassium, magnesium, and iron), high pH and electrical conductivity (EC), low levels of ash and volatile matter, and a high total organic carbon content (between 30 and 70%, depending on pyrolysis conditions like temperature, oxygen levels, and duration). However, the kind of feedstock that is utilized is a major factor in defining the particular characteristics of the biochar (Ronsse et al., 2013). In addition to using scanning electron microscopy, advanced methods for characterizing biochar include measuring and identifying surface functional groups, aromatic chemicals, polycyclic aromatic hydrocarbons, and active surface area (Harris et al., 2013). Thus, a detailed analysis of biochar is necessary prior to its use (Kuppusamy et al., 2016).

4 Functional Characteristics of Biochar

One important characteristic of biochar amendments is its adsorption capability, which is mostly caused by the surface functional groups that are created or changed during pyrolysis based on the feedstock type and temperature. Organic feedstocks lose readily degradable components like cellulose and lignin during pyrolysis, which occurs at about 300 °C. Aromatic chemicals condense and the surface area rises with warmth (Ahmad et al., 2012; Inyang et al., 2016). Zhang and Luo (2014) used FTIR, Boehm titrations, and scanning electron microscopy (SEM) to investigate the surface functional characteristics of biochar made from eucalyptus leaves (EBC) and anaerobically digested garden waste (ADB). According to Ali et al. (2017), they discovered that although the EBC had more functional groups overall, the ADB biochar had more carboxyl, phenolic, and basic functional groups but fewer lactonic groups. Four distinct biochars made from diverse feedstocks—wood, bamboo, rice husk, and rice husk ash—were reported, and their functional characteristics were compared. In comparison to the other biochars, it was shown that wood-based biochar had greater concentrations of phenolic, lactone, and carboxyl groups (Zhang et al., 2014). Wood biochar (charcoal) has a higher aromatic content than biochar made from hydrothermal carbonization (HTC) and low-temperature conversion (LTC) of sewage sludge, according to Qayyum et al. (2012). In comparison to charcoal and LTC biochars, HTC biochar has more decomposable organic groups, according to FTIR examination of the three biochars. Boehm titrations showed that basic functional groups like ketones, pyrones, and chromenes increased whereas acidic functional groups like carboxyl, phenolic, and lactone decreased in pine needle-derived biochar (Ahmad et al., 2013). Biochar is a cost-effective and efficient solution for the sorption and removal of pollutants and toxins from soil and water because of these and other functional groups (Ahmad et al., 2014; Rajapaksha et al., 2016).

5 Effect of Biochar on Plant Physiology and Biochemistry Under Drought Stress

5.1 Effect of Biochar on Seed Germination, Plant Growth, Development, and Yield Traits Under Drought Stress

One of the most detrimental stressors is drought, which severely impairs plant growth, development, seed germination, and total crop yield. When Glycine max was subjected to drought stress, the application of biochar enhanced seed germination, seedling length, sugar, and proline content (Hafeez et al., 2017). By boosting germination, seedling length, biomass, total chlorophyll content, and antioxidant content like CAT, POX, and SOD, biochar derived from *Parthenium hysterophorus* lessened the harmful effects of drought stress in barley (Gul et al., 2023). By enhancing seed germination, seedling growth, dry biomass, leaf area, chlorophyll contents, CO_2 assimilation, water use efficiency, soluble sugar contents, osmolyte accumulation, and leaf K+ concentration, as well as by preventing oxidative damage, biochar treatment of cowpeas reduced the harmful effects of drought stress (Farooq et al., 2021).

All of the soil's biochemical characteristics were adversely impacted by drought stress, while biochar amendments enhanced these characteristics, increasing wheat productivity and production (Zaheer et al., 2021). Following biochar addition, the negative effects of drought were lessened in Vicia faba because it enhanced germination characteristics such as germination energy, germination index, and water usage efficiency. Additionally, biochar reduced the harmful effects of drought and enhanced plant biomass and growth (Nafees et al., 2021). An organic soil additive called biochar is frequently used to improve soil fertility overall during drought stress, raise soil carbon levels, increase organic content, and improve water retention. However, under drought conditions, the application of biochar to wheat increased the number of grains per spike, spike length, thousand grain weight, and biological and economic output. Furthermore, by improving the physiological and biochemical characteristics of wheat plants, biochar lessened the harmful effects of drought (Haider et al., 2020). Due to increased electrolyte leakage, excessive hydrogen peroxide and malondialdehyde generation, and other factors, drought stress had a negative effect on rapeseed growth. Nonetheless, photosynthesis, antioxidant activity, osmolyte levels, growth, and yield were all enhanced by the use of biochar (Khan et al., 2021).

Under drought stress, the application of biochar to tomatoes increased the fruit's nutritional value and lycopene content, as well as the plant's height, root length, fresh and dry weight, and number of fruits per plant (Usman et al., 2023). The application of biochar considerably improved the growth of spinach under drought stress, resulting in increases in the fresh and dry weight of the shoots and roots as well as their length. The effectiveness of the biochar treatment in reducing drought stress was further supported by the higher concentrations of nutrients (nitrogen, phosphorous, and potassium) in the roots and shoots (Alotaibi et al., 2023). By strengthening the soil's structure and boosting its capacity to hold water, biochar reduced the negative effects of drought and improved plant water availability, hence reducing drought stress. In the end, this enhanced the features of plant growth, development, and yield (Zhang et al., 2023). Applying biochar to rapeseed under drought stress increased biomass and yield. In dry conditions, it also increased biomass, the number of pods per plant, and the 1000 seed weight (Khan et al., 2021). Plants under drought stress had lower levels of both protein and oil, but biochar treatment considerably raised both levels in spite of the dryness (Khan et al., 2021). Using biochar considerably decreased the amount of erucic acid, while drought stress caused it to rise. Oleic acid levels also significantly decreased during drought, whereas biochar treatment significantly increased oleic acid concentrations during drought (Khan et al., 2021). Comparing biochar application to control treatments, linoleic acid content increased significantly under both normal and drought stress conditions (Khan et al., 2021).

5.2 Effect of Biochar on Photosynthesis and Gas Exchange Traits Under Drought Stress

In a number of plants, biochar enhanced growth, photosynthetic efficiency, and drought resistance (Fig. 1). Phragmites karka's chlorophyll content, net photosynthetic rate, and water use efficiency were all improved by biochar under drought stress (Abideen et al., 2020). Biochar improved chlorophyll fluorescence characteristics and considerably slowed soil moisture loss in *Pyrus ussuriensis*, hence mitigating drought stress. While increasing the maximal photochemical efficiency of photosystem II (F_v/F_m), electron transport efficiency (ψ_o), and photosynthetic performance index (PI_{a6s}), it decreased initial fluorescence (F_o) and J points in relative variable fluorescence (V_a). These modifications assisted in preventing damage to photosystem II. Furthermore, applying biochar enhanced Rubisco enzyme activity and net photosynthesis (P_n), bolstering plant resistance to drought (Lyu et al., 2016). According to Abideen et al. (2020), the application of biochar optimized carboxylation efficiency and helped to increase biomass production under conditions of water deficit by highlighting the interconnected changes in photosynthetic rate, stomatal conductance (g_s), and sub-stomatal conductance (C_i). Because water shortage conditions decrease photosynthetic attributes and other physiological processes like g_s, C_i, transpiration rate, and chlorophyll content, they have a negative impact on soybean growth and productivity. However, by maintaining cellular activity, the use of biochar helps offset these negative consequences. By significantly increasing photosynthesis, transpiration, Cs, and g_s, the use of biochar helped to lessen the effects of water stress (Nawaz et al., 2023). Using biochar to coat rice seeds has been known as a sustainable way to boost agricultural yields and increase resistance to water scarcity.

Furthermore, through enhancing photosynthetic pigment levels, improving stomatal aperture, and increasing stomatal density, biochar improved photosynthetic performance during drought. According to ultrastructural research, biochar protected cell walls and stabilized chloroplast structure under drought (Zhang et al.,

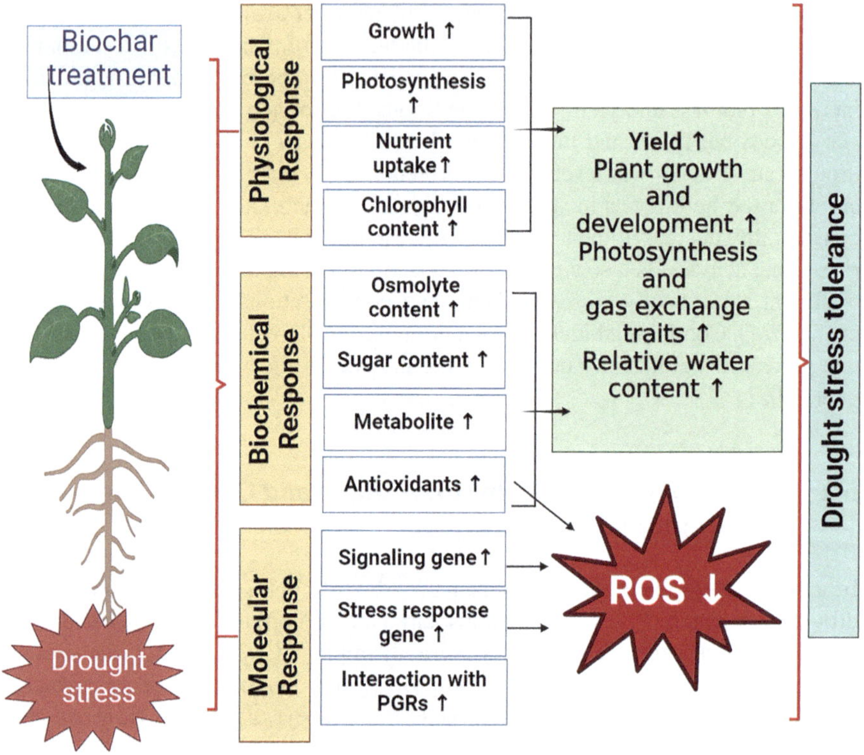

Fig. 1 Schematic representation of biochar mediated stress tolerance in crop plants

2024). According to Alotaibi et al. (2023), the application of biochar to spinach under drought stress led to increased levels of chlorophyll a, chlorophyll b, and total chlorophyll as well as improvements in photosynthetic rate, transpiration rate, and gas exchange characteristics. By enhancing growth, chlorophyll a and b content, photosynthetic rate, transpiration rate, stomatal conductance, and CO_2 concentration, the application of biochar significantly reduced the detrimental effects of drought stress in wheat (Zulfiqar et al., 2022).

Biochar treatment dramatically increased tobacco's growth, biomass, total chlorophyll content, and carotenoids under drought stress (Zaman et al., 2024). In drought-stressed tomato leaves, biochar significantly increased stomatal density and improved chlorophyll content, stomatal conductance, photosynthetic rate, water usage efficiency (WUE), and relative water content, according to Akhtar et al. (2014). In a similar vein, Haider et al. (2015) discovered that applying biochar to drought-stressed maize enhanced leaf RWC, transpiration rate, and osmotic potential in comparison to untreated controls. According to the findings, applying biochar to crops under drought stress increased their leaf area, shoot and root biomass, root volume, and photosynthetic rate (P_n) (Haider et al., 2015).

5.3 Effect of Biochar on Nutrient Uptake and Nutrient Homeostasis Under Drought Stress

Drought stress disturbs nutritional balance in plants, resulting to severe losses in growth and output. But as a useful soil addition, biochar encourages substantial plant growth and aids in reestablishing nutrient equilibrium (Fig. 1). Applications of biochar, for instance, have been demonstrated to improve nitrogen intake. This is probably because it can increase the soil's retention of nitrogen, which improves nutrient availability during drought circumstances (Ibrahim et al., 2020). An improved cation exchange capacity (CEC) is linked to the increase in nitrogen uptake caused by biochar because soils with a greater CEC are better able to absorb NH_4^+ and promote nitrogen utilization. The application of biochar also increases the concentrations of calcium (Ca), magnesium (Mg), and potassium (K) in the soil solution, boosting nutrient availability during drought circumstances, according to Glaser et al. (2002). By altering the pH of the soil, biochar improves the availability of nutrients in the soil. Applying biochar improved nitrogen uptake, which lessened the detrimental effects of drought stress on peanuts (Zhang et al., 2021). Similarly, under drought conditions, Zoghi et al. (2019) discovered that biochar enhances nutrient absorption, most likely by increasing soil water uptake capacity (WUC) and water use efficiency (WUE). According to, the application of biochar improved the leaf area, N and P absorption, and WUE.

Higher rates of biochar application were found to enhance magnesium concentration, with mild drought stress producing the greatest Mg levels (Langeroodi et al., 2019). Similar findings were made by Poormansour et al. (2019), who discovered that increasing the amount of biochar applied to faba bean plants at varying irrigation levels increased the uptake and concentrations of calcium and magnesium in the soil. In conclusion, biochar improved nutrient uptake and CEC, which in turn improved drought tolerance by boosting antioxidant activity and promoting improved plant physiological performance.

5.4 Effect of Biochar on Osmolytes Under Drought Stress

Osmolytes are crucial in helping plants cope with drought stress, as they aid in maintaining cellular function and stability under adverse conditions. However, drought has been shown to disrupt the balance of plant hormones and interfere with the accumulation of osmolytes, thereby impairing the plant's ability to manage stress effectively. This hormonal imbalance, coupled with reduced osmolyte synthesis, can weaken the plant's drought tolerance, affecting growth and overall productivity. Under stress conditions, proline plays a dual role by functioning as both an osmotic regulator and a scavenger of reactive oxygen species (ROS), helping to mitigate damage and maintain cellular homeostasis. In barley plant, biochar application regulated levels of osmolytes such as proline, soluble sugars, sucrose, and starch under drought stress (Hafez et al., 2020). The regulation of these traits, which serve as vital osmoprotectants, may represent a key strategy for barley plants to adapt to and withstand drought conditions effectively (Hafez et al., 2020). Biochar amendment improved the cabbage plant photosynthetic activity, nutrient uptake, and osmotic content such as proline and sucrose, under drought stress (Yildirim et al., 2021). Under drought stress, tobacco plant regulated protein, proline, and soluble sugar content after biochar amendment (Zaman et al., 2024). Biochar supplementation regulated proline content in rice plant after drought stress (Hafez et al. 2021b). Drought stress in *Brassica napus* was mitigated after biochar supplementation as it regulated proline content and reduced production of ROS and malondialdehyde (MDA) (Lalay et al., 2022). The application of exogenous salicylic acid combined with biochar in *Borago officinalis* L. enhances the plant's stress tolerance mechanisms under water deficit conditions by increasing proline and soluble sugar content, which are key factors in improving drought resilience (Taghizadeh Tabari et al., 2022). Application of biochar on *Vicia faba* improved flavonoids, phenols, proline, and glycine betaine content and reduced toxic effect of drought (Nafees et al., 2022). Similarly, in *Nigella sativa*, biochar treatment regulated compatible osmolyte, proline, and soluble carbohydrates content, under drought stress. Therefore, biochar plays crucial role in improving the physiological traits and protective osmolytes of *Nigella sativa* under drought stress (Kiani et al., 2023).

5.5 Effect of Biochar on Antioxidants and ROS Under Drought Stress

Drought stress causes plants to produce more ROS, which can seriously harm important cellular molecules and structures and induce oxidative stress. In water-stressed plants, for instance, higher superoxide dismutase (SOD) activity has been linked to improved photo protection and greater membrane integrity, which aids the plant in overcoming stress. SOD and ascorbate peroxidase (APX) activity were elevated in both control and water-stressed plants treated with biochar, indicating that the application of biochar increases antioxidant activity in plants (Gharred et al., 2022). Applying biochar stabilizes electron transport and increases antioxidant activity, which lessens the negative impacts of drought stress on the plant's photosynthetic system (Chaves et al., 2009). Applying biochar under drought stress increased the AsA/DHAsA ratio and increased the activities of SOD, APX, glutathione reductase (GR), and glutathione peroxidase (GPX), thereby scavenging ROS and preventing the negative effects of drought (Foyer et al., 2009). By encouraging osmotic regulation, preserving hormonal balance, and increasing antioxidant activity, biochar has been shown to improve drought resistance (Mickan et al., 2016). By increasing the activity of catalase (CAT), peroxidase (POD), and SOD in the soil, biochar improves the metabolic performance of wheat plants, supports cell growth, and lowers the formation of reactive oxygen species (ROS). Plant performance under drought stress is greatly enhanced as a result (Zulfiqar et al., 2022). The application of chitosan and biochar to barley plants under drought stress led to a significant increase in the activity of important antioxidant enzymes, such as CAT, POD, and GR. By neutralizing the ROS produced by dryness, this increase in enzyme activity greatly decreased oxidative damage. Thus, the combination treatment improved the plants' physiological stability and shielded them from drought-induced stress, which led to improved growth and drought tolerance (Hafez et al., 2020). Biochar treatment dramatically decreased ROS production, preserved cell membrane integrity, preserved redox equilibrium, and enhanced osmoprotectant accumulation in rice plants under drought stress (Zhang et al., 2024).

Under both well-irrigated and drought stress conditions, the application of biochar to wheat plants increased relative water content and antioxidant levels while decreasing hydrogen peroxide (H_2O_2) levels, electrolyte leakage, proline buildup, and malondialdehyde (MDA) generation (Chowdhury et al., 2024). Likewise, the use of biochar to drought-stressed cabbage increased antioxidant machinery, lowering the levels of MDA and H_2O_2 (Yildirim et al., 2021). In order to mitigate the adverse effects of drought on spinach, biochar treatment may be able to control antioxidant activity during drought stress (Alotaibi et al., 2023). To sum up, biochar increases antioxidant activity, which is essential for lowering ROS and protecting plants from harsh drought stressors. This protective mechanism not only mitigates oxidative damage but also supports overall plant health and growth, resulting in improved resilience and performance under drought conditions (Fig. 1).

6 Effect of Biochar on Gene Expression and Stress-Responsive Genes Under Drought Stress

Applying biochar improves gene expression to help plants cope with drought stress. For instance, when subjected to drought, plants treated with biochar showed enhanced expression of the genes for CAT, APX, and Mn-SOD (Hafez et al., 2021b). These genes were upregulated when biochar was applied during drought stress (Racioppi et al., 2019). Applying biochar to wheat plants promotes both germination and growth by activating the auxin-responsive growth pathway. Gibberellins (GAs) and indole-3-acetic acid (IAA) are hormones that stimulate plant growth, and xyloglucan endotransglucosylase/hydrolase (XTH) genes are essential for controlling cell wall flexibility (Sánchez-Rodríguez et al., 2010). After applying biochar, the Saragolla cultivar's XTH gene expression is increased, which encourages the synthesis of GAs (Racioppi et al., 2019). The growth-promoting pathways of IAA and brassinosteroids (BRs) are triggered by calcium (Ca^{2+}) and ROS-mediated cell signaling, which is activated by changes in soil pH and increased potassium (K^+) availability in biochar-treated soil, according to Viger et al. (2015).

7 Conclusions

A reduction in crop yield results from the substantial disruption of plant growth caused by drought stress, which impacts a number of biochemical, physiological, and molecular pathways. By increasing plant resilience, biochar is a useful soil supplement that lessens these negative impacts. Applying biochar helps plants better withstand drought conditions by enhancing membrane integrity, facilitating improved nutrient absorption, and preserving nutritional balance. Biochar also promotes antioxidant activity and increases photosynthetic efficiency, both of which are critical for preserving hormonal balance and protecting plants from oxidative and osmotic stress during drought. By neutralizing damaging ROS and MDA, this enhanced antioxidant activity helps shield plant cells from injury (Fig. 1). Additionally, biochar promotes increased water use efficiency and stomatal conductance, which enhances plant performance during drought stress. In addition to providing immediate drought relief, biochar helps plants develop long-term drought tolerance by enhancing nutrient availability, maintaining important physiological functions, and strengthening stress-response pathways. Biochar is an important resource for sustainable agricultural operations because of its capacity to boost growth and stress management. This is especially true in areas that are prone to drought, where it may assist guarantee better, more resilient crops and higher yields.

Acknowledgments Authors acknowledge the support by the Strategic Academic Leadership Program of the Southern Federal University ("Priority 2030").

References

Abideen, Z., Koyro, H. W., Huchzermeyer, B., Ansari, R., Zulfiqar, F., & Gul, B. J. P. B. (2020). Ameliorating effects of biochar on photosynthetic efficiency and antioxidant defence of *Phragmites karka* under drought stress. *Plant Biology, 22*(2), 259–266.

Agarwal, P., Parida, S. K., Raghuvanshi, S., Kapoor, S., Khurana, P., Khurana, J. P., & Tyagi, A. K. (2016). Rice improvement through genome-based functional analysis and molecular breeding in India. *Rice, 9*, 1–17.

Ahmad, M., Lee, S. S., Dou, X., Mohan, D., Sung, J. K., Yang, J. E., & Ok, Y. S. (2012). Effects of pyrolysis temperature on soybean stover-and peanut shell-derived biochar properties and TCE adsorption in water. *Bioresource Technology, 118*, 536–544.

Ahmad, M., Lee, S. S., Rajapaksha, A. U., Vithanage, M., Zhang, M., Cho, J. S., et al. (2013). Trichloroethylene adsorption by pine needle biochars produced at various pyrolysis temperatures. *Bioresource Technology, 143*, 615–622.

Ahmad, M., Rajapaksha, A. U., Lim, J. E., Zhang, M., Bolan, N., Mohan, D., et al. (2014). Biochar as a sorbent for contaminant management in soil and water: A review. *Chemosphere, 99*, 19–33.

Akhtar, S. S., Li, G., Andersen, M. N., & Liu, F. (2014). Biochar enhances yield and quality of tomato under reduced irrigation. *Agricultural Water Management, 138*, 37–44.

Alam, P., Balawi, T. A., & Faizan, M. (2022). Salicylic acid's impact on growth, photosynthesis, and antioxidant enzyme activity of Triticum aestivum when exposed to salt. *Molecules, 28*, 100.

Ali, S., Rizwan, M., Qayyum, M. F., Ok, Y. S., Ibrahim, M., Riaz, M., et al. (2017). Biochar soil amendment on alleviation of drought and salt stress in plants: A critical review. *Environmental Science and Pollution Research, 24*, 12700–12712.

Alotaibi, M. O., Ikram, M., Alotaibi, N. M., Hussain, G. S., Ghoneim, A. M., Younis, U., & Danish, S. (2023). Examining the role of AMF-Biochar in the regulation of spinach growth attributes, nutrients concentrations, and antioxidant enzymes in mitigating drought stress. *Plant Stress, 10*, 100205.

Bhat, J. A., Faizan, M., Bhat, M. A., Huang, F., Yu, D., Ahmad, A., Bajguz, A., & Ahmad, P. (2022). Defense interplay of the zinc-oxide nanoparticles and melatonin in alleviating the arsenic stress in soybean (Glycine max L.). *Chemosphere, 288*, 132471.

Chaves, M. M., Flexas, J., & Pinheiro, C. (2009). Photosynthesis under drought and salt stress: Regulation mechanisms from whole plant to cell. *Annals of Botany, 103*(4), 551–560.

Chowdhury, M. S. N., Sani, M. N. H., Siddique, A. B., Hossain, M. S., & Yong, J. W. H. (2024). Synergistic effects of biochar and potassium co-application on growth, physiological attributes, and antioxidant defense mechanisms of wheat under water deficit conditions. *Plant Stress, 12*, 100452.

Cui, G., Zhao, X., Liu, S., Sun, F., Zhang, C., & Xi, Y. (2017). Beneficial effects of melatonin in overcoming drought stress in wheat seedlings. *Plant Physiology and Biochemistry, 118*, 138–149.

Faisal, M., Faizan, M., & Alatar, A. A. (2024). Metallic allies in drought resilience: Unveiling the influence of silver and zinc oxide nanoparticles on enhancing tomato (*Solanum lycopersicum*) resistance through oxidative stress regulation. *Plant Physiology and Biochemistry, 212*, 108722.

Faizan, M., Karabulut, F., Alam, P., Yusuf, M., Tonny, S. H., Adil, M. F., Sehar, S., Ahmed, S. M., & Hayat, S. (2023a). Nanobionics: A sustainable agricultural approach towards understanding plant response to heavy metals, drought, and salt stress. *Nanomaterials, 13*, 974.

Faizan, M., Alam, P., Rajput, V. D., Tonny, S. H., Yusuf, M., Sehar, S., Adil, M. F., & Hayat, S. (2023b). Nanoparticles: An emerging soil crops saviour under drought and heavy metal stresses. *Egyptian Journal of Soil Science, 63*(3), 355–366.

Faizan, M., Ashoka, H., Karabulut, F., Alam, P., Ahmed, S. M., Khan, I., Soysal, S., Hayat, S., & Ahmed, N. (2024). Annexins: A family of calcium binding proteins with variety of roles in plant development and abiotic stress tolerance. *Plant Stress, 14*, 100573.

Farooq, M., Romdhane, L., Rehman, A., Al-Alawi, A. K., Al-Busaidi, W. M., Asad, S. A., & Lee, D. J. (2021). Integration of seed priming and biochar application improves drought tolerance in cowpea. *Journal of Plant Growth Regulation, 40*, 1972–1980.

Foyer, C. H., Bloom, A. J., Queval, G., & Noctor, G. (2009). Photorespiratory metabolism: Genes, mutants, energetics, and redox signaling. *Annual Review of Plant Biology, 60*(1), 455–484.

Geng, S. M., Yan, D. H., Zhang, T. X., Weng, B. S., Zhang, Z. B., & Qin, T. L. (2015). Effects of drought stress on agriculture soil. *Natural Hazards, 75*, 1997–2011.

Gharred, J., Derbali, W., Derbali, I., Badri, M., Abdelly, C., Slama, I., & Koyro, H. W. (2022). Impact of biochar application at water shortage on biochemical and physiological processes in medicago ciliaris. *Plants, 11*(18), 2411.

Glaser, B., Lehmann, J., & Zech, W. (2002). Ameliorating physical and chemical properties of highly weathered soils in the tropics with charcoal – A review. *Biology and Fertility of Soils, 35*, 219–230.

Gorovtsov, A. V., Minkina, T. M., Mandzhieva, S. S., Perelomov, L. V., Soja, G., Zamulina, I. V., Rajput, V. D., Sushkova, S. N., Mohan, D., & Yao, J. (2020). The mechanisms of biochar interactions with microorganisms in soil. *Environmental Geochemistry and Health, 42*(8), 2495–2518. https://doi.org/10.1007/s10653-019-00412-5

Gul, F., Khan, I. U., Rutherford, S., Dai, Z. C., Li, G., & Du, D. L. (2023). Plant growth promoting rhizobacteria and biochar production from Parthenium hysterophorus enhance seed germination and productivity in barley under drought stress. *Frontiers in Plant Science, 14*, 1175097.

Hafeez, Y., Iqbal, S., Jabeen, K., Shahzad, S., Jahan, S., & Rasul, F. (2017). Effect of biochar application on seed germination and seedling growth of *Glycine max* (l.) Merr. Under drought stress. *Pakistan Journal of Botany, 49*(51), 7–13.

Hafez, Y., Attia, K., Alamery, S., Ghazy, A., Al-Doss, A., Ibrahim, E., et al. (2020). Beneficial effects of biochar and chitosan on antioxidative capacity, osmolytes accumulation, and anatomical characters of water-stressed barley plants. *Agronomy, 10*(5), 630.

Hafez, E. M., Omara, A. E. D., Alhumaydhi, F. A., & El-Esawi, M. A. (2021b). Minimizing hazard impacts of soil salinity and water stress on wheat plants by soil application of vermicompost and biochar. *Physiologia Plantarum, 172*(2), 587–602.

Haider, G., Koyro, H. W., Azam, F., Steffens, D., Müller, C., & Kammann, C. (2015). Biochar but not humic acid product amendment affected maize yields via improving plant-soil moisture relations. *Plant and Soil, 395*, 141–157.

Haider, I., Raza, M. A. S., Iqbal, R., Aslam, M. U., Habib-ur-Rahman, M., Raja, S., et al. (2020). Potential effects of biochar application on mitigating the drought stress implications on wheat (*Triticum aestivum* L.) under various growth stages. *Journal of Saudi Chemical Society, 24*(12), 974–981.

Haiying, T. A. N. G., Shubin, W. A. N. G., Ying, L. I. U., Hassan, M. U., Ying, S. O. N. G., Huang, G., et al. (2022). Biochar: A promising soil amendment to mitigate heavy metals toxicity in plants. *Notulae Botanicae Horti Agrobotanici Cluj-Napoca, 50*(3), 12778–12778.

Harris, K., Gaskin, J., Cabrera, M., Miller, W., & Das, K. C. (2013). Characterization and mineralization rates of low temperature peanut hull and pine chip biochars. *Agronomy, 3*(2), 294–312.

Hassan, M. U., Chattha, M. U., Khan, I., Chattha, M. B., Barbanti, L., Aamer, M., et al. (2021). Heat stress in cultivated plants: Nature, impact, mechanisms, and mitigation strategies – A review. *Plant Biosystems – An International Journal Dealing with all Aspects of Plant Biology, 155*(2), 211–234.

Ibrahim, M. M., Tong, C., Hu, K., Zhou, B., Xing, S., & Mao, Y. (2020). Biochar-fertilizer interaction modifies N-sorption, enzyme activities and microbial functional abundance regulating nitrogen retention in rhizosphere soil. *Science of the Total Environment, 739*, 140065.

Inam, A., Javad, S., Naseer, I., Alam, P., Almutairi, Z. M., Faizan, M., Zauq, S., & Shah, A. A. (2024). Efficacy of chitosan loaded zinc oxide nanoparticles in alleviating the drastic effects of drought from corn crop. *Plant Stress, 14*, 100617.

Inyang, M. I., Gao, B., Yao, Y., Xue, Y., Zimmerman, A., Mosa, A., et al. (2016). A review of biochar as a low-cost adsorbent for aqueous heavy metal removal. *Critical Reviews in Environmental Science and Technology, 46*(4), 406–433.

Jatav, H. S., Rajput, V. D., Minkina, T., Singh, S. K., Chejara, S., Gorovtsov, A., Barakhov, A., Bauer, T., Sushkova, S., Mandzhieva, S., Burachevskaya, M., & Kalinitchenko, V. P. (2021). Sustainable approach and safe use of biochar and its possible consequences. *Sustainability, 13*(18).

Jindo, K., Mizumoto, H., Sawada, Y., Sanchez-Monedero, M. A., & Sonoki, T. (2014). Physical and chemical characterization of biochars derived from different agricultural residues. *Biogeosciences, 11*(23), 6613–6621.

Khan, Z., Khan, M. N., Zhang, K., Luo, T., Zhu, K., & Hu, L. (2021). The application of biochar alleviated the adverse effects of drought on the growth, physiology, yield and quality of rapeseed through regulation of soil status and nutrients availability. *Industrial Crops and Products, 171*, 113878.

Kiani, H., Khalesro, S., Mokhatssi-Bidgoli, A., & Sharifi, Z. (2023). Effects of biochar and drought stress on some protective mechanisms of Nigella sativa L. *Iranian Journal of Medicinal and Aromatic Plants Research, 39*(2), 285–302.

Kravchenko, E., Dela Cruz, T. L., Sushkova, S., & Rajput, V. D. (2024). Effect of wood and peanut shell hydrochars on the desiccation cracking characteristics of clayey soils. *Chemosphere, 358*, 142134. https://doi.org/10.1016/j.chemosphere.2024.142134

Kuppusamy, S., Thavamani, P., Megharaj, M., Venkateswarlu, K., & Naidu, R. (2016). Agronomic and remedial benefits and risks of applying biochar to soil: Current knowledge and future research directions. *Environment International, 87*, 1–12.

Lalay, G., Ullah, S., & Ahmed, I. (2022). Physiological and biochemical responses of *Brassica napus* L. to drought-induced stress by the application of biochar and plant growth promoting Rhizobacteria. *Microscopy Research and Technique, 85*(4), 1267–1281.

Langeroodi, A. R. S., Campiglia, E., Mancinelli, R., & Radicetti, E. (2019). Can biochar improve pumpkin productivity and its physiological characteristics under reduced irrigation regimes? *Scientia Horticulturae, 247*, 195–204.

Lashari, M. S., Liu, Y., Li, L., Pan, W., Fu, J., Pan, G., et al. (2013). Effects of amendment of biochar-manure compost in conjunction with pyroligneous solution on soil quality and wheat yield of a salt-stressed cropland from Central China Great Plain. *Field Crops Research, 144*, 113–118.

Latif, M., Bukhari, S. A. H., Alrajhi, A. A., Alotaibi, F. S., Ahmad, M., Shahzad, A. N., et al. (2022). Inducing drought tolerance in wheat through exopolysaccharide-producing rhizobacteria. *Agronomy, 12*(5), 1140.

Liu, Z., Faizan, M., Zheng, L., Cui, L., Han, C., Chen, H., & Yu, F. (2023). Nanoparticles enhance plant resistance to abiotic stresses: A bibliometric statistic. *Agronomy, 13*, 729. (IF = 3.3).

Lyu, S., Du, G., Liu, Z., Zhao, L., & Lyu, D. (2016). Effects of biochar on photosystem function and activities of protective enzymes in *Pyrus ussuriensis* Maxim. under drought stress. *Acta Physiologiae Plantarum, 38*, 1–10.

Ma, S., Wang, Z., Guo, X., Wang, F., Huang, J., Sun, B., & Wang, X. (2021). Sourdough improves the quality of whole-wheat flour products: Mechanisms and challenges – A review. *Food Chemistry, 360*, 130038.

Mahmoud, A. W. M., & Swaefy, H. M. (2020). Comparison between commercial and nano NPK in presence of nano zeolite on sage plant yield and its components under water stress. *Agriculture (Pol'nohospodarstvo), 66*(1), 24–39.

Mariotte, P., Robroek, B. J., Jassey, V. E., & Buttler, A. (2015). Subordinate plants mitigate drought effects on soil ecosystem processes by stimulating fungi. *Functional Ecology, 29*(12), 1578–1586.

Mickan, B. S., Abbott, L. K., Stefanova, K., & Solaiman, Z. M. (2016). Interactions between biochar and mycorrhizal fungi in a water-stressed agricultural soil. *Mycorrhiza, 26*(6), 565–574.

Mohan, D., Pittman, C. U., Jr., & Steele, P. H. (2006). Pyrolysis of wood/biomass for bio-oil: A critical review. *Energy & fuels, 20*(3), 848–889.

Moragues-Saitua, L., Arias-González, A., Blanco, F., Benito-Carnero, G., & Gartzia-Bengoetxea, N. (2023). Effects of biochar and wood ash amendments in the soil-water-plant environment of two temperate forest plantations. *Frontiers in Forests and Global Change, 5*, 878217.

Nafees, M., Ullah, S., & Ahmed, I. (2021). Morphological and elemental evaluation of biochar through analytical techniques and its combined effect along with plant growth promoting rhizobacteria on *Vicia faba* L. under induced drought stress. *Microscopy Research and Technique, 84*(12), 2947–2959.

Nafees, M., Ullah, S., & Ahmed, I. (2022). Modulation of drought adversities in *Vicia faba* by the application of plant growth promoting rhizobacteria and biochar. *Microscopy Research and Technique, 85*(5), 1856–1869.

Nawaz, F., Rafeeq, R., Majeed, S., Ismail, M. S., Ahsan, M., Ahmad, K. S., et al. (2023). Biochar amendment in combination with endophytic bacteria stimulates photosynthetic activity and antioxidant enzymes to improve soybean yield under drought stress. *Journal of Soil Science and Plant Nutrition, 23*(1), 746–760.

Ohashi, M., Kume, T., Yoshifuji, N., Kho, L. K., Nakagawa, M., & Nakashizuka, T. (2015). The effects of an induced short-term drought period on the spatial variations in soil respiration measured around emergent trees in a typical Bornean tropical forest, Malaysia. *Plant and Soil, 387*, 337–349.

Poormansour, S., Razzaghi, F., & Sepaskhah, A. R. (2019). Wheat straw biochar increases potassium concentration, root density, and yield of faba bean in a sandy loam soil. *Communications in Soil Science and Plant Analysis, 50*(15), 1799–1810.

Qayyum, M. F., Steffens, D., Reisenauer, H. P., & Schubert, S. (2012). Kinetics of carbon mineralization of biochars compared with wheat straw in three soils. *Journal of Environmental Quality, 41*(4), 1210–1220.

Qayyum, M. F., Abid, M., Danish, S., Saeed, M. K., & Ali, M. A. (2015). Effects of various biochars on seed germination and carbon mineralization in an alkaline soil. *Pakistan Journal of Agricultural Sciences, 51*, 977–982.

Racioppi, M., Tartaglia, M., De la Rosa, J. M., Marra, M., Lopez-Capel, E., & Rocco, M. (2019). Response of ancient and modern wheat varieties to biochar application: Effect on hormone and gene expression involved in germination and growth. *Agronomy, 10*(1), 5.

Rajapaksha, A. U., Chen, S. S., Tsang, D. C., Zhang, M., Vithanage, M., Mandal, S., et al. (2016). Engineered/designer biochar for contaminant removal/immobilization from soil and water: Potential and implication of biochar modification. *Chemosphere, 148*, 276–291.

Rajput, V. D., Chernikova, N., Minkina, T., Gorovtsov, A., Fedorenko, A., Mandzhieva, S., Bauer, T., Tsitsuashvili, V., Beschetnikov, V., & Wong, M. H. (2023). Biochar and metal-tolerant bacteria in alleviating ZnO nanoparticles toxicity in barley. *Environmental Research, 220*, 115243. https://doi.org/10.1016/j.envres.2023.115243

Rajput, P., Kumar, P., Priya, A. K., Kumari, S., Shiade, S. R. G., Rajput, V. D., Fathi, A., Pradhan, A., Sarfraz, R., Sushkova, S., Mandzhieva, S., Minkina, T., Soldatov, A., Wong, M. H., & Rensing, C. (2024). Nanomaterials and biochar mediated remediation of emerging contaminants. *Science of the Total Environment, 916*, 170064. https://doi.org/10.1016/j.scitotenv.2024.170064

Rath, K. M., & Rousk, J. (2015). Salt effects on the soil microbial decomposer community and their role in organic carbon cycling: A review. *Soil Biology and Biochemistry, 81*, 108–123.

Ronsse, F., Van Hecke, S., Dickinson, D., & Prins, W. (2013). Production and characterization of slow pyrolysis biochar: Influence of feedstock type and pyrolysis conditions. *Gcb Bioenergy, 5*(2), 104–115.

Sánchez-Rodríguez, C., Rubio-Somoza, I., Sibout, R., & Persson, S. (2010). Phytohormones and the cell wall in Arabidopsis during seedling growth. *Trends in Plant Science, 15*(5), 291–301.

Sarker, U., & Oba, S. (2018). Drought stress effects on growth, ROS markers, compatible solutes, phenolics, flavonoids, and antioxidant activity in *Amaranthus tricolor*. *Applied Biochemistry and Biotechnology, 186*, 999–1016.

Singh, J., & Thakur, J. K. (2018). Photosynthesis and abiotic stress in plants. *Biotic and Abiotic Stress Tolerance in Plants*, 27–46.

Singh, M., Saini, R. K., Singh, S., & Sharma, S. P. (2019). Potential of integrating biochar and deficit irrigation strategies for sustaining vegetable production in water-limited regions: A review. *HortScience, 54*(11), 1872–1878.

Song, F., Han, X., Zhu, X., & Herbert, S. J. (2012). Response to water stress of soil enzymes and root exudates from drought and non-drought tolerant corn hybrids at different growth stages. *Canadian Journal of Soil Science, 92*(3), 501–507.

Sultan, H., Li, Y., Ahmed, W., Yixue, M., Shah, A., Faizan, M., Ahmad, A., Abbas, M., Nie, L., & Khan, M. N. (2024). Biochar and nano biochar: Enhancing salt resilience in plants and soil while mitigating greenhouse gas emission: A comprehensive review. *Journal of Environmental Management, 355*, 120448.

Taghizadeh Tabari, Z., Asghari, H. R., Abbasdokht, H., & Babakhanzadeh Sajirani, E. (2022). Salicylic acid and biochar improve drought tolerance in *Borago officinalis* L. by enhancing antioxidant enzymes, leaf proline and soluble sugars. *Journal of Organic Farming of Medicinal Plants, 1*(1), 15–26.

Usman, M., Ahmad, N., Raza, W., Zhao, Z., Abubakar, M., Rehman, S. U., et al. (2023). Impact of biochar on the yield and nutritional quality of tomatoes (*Solanum lycopersicum*) under drought stress. *Journal of the Science of Food and Agriculture, 103*(7), 3479–3488.

Viger, M., Hancock, R. D., Miglietta, F., & Taylor, G. (2015). More plant growth but less plant defence? First global gene expression data for plants grown in soil amended with biochar. *Gcb Bioenergy, 7*(4), 658–672.

Wang, Y., Xia, H., Yao, Z., & Jiang, C. (2018). Effects of biochar and other amendments on the cabbage growth and soil fertility in yellow-brown soil and red soil. *Scientia Agricultura Sinica, 51*, 4306–4315.

Woolf, D., Amonette, J. E., Street-Perrott, F. A., Lehmann, J., & Joseph, S. (2010). Sustainable biochar to mitigate global climate change. *Nature Communications, 1*(1), 56.

Xu, F., Liu, W., Wang, H., Alam, P., Zheng, W., & Faizan, M. (2023). Genome identification of the tea plant (Camella sinensis) ASMT gene family and its expression analysis under abiotic stress. *Genes, 14*, 409.

Yildirim, E., Ekinci, M., & Turan, M. (2021). Impact of biochar in mitigating the negative effect of drought stress on cabbage seedlings. *Journal of Soil Science and Plant Nutrition, 21*(3), 2297–2309.

Zaheer, M. S., Ali, H. H., Soufan, W., Iqbal, R., Habib-ur-Rahman, M., Iqbal, J., et al. (2021). Potential effects of biochar application for improving wheat (Triticum aestivum L.) growth and soil biochemical properties under drought stress conditions. *Land, 10*(11), 1125.

Zaman, Q. U., Rehman, M., Feng, Y., Liu, Z., Murtaza, G., Sultan, K., et al. (2024). Combined application of biochar and peatmoss for mitigation of drought stress in tobacco. *BMC Plant Biology, 24*(1), 862.

Zhang, Y., & Luo, W. (2014). Adsorptive removal of heavy metal from acidic wastewater with biochar produced from anaerobically digested residues: Kinetics and surface complexation modeling. *BioResources, 9*(2).

Zhang, J., Lü, F., Luo, C., Shao, L., & He, P. (2014). Humification characterization of biochar and its potential as a composting amendment. *Journal of Environmental Sciences, 26*(2), 390–397.

Zhang, Q., Shao, M., Jia, X., & Wei, X. (2019). Changes in soil physical and chemical properties after short drought stress in semi-humid forests. *Geoderma, 338*, 170–177.

Zhang, Y., Wang, J., & Feng, Y. (2021). The effects of biochar addition on soil physicochemical properties: A review. *Catena, 202*, 105284.

Zhang, W., Wei, J., Guo, L., Fang, H., Liu, X., Liang, K., et al. (2023). Effects of two biochar types on mitigating drought and salt stress in tomato seedlings. *Agronomy, 13*(4), 1039.

Zhang, K., Han, X., Fu, Y., Khan, Z., Zhang, B., Bi, J., et al. (2024). Biochar coating promoted rice growth under drought stress through modulating photosynthetic apparatus, chloroplast ultrastructure, stomatal traits and ROS homeostasis. *Plant Physiology and Biochemistry, 109145.*

Zhao, Z. Y., Wang, P. Y., Xiong, X. B., Wang, Y. B., Zhou, R., Tao, H. Y., et al. (2022). Environmental risk of multi-year polythene film mulching and its green solution in arid irrigation region. *Journal of Hazardous Materials, 435*, 128981.

Zhou, J., Wang, X., Jiao, Y., Qin, Y., Liu, X., He, K., et al. (2007). Global genome expression analysis of rice in response to drought and high-salinity stresses in shoot, flag leaf, and panicle. *Plant Molecular Biology, 63*, 591–608.

Zoghi, Z., Hosseini, S. M., Kouchaksaraei, M. T., Kooch, Y., & Guidi, L. (2019). The effect of biochar amendment on the growth, morphology and physiology of *Quercus castaneifolia* seedlings under water-deficit stress. *European Journal of Forest Research, 138*, 967–979.

Zulfiqar, B., Raza, M. A. S., Saleem, M. F., Aslam, M. U., Iqbal, R., Muhammad, F., et al. (2022). Biochar enhances wheat crop productivity by mitigating the effects of drought: Insights into physiological and antioxidant defense mechanisms. *PLoS One, 17*(4), e0267819.

Application of Biochar for Salinity Stress Management of Crop Plants

Akbar Hossain, Annika Jahan Aonti, Md. Mobinur Rahman, Md. Parvez Kabir, Bulbul Ahmed, and Md. Israfil Haq

Abstract

Salinity is a critical abiotic component that impacts crop yield and is a primary risk to global food security, particularly in arid and semiarid areas. One solution against this increasingly global issue that can be accomplished with one beneficial technique is biochar application. Biochar is a carbon-rich product produced from the pyrolysis of biomass that improves soil health and increases crop production. Interaction of soil microbes and roots and remediation on physicochemical properties of soil improved by biochar amendment. In some instances, reducing the requisite fertilizer application by as high as 25%, thereby optimizing bulk density and increasing saturated hydraulic conductivity (Ks) under saline conditions to enhance crop yield. This may be attributed to the high adsorption capacity of biochar that prevented transient Na^+ binding, decreased osmotic stress (because of increased soil moisture content), and the release of mineral nutrients (notably K^+, Ca^{++}, and Mg^{++}) into the solution. Their research results indicated that biochar can enhance soil quality and plant production in saline conditions. Future studies on biochar amendments are expected to eluci-

A. Hossain (✉) · A. J. Aonti · M. P. Kabir · B. Ahmed
Soil Science Division, Bangladesh Wheat and Maize Research Institute, Dinajpur, Bangladesh

M. M. Rahman
Maize Breeding Division, Bangladesh Wheat and Maize Research Institute, Dinajpur, Bangladesh

M. I. Haq
Soil Science Division, Bangladesh Wheat and Maize Research Institute, Dinajpur, Bangladesh

Department of Agricultural Extension (DAE), Langadu, Rangamati, Bangladesh

S. Singh, S. Srivastava (eds.), *Biochar: A Precious Resource from Biological Waste*,
https://doi.org/10.1007/978-981-95-0425-1_9

date the interplay of biochar with soil components, enhance remediation efforts for soil, and boost crop yields with salt stress.

Keywords

Biochar · Salinity · Soil health · Crop productivity · Soil remediation

1 Introduction

Salinity stress is an important environmental problem that negatively impacts agricultural productivity worldwide. Soil salinity is a global challenge and is damaging approximately 950 million hectares of land, of which about 250 million hectares are irrigated, creating an estimated economic loss of $27.3 billion annually (Ayub et al., 2020). This stress impairs physiological functions such as nutrient uptake, photosynthesis, growth, and yields of plants, which are detrimental to plant growth and development (Munns & Tester, 2008; Gupta & Huang, 2014). Additionally, due to salt stress, plants experience oxidative stress due to the generation of reactive oxygen species (ROS) (Cruz de Carvalho, 2008; Abbasi et al., 2015). Salinity stress also adversely impacts soil properties; for instance, it depletes organic matter content and the C:N ratio (Morrissey et al., 2014). High salt concentrations inhibit soil microbial activity and interfere with soil chemical and physical properties, reducing soil productivity (Yan et al., 2015). Soil salinization is a significant and profound interruption to carbon storage (Wong et al., 2008) during the very long periods in which carbon occurs.

Both organic and inorganic amendments efficiently mitigate the soil salinity by enhancing soil physicochemical qualities and supplying critical nutrients. Organic resources such as compost, green manures, and food industry wastes have been proven to be effective in alleviating the saline problems of soils and improving the carbon stocks and fertility of the applied organic resources. Biochar, an organic amendment, can be incorporated into soil to improve soil quality (Pandian et al., 2024). Pyrolysis is a high-temperature process (up to 700 °C) in which organic material is degraded in environments with very low oxygen levels to produce a high-carbon (high C) residue or char Kwapinski et al., 2010. Biochar application on saline soil ameliorates salinity effects in the crop by increasing nutrient availability, soil structure, and water retention (Liang et al., 2005; Laird et al., 2010). Furthermore, biochar has been demonstrated to increase the activity of microbes in the rhizosphere, which can improve plant growth and stress tolerance. (Lehmann et al., 2011).

Biochar application is seen as a win-win approach to achieve multiple goals for sustainable agriculture. Many studies show that biochar application to salt-stressed plants improved photosynthesis, nutrient uptake, plant growth, and biomass (Akhtar et al., 2015; Ali et al., 2017; Haider et al., 2015; Kim et al., 2016). It has been shown that biochar also plays an important role in enhancing soil physicochemical properties, including pH, cation exchange capacity (CEC), soil structure, water holding capacity (WHC), and surface area (Chaganti & Crohn, 2015; Hammer et al., 2015; Bamminger et al., 2016; Lim et al., 2016).

Applying biochar reduces Na+ deposition (Chaganti et al., 2015; Diacono & Montemurro, 2015; Sun et al., 2016), avoiding salinity stress and improving K^+ deposition in plants. According to Kim et al. (2016), it further promotes mineral uptake ability and regulates plant stomatal conductance and phytohormones (Ali et al., 2017). Salt stress induces a dose-dependent increase of phosphorus (P) in maize tissues supplemented with biochar (Kim et al., 2016).

Other studies have shown that biochar alleviates plant salt stress by reducing phytohormone synthesis (Akhtar et al., 2015; Lashari et al., 2015). Biochar can enhance the activity of antioxidant enzymes and promote plant growth and biomass under salt stress by reducing oxidative stress. Reports show that the activity of glutathione reductase (GR) and ascorbate peroxidase (APX) in maize under salt stress is lower when biochar is applied compared to the control (Kim et al., 2016). In addition, biochar and endophytic bacteria co-applicate inhibit xylem abscisic acid (ABA) secretion in wheat and maize under saline conditions (Akhtar et al., 2015).

Biochar amendment is an environmentally friendly, sustainable method to ameliorate salinity stresses on crop plants. Thus, this chapter aims to summarize the current knowledge on mechanisms through which biochar enhances plants' ability to withstand salinity stresses and the most recent approaches to how biochar amendment affects soils. However, more research is needed to fully understand how biochar alleviates salt stress and develop best practices for using it in various agricultural systems.

2 Consequences of Salinity Stress on Soil Health and Crop Productivity

Soil salinity, defined as the presence of soluble salts in the soil, represents one of the most significant challenges for the future of croplands globally and threatens agricultural productivity and environmental sustainability (Shahid et al., 2018). High salinity values are naturally found in arid and semiarid soils; however, when salt accumulates in the zone of crop roots at concentration levels that affect crop yield, salinity problems are developed (Francois & Maas, 1999). The global extent of salt-affected soils is estimated to be 935 million hectares (M ha) (Ivushkin et al., 2019), with >70% of this area distributed in arid and semiarid regions, including the USA, China, Africa, India, Pakistan, Australia, and the Middle East (Basak et al., 2022). Salinity affects more than 424 million hectares of surface soil (0–30 cm) and 833 million hectares of subsoil (30–100 cm) of the world's total land area. On a global basis, approximately 10 ha of arable land is lost annually every minute due to salinization, with more than two-thirds of these areas located in arid and semiarid climate zones (Fig. 1).

Sodium bicarbonates and carbonates play a crucial role in the formation of soil sodicity. The sodium adsorption ratio (SAR) and exchangeable sodium percentage (ESP) are the best measures of risk where salt is concerned in soil and soil mixtures

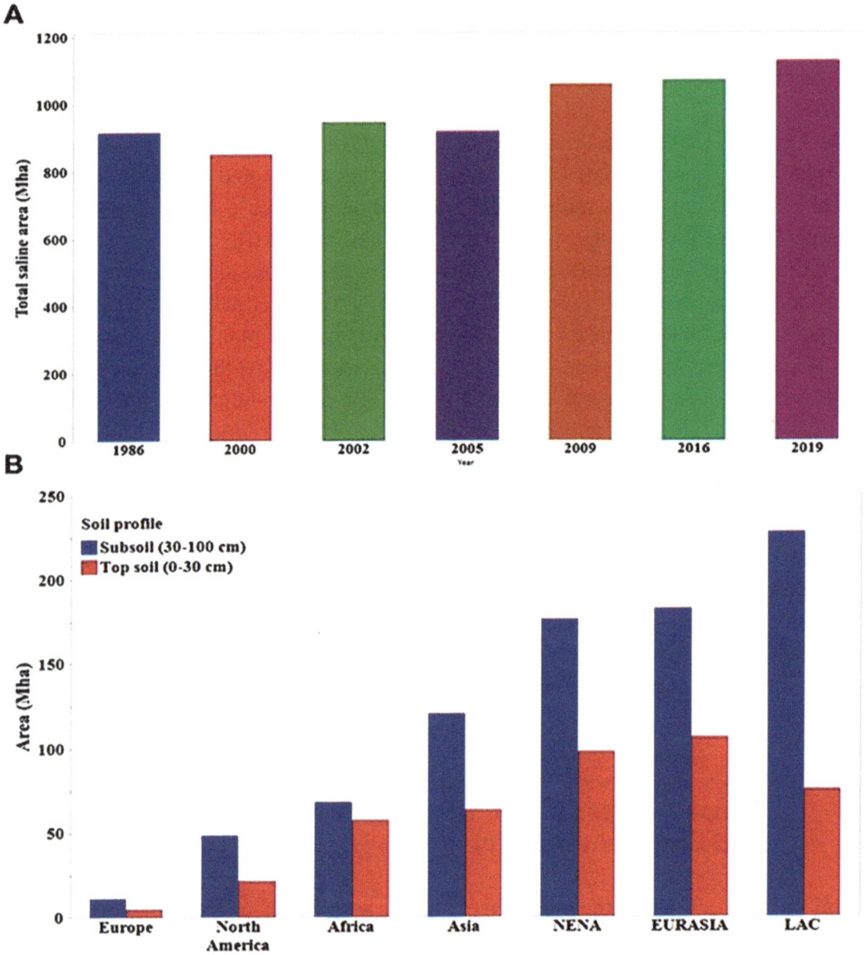

Fig. 1 (**a**) Annual distribution of the global total saline area; (**b**) global distribution of salt-affected area at topsoil and subsoil. (Data sources: Ivushkin et al. (2019), and FAO (2021))

in that order. Salt-affected soil is classified based on the soil's pH, acidity, ESP, SAR, and EC, as shown in Table 1.

Salinity is the measure of dissolved inorganic solutes (Na^+, Mg^{++}, Ca^{++}, K^+, Cl^-, SO_4^-, HCO_3^-, NO_3, and CO_3^-) in water and soil (Bernstein et al., 1974; Tanji, 1990). While sodium (Na^+), chloride (Cl^-), calcium (Ca^{+2}), magnesium (Mg^{+2}), and some other ions are essential for plant nutrition, excess Na^+ and/or Cl^- levels in salt-affected soils can induce stress in economically important crops. The excessive addition of other nutrients (Ca, K, Mg, etc.) can also cause an imbalance of nutrients, resulting in a decrease in the yield of crops (Wang et al., 2022). The activity of enzymes, soil respiration, soil microbial biomass, and bacterial growth rates can be reduced by salinity, which may ultimately influence biogeochemical cycling and

Table 1 Types of salt-affected soil (Stavi et al., 2021)

Types of soil	pH of soil	Soil EC (dsm^{-1})	Soil ESP	Soil SAR
Saline	< 8.5	>4.0	<15	<13
Sodic	>8.5	Variable	>15	>13
Saline-sodic	> 8.5	>4.0	Variable	>13

soil fertility (Wu et al., 2022). In contaminated areas, soil's salinity stress on plants may intensify, and nutrients or water in the soil that farmed plants need may not be available (Li et al., 2023). Salt-affected soils influenced by heavy metal and/or organic contaminants have shown extra stress (Wu et al., 2022). Soil salts disturb functional activities in soil and plants. The presence of high concentrations (values) of salt in the soil adversely affects the physicochemical (e.g., bulk density, infiltration, aeration, soil water potential, soil aggregates, and soil fertility) and biological (e.g., soil enzyme activity, microbial activity, and biodiversity) attributes. Biochemical, physiological, and molecular irregularities develop in the crops planted in salt-affected soils.

Soil salinity presents various challenges to crop production. It is not limited to coastal areas but can also be attributed to other factors, including poor agronomic practices or using recycled water to irrigate plants that might contain high concentrations of salts (Kumar & Sharma, 2020). The impairing resilience of saline soils due to their high levels of solutes also affects crop plants, resulting in production losses (Mbarki et al., 2018). When the soil and water have a high salt concentration, water transportation from the soil to the roots is affected, and water consumption efficiency also decreases (Yuan et al., 2018). When soil solutes' osmotic pressure is higher than root cells, the soil withdraws water from the roots (Munns, 2005). It causes water stress by elevating osmotic pressure in plants and has the potential to pollute the groundwater and freshwater systems (Greene et al., 2016). Soil salinity affects crop yield, appropriateness, nutrient availability, nutrient uptake, and soil microbial activity (Munns & Gilliham, 2015; Yuan et al., 2018). Soil ionic and osmotic stresses can change the rhizosphere microbial community, composition, and enzyme activity.

It is well known that nutrient deficiency and/or toxicity can arise as a consequence of osmotic stress, limited water (Rengasamy, 2006), ionic stress (FAO and ITPS, 2015), and changes in the ionic balance of the cell (Qadir et al., 2008; Singh et al., 2015; Hazman et al., 2016). Beyond plant nutrition, soil salinity can impact microbial activity, soil nutrient availability, soil organism connections, and soil fertility. Global soil salinization severely challenges ecosystem resilience and agricultural sustainability with widespread effects. This results in negative impacts like lower agricultural productivity, soil structure and fertility degradation, decline in species diversity, and disruption of the ecological balance (Haj-Amor et al., 2022) (Fig. 2). Soil salinity can affect microbial activity and processes and soil carbon (C) and nitrogen (N) availability, leading to changes in nitrous oxide (N_2O), methane (CH_4), greenhouse gas carbon dioxide (CO_2), and soil organic carbon (SOC) emissions (Shahariar et al., 2021).

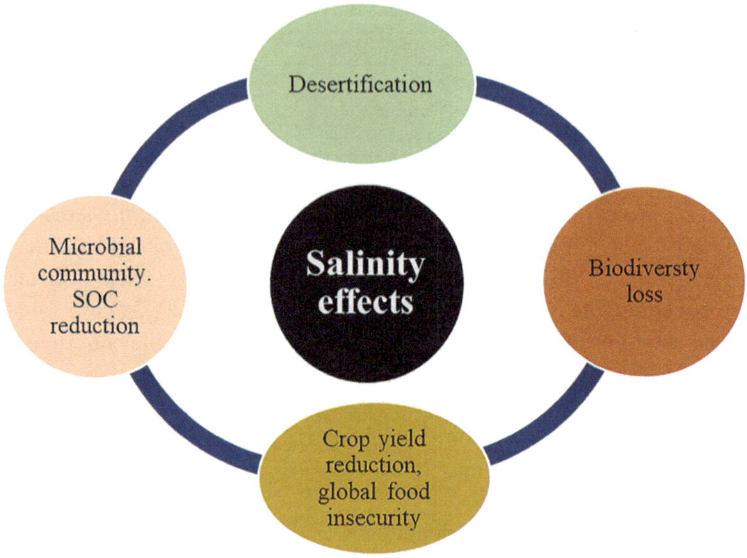

Fig. 2 The overall consequences of salinity

According to Vasantha et al. (2010), osmotic stress, nutritional problems, and toxicity are problems related to saline soils, which are usually faced by crops grown on saline soils. Under these conditions, crop yields can be dramatically reduced (Shrivastava & Kumar, 2015). Some aspects of climate change (e.g., increasing sea levels, increasing air temperatures, and greater evaporation) will increase soil and subsurface water salinities in many places. In many countries, the result will be further decline in crop yields and agricultural productivity. Soil salinity exacerbates the problem of desertification. Soil salinity has resulted in 350 million hectares worldwide that are degraded and at risk of decertification (Ivushkin et al., 2019). The problem has many contributing factors, such as improper irrigation and drainage methods, poor land-use management practices, and climate change (Shrivastava & Kumar, 2015). According to Raj (2020), desertification alters and destroys the important vegetation to biodiversity and many ecosystem services (ESs), such as biomass, dry matter dynamics, litterfall, and the global carbon cycle.

Salinity harms plants at every life cycle stage, but the germination and seedling stages are susceptible to salinity. Salinity has an adverse effect on seed germination by causing a decrease in gibberellic acid levels, an increase in abscisic acid levels, membrane permeability changes, and/or reduced water uptake in seeds (Lee & Luan, 2012). High Na+ and Cl- ion accumulation from soil absorption results in plant cytotoxicity, causing leaf burning, its growth inhibition, and eventually death of the plant. In addition, higher Na+ concentrations deplete other elements (K+, Ca2+, and Mg2+) and can also compete with cations and result in nutrient deficiency (Atta et al., 2021). Most importantly, salinity decreases agricultural yield and production by altering several morphophysiological and biochemical processes (Table 2).

Table 2 Yield reduction due to salinity in several important crops

Crops	Yield reduction (%)	References
Maize	50–60	Zorb et al. (2018)
Wheat	10–50	Panta et al. (2014), Zorb et al. (2018), and Yadav et al. (2020)
Rice	30–50	Eynard et al. (2005)
Brassica spp.	30–70	Pavlovic et al. (2019) and Singh et al. (2021)
Sugarcane	5–60	Plaut et al. (2000), Rao et al. (2015), and Kumar et al. (2023)
Groundnut	25–50	Otitoloju (2014) and Zorb et al. (2018)
Sorghum	15–50	Daniells et al. (2001) and Kenneth and Neeltje (2002)
Barley	10–60	Abd El-Monem et al. (2013)
Bajra	5–25	Yadav et al. (2020)
Cotton	10–20	Zorb et al. (2018)

3 Concepts of Biochar and Its Potential Effects on Soil Health and Crop Productivity

3.1 Effects on Soil Health

Biochar is a carbon-rich material obtained from the thermal decomposition of organic matter through pyrolysis under limited oxygen conditions (Rahman et al., 2020). The feedstock for biochar production includes agricultural residues, wood, and other organic waste, making it a sustainable option for enhancing soil properties (Joseph, 2012). The unique properties of biochar, such as high porosity, large surface area, and chemical stability, also facilitate potential benefits to improve soil health under problematic soil salinity conditions (Agegnehu et al., 2016).

Biochar has been shown to have positive effects on various aspects of soil health. One of its best features is that it improves the soil structure. The addition of biochar improves soil porosity, water holding capacity, and aggregate stability, which are crucial for plant growth, especially in saline soils where these properties are often compromised (Jeffery et al., 2011). Biochar has high porosity, and its surface area is able to store more water, thus increasing the soil moisture content and reducing water stress in plants (Steiner et al., 2007).

In addition to physical properties, biochar can also alter soil chemical constraints. Its alkaline nature helps buffer acidic soils, leading to improved nutrient availability (Rahman et al., 2020). The presence of biochar enhances the cation exchange capacity (CEC), which allows soils to retain essential nutrients such as calcium, magnesium, and potassium more effectively, improving plant uptake. This phenomenon is particularly important in sodic environments, where sodium ions can disrupt nutritional balance (Jeffery et al., 2011).

However, a marked increase in soil microbial activity is caused by biochar. Biochar provides a stable biochemical environment and substrate for microbes, which leads to increases in microbial diversity and abundance (Rondon et al., 2007). This phenomenon is even more important for beneficial microorganisms, such as mycorrhizal fungi and diazotrophic bacteria responsible for nutrient metabolism and nitrogen fixation, which consequently can lead to greater soil fertility, increasing the resistance of plants subjected to stressful situations (Rondon et al., 2007).

The long-term stability of biochar in soils is another advantage, as it remains effective over extended periods, making it a cost-effective strategy for improving soil health (Joseph, 2012). Additionally, the ability of biochar to immobilize heavy metals and toxic compounds can be beneficial in mitigating the negative effects of soil contamination, which is often exacerbated by salinity (Beesley et al., 2011).

Thus, biochar has great potential for enhancing soil health under saline conditions. Its ability to improve soil structure, nutrient retention, microbial activity, and detoxification processes makes it a valuable amendment for sustainable soil health management.

3.2 Effects of Biochar on Crop Productivity

Biochar is a carbon-rich solid material obtained from the thermochemical conversion of organic materials in a limited or deficient-oxygen environment (Zheng et al., 2016; Guo, 2020) that can enhance soil fertility and has physical and chemical characteristics suitable for long-term carbon storage in the environment (Ibarrola et al., 2013).

Biochar is the final form of the carbonization of feedstocks, including crop residues, tree biomass, paper waste, and rice husks (Escalante-Rebolledo et al., 2016; Adeyemi & Idowu, 2017). While any organic habitat can be used for the production of biochar, plant-based habitats are particularly conducive, especially those containing large amounts of lignocellulosic terrains, the major constituents of terrestrial plants in addition to water (Quesada-Kimsey, 2012). Biochar is one of the few carbon-rich products applied to soils for carbon sequestration (Steiner, 2016), and the everyday products that contain carbon include activated carbon. The physical, chemical, and biological characteristics of soil can combine to a certain extent, which can positively impact soil fertility and lead to crop growth and yield (Silva et al., 2017).

Meta-analyses suggest that adding biochar could increase aboveground plant productivity by 10% (Liu et al., 2013; Biederman & Harpole, 2013). As discussed in the previous sections, biochar also enhances the growth of crops by modifying soil physical, biological, and chemical parameters. That said, the results from applying biochar are not always predictable. Jeffery et al. (2011) reported that biochar addition altered plant productivity (crop yield and aboveground biomass) by 28–39% in soils. Biochar application to soils has been demonstrated to considerably improve the yields of many plant species across diverse environments (Lehmann & Joseph, 2015). In 2013, Crane-Droesch et al. found a potential response of crops to biochar treatment over a large area of sub-Saharan Africa, as well as parts of South America, Southeast Asia, and southeastern North America. The better aggregation of soil was responsible for more nutrient retention, and the retention of soil water-holding ability of biochar might be the reason for the increased crop yields in these highly weathered and poor nutrient soils. According to Yamato et al. (2006), under fertilized conditions, applying biochar improves soil pH, CEC, and nutrient availability, thereby increasing the yields of maize, cowpea, and peanuts. Uzoma et al. (2011) also found that the most recommendable maize grain production was obtained in 15 and 20 t/ha biochar applications (150% and 98% increases) because of the improvement of soil's physical and chemical properties.

Collectively, these trends suggest a moderate to negative yield response in most of the world's top grain-producing nations. At the same time, results from studies assessing the effect of biochar amendment on plant productivity in peat soils have indicated unfavorable consequences (Crane-Droesch et al., 2013). Dramatic reductions in crop yields in biochar-treated soils have been linked to dramatic increases in soil C/N ratios, leading to nitrogen immobilization (Bridle & Pritchard, 2004; Chan et al., 2008). According to Kloss et al. (2014), the micronutrients due to biochar might also lead to reduced crop development. For example, in barley and mustard, yields were reduced by 68% following biochar application (Kloss et al., 2014). This was due to significant reductions in the amounts of Cu, Fe, Mn, and Zn available, but increased levels of Mo were found in the plant tissues. The effectiveness of biochar as a means to enhance plant production is influenced by climate, soil characteristics, type of crop grown, and experimental design (Liu et al., 2013; Wang et al., 2012). Besides the interactions of soil biotic and abiotic elements with biochar occurring during soil application, differences in responses can also be elucidated regarding biochar feedstock and pyrolysis conditions (Sohi et al., 2009). In pot experiments, positive crop productivity has been observed more often than in field experiments, in acidic soils than in neutral soils, and in sandy soils than in loam and silt soils (Crane-Droesch et al., 2013; Jeffery et al., 2011). Moreover, in dryland soils, adding biochar in cropping systems increases yields by 11% (Liu et al., 2013). Chicken litter has the most significant positive effect (28%) on yield in terms of the effects of biochar sources compared with biosolids, the only feedstock that has a discernible adverse effect (Jeffery et al., 2011).

Combining wastes with biochar rather than producing biochar from wastes may also prove equally or more advantageous for nutrient retention and reduced risk of environmental contamination (Awasthi et al., 2018). A growing number of studies have focused on the effects of biochar amendment on salt-affected soils that are common in arid regions (Saifullah et al., 2018). For instance, biochar addition mitigates salinity stress damage in wheat, resulting in greater leaf water potential and reduced proline, soluble sugars, and superoxide dismutase (Kanwal et al., 2018). According to Hammer et al. (2015), the plants growing in biochar-amended soils may display commensurate decreases in Na and increased K. The soils may also show lower electrical conductivity, indicating Na uptake. Hammer et al. (2015) indicate the potential for "joint management in agricultural systems" with symbiotic microbe with biochar interactions. While these benefits are potentially attractive, they rely on appropriate feedstock sources and processing methods, since elevated biochar application could lead to high soil salinity and sodicity (Saifullah et al., 2018). Besides, while current prices for most of the black carbons (char-derived products) would make large-scale applications uneconomical, the application rates where each described mechanism should be beneficial would depend on biomass feedstock availability and cheap production methods (Saifullah et al., 2018).

4 Prospects of the Use of Biochar for Ameliorating Soil Salinity

Worldwide, agricultural sustainability is now threatened by soil salinity, especially in arid and semiarid areas where the accumulation of salt within soils is behind a specific threshold that inhibits plant growth and development (Rahman et al., 2020). Soil salinity has adverse effects on plant water uptake, nutrient availability and soil structure, which reduces agricultural outputs and results in the salinization of arable land (Munns & Tester, 2008). Addressing soil salinity is crucial for sustaining agricultural productivity, and biochar, a carbon-rich material obtained from the pyrolysis of organic materials, has potential for effective salinity management in salt-affected soils.

4.1 Better Soil Physical Properties

Improving soil physical properties is one of the major mechanisms by which biochar ameliorates saline soils (Rahman et al., 2020). Biochar is a porous material that increases soil porosity and water-holding capacity and reduces bulk density (Lehmann et al., 2011). These changes help improve soil productivity and maintain nutrient availability in soil, which is a key limiting factor for saline soils. In addition, the water-retention properties of biochar allow more leaching to wash salts through the rhizosphere to reduce plant salt stress (Karhu et al., 2011).

4.2 Improvements in Soil Chemical Properties

The use of biochar enhances soil chemical properties, which is crucial for reducing the effects of salinity (Rahman et al., 2020). Biochar can increase soil pH, especially in acidic soils, and its alkaline nature helps neutralize acidic components that are often associated with saline soils (Cheng et al., 2006). In addition, the cation exchange capacity (CEC) of soil is high with biochar, and biochar can retain essential nutrients for a relatively long period, which increases plant availability even under saline conditions, resulting in a potential yield loss reduction (Yuan & Xu, 2011). This gradual release of nutrients to plants ensures a stable nutrient profile in the soil for proper plant growth, especially under saline conditions.

4.3 Enhancing Microbial Activity

The activity of soil microorganisms plays a crucial role in maintaining soil health and fertility. Consequently, a decrease in both the variety and function of microorganisms exacerbates the problems associated with salinity. For example, biochar has been suggested to increase microbial activity via physicochemical mechanisms and/or through providing a habitat or energy source for soil microbes (Lehmann et al., 2011). Biochar application has been found to increase microbial activity, leading to faster degradation of organic matter, which in turn is instrumental for the nutrient recycling needed to maintain soil fertility under saline conditions (Rahman et al., 2020). Biochar not only affects the increase in soil porosity and pore size distribution with a positive impact on improving soil aggregation by shifting microbial communities but is also associated with beneficial microbes that can produce biofilms or exopolysaccharides during particle aggregation, thus resulting in salinity amelioration in salt-affected soil (Ali et al., 2017).

4.4 A Lower Sodium Adsorption Ratio (SAR)

The SAR is a significant index of the degree of soil saltiness, and too many high-level SAR values can destroy not only agricultural soils but also prevent plant growth. Biochar application can reduce soil SAR because of its effects on the exchange between sodium ions and calcium or magnesium in soil (Liu et al., 2016). This ion exchange process is favorable in terms of minimizing sodium (Na^+)-associated problems that cause soil structure degradation and hence help ameliorate conditions for plant growth. Similarly, biochar has the potential to adsorb sodium, which may reduce soluble Na^+ activity, leading to a decrease in total salinity (Busscher et al., 2010).

4.5 Long-Term Benefits and Sustainability

The stability and persistence of biochar in the soil maintain its usability for several years, which ultimately benefits soil salinity management (Rahman et al., 2020). While other soil amendments must be reapplied frequently, slow degradation rates of biochar lead it to remain in soils for long periods, improving many physical properties, continually improving soil properties and mitigating salinity, which also acts as a mechanism causing expansion against salinity. This, in turn, indicates that biochar is an eco-sustainable solution for the long-term mitigation of soil salinity (Lehmann & Joseph, 2015). Additionally, biochar production can be integrated into waste management and is an inexpensive alternative to sodic soil reclamation (Verheijen et al., 2010).

5 Conclusion

In conclusion, many reports have shown that biochar has the potential to reduce soil salinity, which, in terms of the physical and chemical properties of soils, enhances microbial activity, resulting in SAR reduction combined with growth enhancement features such as long-term sustainability. Although there may be variation in the required biochar application rates and methods with respect to the soil type and cropping system, the overall body of evidence reviewed here suggests that this is a promising new strategy for mitigating salt stress to plants caused by soils exerting salinity pressure on crops from saline environments.

References

Abbasi, G. H., Akhtar, J., Ahmad, R., Jamil, M., Anwar-ul-Haq, M., Ali, S., & Ijaz, M. (2015). Potassium application mitigates salt stress differentially at different growth stages in tolerant and sensitive maize hybrids. *Plant Growth Regulation, 76*, 111–125.

Abd El-Monem, A. A., El-Habbasha, S. F., & Hozayn, M. (2013). Mitigation salinity stress effects on barley (*Hordeum vulgare* L.) growth, yield and some physiological aspects by hemin. *Journal of Applied Sciences Research, 9*(3), 2411–2421.

Adeyemi, T., & Idowu, O. (2017). Biochar: Promoting crop yield, improving soil fertility, mitigating climate change and restoring polluted soils. *World News of Natural Sciences, 8*(1), 27–36.

Agegnehu, G., Bass, A. M., Nelson, P. N., & Bird, M. I. (2016). Benefits of biochar, compost and biochar–compost for soil quality, maize yield and greenhouse gas emissions in a tropical agricultural soil. *Science of the Total Environment, 543*, 295–306.

Akhtar, S. S., Andersen, M. N., & Liu, F. (2015). Biochar mitigates salinity stress in potato. *Journal of Agronomy and Crop Science, 201*(5), 368–378.

Ali, S., Rizwan, M., Qayyum, M. F., Ok, Y. S., Ibrahim, M., Riaz, M., et al. (2017). Biochar soil amendment on alleviation of drought and salt stress in plants: A critical review. *Environmental Science and Pollution Research, 24*, 12700–12712.

Atta, K., Pal, A. K., & Jana, K. (2021). Effects of salinity, drought and heavy metal stress during seed germination stage in ricebean [Vigna umbellata (Thunb.) Ohwi and Ohashi]. *Plant Physiology Reports, 26*, 109–115. https://doi.org/10.1007/s40502-020-00542-4

Awasthi, M. K., Awasthi, S. K., Wang, Q., Wang, Z., Lahori, A. H., Ren, X., Chen, H., Wang, M., Zhao, J., & Zhang, Z. (2018). Influence of biochar on volatile fatty acids accumulation and microbial community succession during biosolids composting. *Bioresource Technology, 251*, 158–164.

Ayub, M. A., Ahmad, H. R., Ali, M., Rizwan, M., Ali, S., Rehman, M. Z. u., & Waris, A. A. (2020). Salinity and its tolerance strategies in plants. In *Plant life under changing environment* (pp. 47–76). Academic Press.

Bamminger, C., Poll, C., Sixt, C., Högy, P., Wüst, D., Kandeler, E., & Marhan, S. (2016). Short-term response of soil microorganisms to biochar addition in a temperate agroecosystem under soil warming. *Agriculture, Ecosystems & Environment, 233*, 308–317.

Beesley, L., Moreno-Jiménez, E., Gomez-Eyles, J. L., Harris, E., Robinson, B., & Sizmur, T. (2011). A review of biochars' potential role in the remediation, revegetation and restoration of contaminated soils. *Environmental Pollution, 159*(12), 3269–3282.

Bernstein, L., Francois, L. E., & Clark, R. A. (1974). Interactive effects of salinity and fertility on yields of grains and vegetables. *Agronomy Journal, 66*, 412–421.

Biederman, L. A., & Harpole, W. S. (2013). Biochar and its effects on plant productivity and nutrient cycling: A meta-analysis. *GCB Bioenergy, 5*, 202–214. https://doi.org/10.1111/gcbb.12037

Bridle, T., & Pritchard, D. (2004). Energy and nutrient recovery from sewage sludge via pyrolysis. *Water Science and Technology, 50*, 169–175.
Busscher, W. J., Novak, J. M., Evans, D. E., Watts, D. W., Niandou, M. A. S., & Ahmedna, M. (2010). Influence of pecan biochar on physical properties of a Norfolk loamy sand. *Soil Science, 175*(1), 10–14.
Chaganti, V. N., & Crohn, D. M. (2015). Evaluating the relative contribution of physiochemical and biological factors in ameliorating a saline–sodic soil amended with composts and biochar and leached with reclaimed water. *Geoderma, 259*, 45–55.
Chaganti, V. N., Crohn, D. M., & Šimůnek, J. (2015). Leaching and reclamation of a biochar and compost amended saline–sodic soil with moderate SAR reclaimed water. *Agricultural Water Management, 158*, 255–265.
Chan, K., Van Zwieten, L., Meszaros, I., Downie, A., & Joseph, S. (2008). Agronomic values of greenwaste biochar as a soil amendment. *Soil Research, 45*, 629–634.
Cheng, C. H., Lehmann, J., & Engelhard, M. H. (2006). Natural oxidation of black carbon in soils: Changes in molecular form and surface charge along a climosequence. *Geochimica et Cosmochimica Acta, 70*(4), 717–729.
Crane-Droesch, A., Abiven, S., Jeffery, S., & Torn, M. S. (2013). Heterogeneous global crop yield response to biochar: A meta-regression analysis. *Environmental Research Letters, 8*, 044049.
Cruz de Carvalho, M. H. (2008). Drought stress and reactive oxygen species: Production, scavenging and signalling. *Plant Signalling & Behavior, 3*(3), 156–165.
Daniells, I. G., Holland, J. F., Young, R. R., Alston, C. L., & Bernardi, A. L. (2001). Relationship between yield of grain sorghum (Sorghum bicolor) and soil salinity under field conditions. *Australian Journal of Experimental Agriculture, 41*, 211–217. https://doi.org/10.1071/EA00084
Diacono, M., & Montemurro, F. (2015). Effectiveness of organic wastes as fertilizers and amendments in salt-affected soils. *Agriculture, 5*(2), 221–230.
Escalante-Rebolledo, A., Pérez-López, G., Hidalgo-Moreno, C., López-Collado, J., Campos-Alves, J., Valtierra-Pacheco, E., & Etchevers-Barra, J. D. (2016). Biocarbón (biochar) I: Naturaleza, historia, fabricación y uso en el suelo. *Terra Latinoamericana, 34*(3), 367–382.
Eynard, A., Lal, R., & Wiebe, K. (2005). Crop response in salt-affected soils. *Journal of Sustainable Agriculture, 27*, 5–50. https://doi.org/10.1300/J064v27n01-03
FAO. (2021). *Global map of salt-affected soils*. Available at: https://www.fao.org/3/cb7247en/cb7247en.pdf
FAO and ITPS. (2015). *Status of the world's soil resources (SWSR)—Main report*. Food and Agriculture Organization of the United Nations.
Francois, L. E., & Maas, E. V. (1999). *Crop response and management of salt-affected soils* (pp. 169–192). U.S. Department of Agriculture.
Guo, M. (2020). The 3R principles for applying biochar to improve soil health. *Soil Systems, 4*(9), 1–16. https://doi.org/10.3390/soilsystems4010009
Gupta, B., & Huang, B. (2014). Mechanism of salinity tolerance in plants: Physiological, biochemical, and molecular characterization. *International Journal of Genomics, 2014*, 701596.
Haider, G., Koyro, H. W., Azam, F., Steffens, D., Müller, C., & Kammann, C. (2015). Biochar but not humic acid product amendment affected maize yields by improving plant–soil moisture relations. *Plant and Soil, 395*, 141–157.
Haj-Amor, Z., Araya, T., Kim, D. G., Bouri, S., Lee, J., Ghiloufi, W., & Lal, R. (2022). Soil salinity and its associated effects on soil microorganisms, greenhouse gas emissions, crop yield, biodiversity and desertification: A review. *Science of the Total Environment, 843*, 156946.
Hammer, E. C., Forstreuter, M., Rillig, M. C., & Kohler, J. (2015). Biochar increases arbuscular mycorrhizal plant growth enhancement and ameliorates salinity stress. *Applied Soil Ecology, 96*, 114–121.
Hazman, M., Hause, B., Eiche, E., Riemann, M., & Nick, P. (2016). Different forms of osmotic stress evoke qualitatively different responses in rice. *Journal of Plant Physiology, 202*, 45–56.
Ibarrola, R., Evar, B., & Reay, D. (2013). *Comercialización de Biocarbón (biochar) en México.* Definición del contexto para un programa de investigación multidisciplinario. (Ed.). Universidad Edimburgo, 28 p.

Ivushkin, K., Bartholomeus, H., Bregt, A. K., Pulatov, A., Kempen, B., & De Sousa, L. (2019). Global mapping of soil salinity change. *Remote Sensing of Environment, 231*, 111260. https://doi.org/10.1016/j.rse.2019.111260

Jeffery, S., Verheijen, F. G. A., Velde, M. V. D., & Bastos, A. C. (2011). A quantitative review of the effects of biochar application to soils on crop productivity using meta-analysis. *Agriculture, Ecosystems and Environment, 144*, 175–187. https://doi.org/10.1016/j.agee.2011.08.015

Joseph, S. (2012). *Biochar for environmental management: Science and technology*. Routledge.

Kanwal, S., Ilyas, N., Shabir, S., Saeed, M., Gul, R., Zahoor, M., Batool, N., & Mazhar, R. (2018). Application of biochar in mitigation of negative effects of salinity stress in wheat (*Triticum aestivum* L.). *Journal of Plant Nutrition, 41*(4), 526–538.

Karhu, K., Mattila, T., Bergström, I., & Regina, K. (2011). Biochar addition to agricultural soil increased CH4 uptake and water holding capacity – Results from a short-term pilot field study. *Agriculture, Ecosystems & Environment, 140*(1–2), 309–313. https://doi.org/10.1016/j.agee.2010.12.005

Kenneth, K., & Neeltje, C. (2002). *Agricultural drainage water management in arid and semiarid areas* (p. 61). FAO Irrigation and Drainage Paper.

Kim, H. S., Kim, K. R., Yang, J. E., Ok, Y. S., Owens, G., Nehls, T., & Kim, K. H. (2016). Effect of biochar on reclaimed tidal land soil properties and maize (*Zea mays* L.) response. *Chemosphere, 142*, 153–159.

Kloss, S., Zehetner, F., Wimmer, B., Buecker, J., Rempt, F., & Soja, G. (2014). Biochar application to temperate soils: Effects on soil fertility and crop growth under greenhouse conditions. *Journal of Plant Nutrition and Soil Science, 177*, 3–15.

Kumar, P., & Sharma, P. K. (2020). Soil salinity and food security in India. *Frontiers in Sustainable Food Systems, 4*, 533781. https://doi.org/10.3389/fsufs.2020.533781

Kumar, R., Dhansu, P., Kulshreshtha, N., Meena, M. R., Kumaraswamy, M. H., Appunu, C., et al. (2023). Identification of salinity-tolerant stable sugarcane cultivars via AMMI, GGE and other stability parameters under multiple environments of salinity stress. *Sustainability, 15*, 1119. https://doi.org/10.3390/su15021119

Kwapinski, W., Byrne, C. M., Kryachko, E., Wolfram, P., Adley, C., Leahy, J. J., & Hayes, M. H. (2010). Biochar from biomass and waste. *Waste and Biomass Valorization, 1*, 177–189.

Laird, D. A., Fleming, P., Davis, D. D., Horton, R., Wang, B., & Karlen, D. L. (2010). Impact of biochar amendments on the quality of a typical Midwestern agricultural soil. *Geoderma, 158*(3–4), 443–449.

Lashari, M. S., Ye, Y., Ji, H., Li, L., Kibue, G. W., Lu, H., & Pan, G. (2015). Biochar–manure compost in conjunction with pyroligneous solution alleviated salt stress and improved leaf bioactivity of maize in a saline soil from Central China: A 2-year field experiment. *Journal of the Science of Food and Agriculture, 95*(6), 1321–1327.

Lee, S. C., & Luan, S. (2012). ABA signal transduction at the crossroad of biotic and abiotic stress responses. *Plant Cell Environment, 35*(1), 53–60. https://doi.org/10.1111/j.1365-3040.2011.02426

Lehmann, J., & Joseph, S. (2015). *Biochar for environmental management: Science, technology and implementation*. Routledge.

Lehmann, J., Rillig, M. C., Thies, J., Masiello, C. A., Hockaday, W. C., & Crowley, D. (2011). Biochar effects on soil biota – A review. *Soil Biology and Biochemistry, 43*(9), 1812–1836.

Li, X., Wang, Y., Guo, P., Zhang, Z., Cui, X., Hao, B., & Guo, W. (2023). Arbuscular mycorrhizal fungi facilitate Astragalus adsurgens growth and stress tolerance in cadmium and lead contaminated saline soil by regulating rhizosphere bacterial community. *Applied Soil Ecology, 187*, 104842.

Liang, Y., Si, J., Nikolic, M., Peng, Y., Chen, W., & Jiang, Y. (2005). Organic manure stimulates biological activity and barley growth in soil subject to secondary salinization. *Soil Biology and Biochemistry, 37*(6), 1185–1195.

Lim, T. J., Spokas, K. A., Feyereisen, G., & Novak, J. M. (2016). Predicting the impact of biochar additions on soil hydraulic properties. *Chemosphere, 142*, 136–144.

Liu, X., Zhang, A., Ji, C., Joseph, S., Bian, R., Li, L., Pan, G., & Paz-Ferreiro, J. (2013). Biochar's effect on crop productivity and the dependence on experimental conditions—A meta-analysis of literature data. *Plant and Soil, 373*, 583–594.

Liu, Z., Dugan, B., Masiello, C. A., Barnes, R. T., Gallagher, M. E., & Gonnermann, H. (2016). Impacts of biochar concentration and particle size on hydraulic conductivity and DOC leaching of biochar–sand mixtures. *Journal of Hydrology, 533*, 461–472.

Mbarki, S., Cerdà, A., Zivcak, M., Brestic, M., Rabhi, M., Mezni, M., Jedidi, N., Abdelly, C., & Pascual, J. A. (2018). Alfalfa crops amended with MSW compost can compensate the effect of salty water irrigation depending on the soil texture. *Process Safety and Environmental Protection, 115*, 8–16.

Morrissey, E. M., Gillespie, J. L., Morina, J. C., & Franklin, R. B. (2014). Salinity affects microbial activity and soil organic matter content in tidal wetlands. *Global Change Biology, 20*(4), 1351–1362.

Munns, R. (2005). Genes and salt tolerance: Bringing them together. *New Phytologist, 167*(3), 645–663.

Munns, R., & Gilliham, M. (2015). Salinity tolerance of crops – What is the cost? *New Phytologist, 208*(3), 668–673.

Munns, R., & Tester, M. (2008). Mechanisms of salinity tolerance. *Annual Review of Plant Biology, 59*, 651–681.

Otitoloju, K. (2014). Growth, yield and seed nutritional composition of groundnut (Arachis hypogaea LINN) under elevated level of soil salinity. *Molecular Soil Biology, 5*(5). https://doi.org/10.5376/msb.2014.05.0005

Pandian, K., Vijayakumar, S., Mustaffa, M. R. A. F., Subramanian, P., & Chitraputhirapillai, S. (2024). Biochar–a sustainable soil conditioner for improving soil health, crop production and environment under changing climate: A review. *Frontiers in Soil Science, 4*, 1376159.

Panta, S., Flowers, T., Lane, P. A., Doyle, R., Haros, G., & Shabala, S. N. (2014). Halophyte agriculture: Success stories. *Environmental and Experimental Botany, 107*, 71–83. https://doi.org/10.1016/j.envexpbot.2014.05.006

Pavlovic, I., Mlinarić, S., Tarkowská, D., Oklestkova, J., Novák, O., Lepedus, H., et al. (2019). Early Brassica crops responses to salinity stress: A comparative analysis between chinese cabbage, white cabbage, and kale. *Frontiers in Plant Science, 10*. https://doi.org/10.3389/fpls.2019.00450

Plaut, Z., Meinzer, F. C., & Federman, E. (2000). Leaf development, transpiration and ion uptake and distribution in sugarcane cultivars grown under salinity. *Plant and Soil, 218*, 59–69. https://doi.org/10.1023/A:1014996229436

Qadir, M., Tubeileh, A., Akhtar, J., Larbi, A., Minhas, P. S., & Khan, M. A. (2008). Productivity enhancement of salt –affected environments through crop diversification. *Land Degradation & Development, 19*, 429–453.

Quesada-Kimsey, J. (2012). La carbonización de residuos biomásicos: una exploración con perspectivas emocionantes. *Tecnología en Marcha, 25*(5), 14–21. https://doi.org/10.18845/tm.v25i5.465

Rahman, G. M., Rahman, M. M., Alam, M. S., Kamal, M. Z., Mashuk, H. A., Datta, R., & Meena, R. S. (2020). Biochar and organic amendments for sustainable soil carbon and soil health. In *Carbon and nitrogen cycling in soil* (pp. 45–85). Springer Singapore.

Rao, V. P., Sengar, R. S., Singh, S., & Sharma, V. (2015). Molecular and metabolic perspectives of sugarcane under salinity stress pressure. *Progressive Agriculture, 15*, 77–84.

Rengasamy, P. (2006). World salinization with emphasis on Australia. *Journal of Experimental Botany, 57*, 1017–1023.

Rondon, M. A., Lehmann, J., Ramírez, J., & Hurtado, M. (2007). Biological nitrogen fixation by common beans (Phaseolus vulgaris L.) increases with biochar additions. *Biology and Fertility of Soils, 43*, 699–708.

Saifullah, D. S., Naeem, A., Iqbal, M., Farooq, M. A., Bibi, S., & Rengel, Z. (2018). Opportunities and challenges in the use of mineral nutrition for minimizing arsenic toxicity and accumulation in rice: A critical review. *Chemosphere, 194*, 171–188.

Shahariar, S., Farrell, R., Soolanayakanahally, R., & Bedard-Haughn, A. (2021). Elevated salinity and water table drawdown significantly affect greenhouse gas emissions in soils from contrasting land-use practices in the prairie pothole region. *Biogeochemistry, 155*(1), 127–146.

Shrivastava, P., & Kumar, R. (2015). Soil salinity: A serious environmental issue and plant growth promoting bacteria as one of the tools for its alleviation. *Saudi Journal of Biological Sciences, 22*(2), 123–131.

Silva, I. C. B., Fernandes, L. A., Colen, F., & Sampaio, R. A. (2017). Growth and production of common bean fertilized with biochar. *Ciência Rural, 47*(11), 1–8.

Singh, M., Kumar, J., & Singh, S. (2015). Roles of osmoprotectants in improving salinity and drought tolerance in plants. *Reviews in Environmental Science and Bio/Technology, 14*, 407–426.

Singh, J., Chander Sharma, P., & Singh, V. (2021). Breeding mustard (*Brassica juncea*) for salt tolerance: Problems and prospects. *Brassica Breeding and Biotechnology, 3*, 69–85. https://doi.org/10.5772/intechopen.94551

Sohi, S., Lopez-Capel, E., Krull, E., & Bol, R. (2009). *Biochar, climate change and soil: A review to guide future research*. CSIRO.

Steiner, C. (2016). Considerations in biochar characterization. In *Agricultural and environmental applications of biochar: Advances and barriers*. SSSA Special Publication. 63(1):87–99. https://doi.org/10.2136/sssaspecpub63.2014.0038.5

Steiner, C., Teixeira, W. G., Lehmann, J., Nehls, T., de Macêdo, J. L. V., Blum, W. E., & Zech, W. (2007). Long term effects of manure, charcoal and mineral fertilization on crop production and fertility on a highly weathered central Amazonian upland soil. *Plant and Soil, 291*, 275–290.

Sun, J., He, F., Shao, H., Zhang, Z., & Xu, G. (2016). Effects of biochar application on *Suaeda salsa* growth and saline soil properties. *Environmental Earth Sciences, 75*, 1–6.

Tanji, K. K. (1990). Nature and extent of agricultural salinity. In K. K. Tanji (Ed.), Manuals and reports on engineering practices no. 71. *Agricultural salinity assessment and management* (pp. 1–17). American Society of Civil Engineers.

Uzoma, K., Inoue, M., Andry, H., Fujimaki, H., Zahoor, A., & Nishihara, E. (2011). Effect of cow manure biochar on maize productivity under sandy soil condition. *Soil Use and Management, 27*, 205–212.

Vasantha, S., Venkataramana, S., Gururaja Rao, P. N., & Gomathi, R. (2010). Long term salinity effect on growth, photosynthesis and osmotic characteristics in sugarcane. *Sugar Tech, 12*, 5–8.

Verheijen, F., Jeffery, S., Bastos, A. C., Van der Velde, M., & Diafas, I. (2010). *Biochar application to soils – A critical scientific review of effects on soil properties, processes, and functions*. European Commission, Joint Research Centre, Institute for Environment and Sustainability.

Wang, J., Pan, X., Liu, Y., Zhang, X., & Xiong, Z. (2012). Effects of biochar amendment in two soils on greenhouse gas emissions and crop production. *Plant and Soil, 360*, 287–298.

Wang, Y., Long, Q., Li, Y., Kang, F., Fan, Z., Xiong, H., Zhao, H., Luo, Y., Guo, R., He, X., et al. (2022). Mitigating magnesium deficiency for sustainable citrus production: A case study in Southwest China. *Scientia Horticulturae, 295*, 110832.

Wong, V. N., Dalal, R. C., & Greene, R. S. (2008). Salinity and sodicity effects on respiration and microbial biomass of soil. *Biology and Fertility of Soils, 44*, 943–953.

Wu, L., Yue, W., Zheng, N., Guo, M., & Teng, Y. (2022). Assessing the impact of different salinities on the desorption of Cd, Cu and Zn in soils with combined pollution. *Science of the Total Environment, 836*, 155725.

Yadav, T., Kumar, A., Yadav, R. K., Yadav, G., Kumar, R., & Kushwaha, M. (2020). Salicylic acid and thiourea mitigate the salinity and drought stress on physiological traits governing yield in pearl millet- wheat. *Saudi Journal of Biological Sciences, 27*(8), 2010–2017. https://doi.org/10.1016/j.sjbs.2020.06.030

Yamato, M., Okimori, Y., Wibowo, I. F., Anshori, S., & Ogawa, M. (2006). Effects of the application of charred bark of Acacia mangium on the yield of maize, cowpea and peanut, and soil chemical properties in South Sumatra, Indonesia. *Soil Science & Plant Nutrition, 52*, 489–495.

Yan, N., Marschner, P., Cao, W., Zuo, C., & Qin, W. (2015). Influence of salinity and water content on soil microorganisms. *International Soil and Water Conservation Research, 3*(4), 316–323.

Yuan, J. H., & Xu, R. K. (2011). The amelioration effects of low temperature biochar generated from nine crop residues on an acidic Ultisol. *Soil Use and Management, 27*(1), 110–115.

Yuan, C. F., Feng, S. Y., Wang, J., Huo, Z. L., & Ji, Q. Y. (2018). Effects of irrigation water salinity on soil salt content distribution, soil physical properties and water use efficiency of maize for seed production in arid Northwest China. *International Journal of Agriculture & Biological Engineering, 11*(3), 137–145.

Zheng, W., Holm, N., & Spokas, K. A. (2016). Research and application of biochar in North America. In Z. Guo, M. S. He, & M. Uchimiya (Eds.), *Agricultural and environmental applications of biochar: Advances and barriers*. SSSA Special Publication. 63(1):475–494. https://doi.org/10.2136/sssaspecpub63.2014.0053

Zorb, C., Geilfus, C. M., & Dietz, K. J. (2018). Salinity and crop yield. *Plant Biology, 21*, 31–38. https://doi.org/10.1111/plb.12884

Shahid, S.A., Zaman, M. and Heng, L., 2018. Salinity and sodicity adaptation and mitigation options. In Guideline for salinity assessment, mitigation and adaptation using nuclear and related techniques (pp. 55-89). Cham: Springer International Publishing. https://doi.org/10.1007/978-3-319-96190-3_3

Basak, N., Rai, A.K., Barman, A., Mandal, S., Sundha, P., Bedwal, S., Kumar, S., Yadav, R.K. and Sharma, P.C., 2022. Salt affected soils: Global perspectives. In Soil health and environmental sustainability: Application of geospatial technology (pp. 107-129). Cham: Springer International Publishing. https://doi.org/10.1007/978-3-031-09270-1_6

Greene, R., Timms, W., Rengasamy, P., Arshad, M. and Cresswell, R., 2016. Soil and aquifer salinization: Toward an integrated approach for salinity management of groundwater. In Integrated groundwater management: Concepts, approaches and challenges (pp. 377-412). Cham: Springer International Publishing. DOI: 10.1007/978-3-319-23576-9_29

Raj AD (2020) Modelling climate change impact on surface runoff and sediment yield in a watershed of Shivalik region (M.Sc. Thesis, Academy of Climate Change Education and Research, Kerala Agricultural University, Vellanikkara).

Biochar as a Treasure Trove of Essential Nutrients: Combating Elemental Deficiency in Crops

Bipradeep Mondal, Jayanta Kumar Biswas, Nirmali Bordoloi, Debojyoti Moulick, and Jayjit Majumdar

Abstract

The burgeoning human population has placed immense strain on agricultural resources, leading to widespread elemental deficiencies in crops. This chapter explores the potential of biochar, a carbon-rich matter produced by biomass pyrolysis, as a solution to this escalating problem. Biochar is a treasure trove of essential nutrients, possessing a unique capacity to store and gradually release them into the soil, promoting plant growth and development. It can therefore be utilized as a multifaceted tool to improve soil health, reduce greenhouse gas (GHG) emissions, and rejuvenate plant nutrition and productivity organically. The chapter delves into the mechanisms by which biochar enhances nutrient availability in soil. Further, a relation has been established to show how biochar's porous structure helps it serve as a reservoir for essential elements such as nitrogen, phosphorus, potassium, calcium, magnesium, sulfur, iron, and zinc, while its charged surfaces facilitate the adsorption and slow release of these nutrients. This sustained release ensures a steady supply of nutrients for crops throughout

B. Mondal
Department of Ecological Studies, University of Kalyani, Kalyani, Nadia, West Bengal, India

Department of Environmental Sciences, Central University of Jharkhand, Ranchi, Jharkhand, India

J. K. Biswas · J. Majumdar (✉)
Department of Ecological Studies, University of Kalyani, Kalyani, Nadia, West Bengal, India

N. Bordoloi
Department of Environmental Sciences, Central University of Jharkhand, Ranchi, Jharkhand, India

D. Moulick
Department of Environmental Science, University of Kalyani, Kalyani, Nadia, West Bengal, India

S. Singh, S. Srivastava (eds.), *Biochar: A Precious Resource from Biological Waste*,
https://doi.org/10.1007/978-981-95-0425-1_10

their growth cycle, unlike traditional fertilizers that can be rapidly leached from the soil. The chapter analyzes the impact of biochar application on various aspects of plant health and productivity. It discusses how biochar can improve nutrient uptake efficiency, leading to stronger and more resilient crops. Additionally, the chapter also portrays the indirect effects of biochar in enhancing nutrient availability by fostering beneficial soil microbial communities that play a crucial role in nutrient cycling and decomposition. Nevertheless, the environmental benefits of using biochar for management of nutrients, GHG emissions, and carbon storage can contribute to climate change mitigation strategies. Additionally, biochar application can help reduce soil erosion and improve water retention capacity, leading to more sustainable agricultural practices.

Keywords

Biochar · Nutrient · Nutrient assimilation · Nutrient immobilization · Crop productivity · Climate change

1 Introduction

Biochar, a stable carbon-rich material produced via biomass pyrolysis, presents a potential solution for mitigating climate change and enhancing soil quality. The escalating concentration of atmospheric greenhouse gases necessitates innovative strategies to reduce climate change. Biochar, a stable carbon-rich product derived from biomass pyrolysis, has emerged as a promising carbon sequestration and soil improvement approach. Biochar is synthesized through the thermal decomposition of biomass under anaerobic conditions. The pyrolysis process, influenced by temperature, heating rate, and feedstock composition, yields biochar with varying properties. A wide range of feedstocks, including agricultural residues, forest byproducts, and energy crops, can be employed for biochar production. The physicochemical characteristics of biochar, such as surface area, porosity, pH, and elemental composition, significantly impact its suitability for different applications. These properties are influenced by the pyrolysis conditions and feedstock characteristics. Understanding the relationship between biochar properties and its performance is crucial for optimizing its utilization. Its application in agriculture offers a multifaceted approach to enhancing soil fertility, improving crop productivity, and mitigating climate change. By acting as a nutrient reservoir, improving soil structure, and fostering beneficial microbial activity, biochar can significantly contribute to agricultural systems' long-term health and sustainability. The application of biochar in soil nutrient management is underpinned by its unique physicochemical properties. Its porous structure provides an extensive surface area for nutrient adsorption, enabling it to retain essential elements such as nitrogen, phosphorus, and potassium. This retention capacity helps to prevent nutrient leaching, a critical issue in many agricultural systems. Moreover, biochar's pH can influence nutrient availability. For soils with low pH, biochar can increase pH, thereby enhancing the availability of phosphorus and micronutrients. Beyond nutrient retention, biochar plays a crucial role in improving soil structure. Its addition can increase soil aggregation, leading to enhanced water infiltration, aeration, and root penetration. These improvements in soil physical properties contribute to better nutrient uptake by plants. Furthermore, biochar can act as a slow-release fertilizer, gradually supplying nutrients to crops over time. This sustained nutrient release reduces the need for frequent fertilizer applications, thereby decreasing production costs and minimizing environmental impacts. The interaction between biochar and soil microorganisms is another key aspect of its role in nutrient management. Biochar can provide a habitat for beneficial microorganisms, which play a vital role in nutrient cycling. By stimulating microbial activity, biochar can enhance nutrient mineralization, making nutrients more accessible to plants. Additionally, some studies have shown that biochar can promote the growth of nitrogen-fixing bacteria, which can contribute to soil nitrogen enrichment. While the potential benefits of biochar in soil nutrient management are significant, it is essential to consider the factors influencing its effectiveness. The type of biomass used for biochar production, pyrolysis conditions, and soil properties can all impact biochar's performance. Additionally, the long-term effects of biochar application on soil fertility and crop productivity require further investigation. Biochar is a promising tool for enhancing soil nutrient management and promoting sustainable agriculture. Its ability to improve nutrient retention, soil structure, and microbial activity offers a multifaceted approach to addressing the challenges associated with conventional agricultural practices. However, further research is needed to optimize biochar production and application for specific soil and crop conditions. We can contribute to developing resilient and productive agricultural landscapes by integrating biochar into sustainable farming systems.

2 Manifestation of Deficiency of Essential Nutrients in Crops

Plants, like humans, require a balanced diet of essential nutrients for optimal growth and development. These nutrients are absorbed from the soil and utilized for various physiological processes. Plants require a total of 17 essential elements for optimal growth and development. These elements comprise 14 minerals, categorized as macronutrients such as nitrogen, phosphorus, potassium, magnesium, calcium, and sulfur (1–15 mg/g plant dry weight) and micronutrients such as chlorine, iron, boron, manganese, zinc, copper, nickel, and molybdenum (100–1000 times less than macronutrients), as well as the non-mineral elements carbon, hydrogen, and oxygen (Mengel et al., 2001; Nath & Tuteja, 2016). Their deficiency induces characteristic symptoms that serve as diagnostic indicators. Nutrient deficiencies can significantly reduce crop yield, quality, and plant vigor. While the specific symptoms vary depending on the nutrient, crop species, and severity of deficiency, some general indicators include chlorosis (yellowing of leaves), necrosis (tissue death), stunted growth, and reduced yield. Early detection is crucial for effective management. Macronutrients are required in relatively large quantities by plants. Nitrogen

(N) is essential for protein synthesis, chlorophyll formation, and overall plant growth. Its deficiency manifests as general chlorosis, often starting with older leaves, stunted growth, and reduced yield. Phosphorus (P) is crucial for energy transfer, root development, and flowering. Phosphorus-deficient plants exhibit dark green or purplish leaves, stunted growth, and poor root development. Potassium (K) is involved in protein synthesis, enzyme activation, and water regulation. Potassium deficiency leads to leaf margin burn, weak stems, and reduced disease resistance. Calcium (Ca) is essential for cell wall formation, and its deficiency causes blossom end rot in fruits and leaf margin necrosis. Magnesium (Mg) is a component of chlorophyll, and its deficiency results in interveinal chlorosis, particularly in older leaves. Sulfur (S) is involved in protein synthesis and chlorophyll formation, and its deficiency causes general chlorosis, like nitrogen but often affecting younger leaves. Micronutrients are required in trace amounts but are equally vital for plant health. Iron (Fe) is essential for chlorophyll formation, and its deficiency causes interveinal chlorosis, primarily in young leaves. Zinc (Zn) is involved in enzyme activation and plant growth, and its deficiency results in small, chlorotic leaves with shortened internodes. Manganese (Mn) is essential for photosynthesis and respiration, and its deficiency causes interveinal chlorosis with brown spots. Copper (Cu) is involved in enzyme activity and chlorophyll synthesis, and its deficiency leads to whitish or gray spots on leaves and dieback of shoots. Boron (B) is crucial for cell wall formation and pollen development, and its deficiency causes thickened, dark green leaves, blossom end rot, and heart rot in root crops. Molybdenum (Mo) is involved in nitrogen metabolism, and its deficiency results in chlorosis and leaf curling, like nitrogen deficiency.

Accurate diagnosis of nutrient deficiencies often requires soil and plant tissue analysis. Visual symptoms alone might be misleading. Consulting with a local agricultural extension service can provide valuable guidance. Prevention and management of nutrient deficiencies involve regular soil testing, balanced fertilization based on crop needs and soil analysis, crop rotation, proper irrigation, and the use of organic amendments to improve soil fertility. By understanding the signs of nutrient deficiency and implementing appropriate management practices, farmers can ensure optimal crop growth, yield, and quality.

2.1 Factors Responsible for Nutrient Deficiency in Crops

Higher plants, rooted in place, have evolved sophisticated mechanisms to acquire essential nutrients. They possess intricate sensing and signaling networks that detect fluctuations in soil nutrient availability. The root system, the primary interface with the soil, undergoes adaptive changes in architecture in response to differing nutrient concentrations. This architectural plasticity maximizes nutrient uptake by increasing the root's surface area and volume.

2.1.1 Nitrogen Acquisition, Assimilation, and Deficiency

Nitrogen (N) is a critical nutrient for plant growth and agricultural productivity, but its excessive use and mismanagement can lead to significant losses from the soil. Nitrogen constitutes a substantial component of plant biomass, contributing approximately 1.5–8% of total dry matter and a significant 16% of total plant protein. Despite the atmosphere's nitrogen-rich composition, most higher plants are unable to directly assimilate this elemental form. Instead, they rely on nitrogen-fixing bacteria and associated enzymes to convert atmospheric nitrogen into usable forms. Inorganic nitrogen compounds, primarily nitrate (NO_3^-) and ammonium (NH_4^+) are subsequently absorbed by plants through specialized transporter proteins (Frink et al., 1999; Nath & Tuteja, 2016). Nitrogen deficiency in plants occurs due to low organic matter and nitrogen content in the soil. Apart from that, lack of replenishment of N, on crop harvest and removal; leaching and runoff of nitrogen due to excessive rain and irrigation; the abundance of denitrifying microbiota in soil; and overgrazing and immobilization of nitrogen adds to the nitrogen limitation for the crops/plants (Eickhout et al., 2006; Udverdi et al., 2015). Nitrate, the predominant inorganic form of nitrogen, is highly soluble and readily moves with water through the soil profile. Excessive rainfall or irrigation can leach nitrate below the root zone, rendering it unavailable to plants and potentially contaminating groundwater. Leaching is particularly problematic in sandy soils with low organic matter content. Under anaerobic conditions, often prevalent in waterlogged soils or compacted layers, denitrifying bacteria convert nitrate into gaseous nitrogen (N_2) and nitrous oxide (N_2O). These gases escape into the atmosphere, representing a significant loss of nitrogen from the soil. Denitrification is exacerbated by high organic matter content, low-oxygen levels, and warm temperatures (Firestone, 1982; Zumft, 1997; Martínez-Espinosa et al., 2020; Yao et al., 2024). Ammonium-based fertilizers, when applied to the soil surface, can undergo volatilization, converting to ammonia gas (NH_3) and escaping into the atmosphere. This loss is influenced by soil pH, temperature, and moisture conditions. High pH, high temperatures, and low soil moisture favor ammonia volatilization (De Vries & Bardgett, 2012). During the low nitrogen period, the decomposition rate increases due to the mining of nitrogen by microbiota to utilize labile nitrogen content from recalcitrant organic matter; this is needed to be replenished by the application of fertilizers. Otherwise, the soil will go further low in nitrogen content (Craine et al., 2007).

Nitrogen leaching from soil is a significant environmental concern that can lead to groundwater and surface water pollution. Several factors contribute to nitrogen leaching, including nitrogen deposition, nitrogen fertilization, soil properties, climate conditions, and agricultural practices. Studies have shown that elevated nitrogen deposition can cause nitrogen leaching from the soil in the form of NO_3^-, leading to the loss of essential nutrients such as K^+, Mg^{2+}, and Ca^{2+} (Liu et al., 2019). The application of nitrogen fertilizers can increase nitrate leaching, with higher rates of fertilizer application resulting in more nitrogen losses (Tripolskaja & Verbylienė, 2014). Additionally, the excess nitrogen from fertilizers can easily move downward with water, primarily as NO_3^--N, representing a substantial portion of total dissolved nitrogen loss (Zhang et al., 2021a, b, c). Climate factors such as warming, soil moisture content, and precipitation can influence nitrogen fluxes in soil, affecting nitrogen content and leaching rates (Liu et al., 2021; Ju, 2023). Furthermore, agricultural practices such as irrigation and nitrogen fertilizer application rate can impact soil nitrogen leaching, with higher irrigation amounts leading to more leaching losses (Zhang et al., 2021a, b, c; Zhang et al., 2023). The type of nitrogen fertilizer used also plays a role, as organic fertilizers can result in higher leaching rates compared to chemical fertilizers (Hina, 2024). Moreover, soil properties such as texture can influence nitrogen leaching, with higher leaching observed in silt and peat soils compared to sandy soils (Yläranta et al., 1993). Studies have also shown that changes in soil organic carbon and nitrogen content are crucial in reducing the risk of nitrogen leaching, emphasizing the importance of organic material inputs and combined organic-inorganic fertilizer applications (Guo et al., 2014). Nitrous oxide (N_2O) emission from soil is a significant contributor to greenhouse gas concentrations in the atmosphere, impacting global climate change. Understanding the processes and controls of N_2O emissions from soil is crucial for developing effective mitigation strategies (Butterbach-Bahl et al., 2013). Research has shown that soil N_2O emissions are closely linked to pedo-climatic conditions, such as rainfall, environmental variables such as temperature and soil temperature, moisture, as well as management factors like tillage practices and crop residue management (Trozzo et al., 2020; Ju, 2023). Additionally, the type of crop grown and the specific soil management techniques employed can also influence N_2O emissions (Rashti et al., 2015). Moreover, the availability of nutrients in the soil, particularly nitrogen, plays a crucial role in regulating N_2O emissions. Excessive nitrogen fertilization can disrupt the balance between soil nitrogen supply and plant nitrogen utilization, leading to increased N_2O emissions (Zhang et al., 2021a, b, c). Strategies such as incorporating biochar into soil have been proposed to mitigate N_2O emissions by reducing the moisture sensitivity of soil N_2O production (Fidel et al., 2018). The Intergovernmental Panel on Climate Change (IPCC) has developed frameworks to quantify the impacts of nitrogen availability and environmental factors on N_2O emissions from agricultural soils (Shang et al., 2020). These frameworks utilize emission factors to estimate N_2O emissions at regional scales, highlighting the importance of considering the interactions between soil properties, management practices, and environmental conditions in emission estimations (Shang et al., 2020). Studies have also shown that the type of soil, the pH levels, and the presence of specific elements like metal ions can influence N_2O emissions from soils (Balashov et al., 2021; Lin et al., 2024). In agricultural settings, the type of crop and management practices can significantly impact N_2O emissions from soil. For example, intercropping systems like maize-legume rotations have been found to affect N_2O emissions due to differences in soil conditions and nutrient availability (Huang et al., 2014). Similarly, pasture management practices can influence greenhouse gas emissions, with faster nitrogen cycling leading to increased N_2O emissions (Cardoso et al., 2022). Moreover, the use of manure as a fertilizer can contribute to N_2O emissions from soils, emphasizing the need for effective control and mitigation strategies in agricultural practices (Júnior et al., 2021). Additionally, the presence of peatlands and groundwater levels can impact N_2O emissions, indicating the importance of considering local environmental conditions in emission assessments (Nardi et al., 2021). Furthermore, the contribution of the nongrowing season to annual N_2O emissions in regions like the continuous permafrost area underscores the need to account for seasonal variations in emission estimates (Gao et al., 2020b).

N_2O emission from soil is a complex phenomenon influenced by a multitude of factors including soil properties, management practices, environmental conditions, and crop types. Understanding the processes and controls of N_2O emissions is essential for developing effective mitigation strategies to reduce greenhouse gas emissions from agricultural and natural soils. Considering the interactions among soil conditions, management practices, and environmental variables, researchers can work towards sustainable soil management practices that minimize N_2O emissions and contribute to global efforts to combat climate change.

2.1.2 Phosphorus Acquisition, Assimilation, and Deficiency

Phosphorus is an essential nutrient for plant growth and development, playing a crucial role in various physiological processes within plants. However, the availability of phosphorus in soils is often limited, leading to phosphorus deficiency in plants. Orthophosphate is the primary available form of phosphorus that plants can assimilate, but in many natural and agricultural ecosystems, phosphate levels are suboptimal (López-Arredondo et al., 2014). When phosphate fertilizers are applied to soils, they are quickly immobilized due to fixation and microbial activity, further exacerbating the challenge of phosphorus acquisition by plants. Root architecture is a critical factor influencing phosphorus acquisition by plants. Studies have shown that root architectural plasticity, such as changes in root growth angle, can significantly impact the ability of plants to acquire immobile nutrients like phosphorus (Bonser et al., 1996). Additionally, the formation of aerenchyma in roots has been identified as an adaptation to low phosphorus availability, reducing the respiratory and phosphorus requirements for soil exploration by roots (Fan et al., 2003). This adaptation allows plants to optimize their phosphorus uptake efficiency in phosphorus-stressed environments. Furthermore, the symbiotic relationship between plants and phosphate-solubilizing microorganisms plays a vital role in phosphorus solubilization and uptake.

Soil phosphorus loss mechanisms are primarily influenced by erosion processes and runoff dynamics. Research indicates that rainfall-induced runoff significantly contributes to phosphorus loss, with both particulate phosphorus (PP) and soluble phosphorus (SP) being transported away from agricultural lands. The erosion of soil, particularly in sloped terrains, enhances the loss of phosphorus, as steeper slopes increase the erosive capacity of runoff, leading to greater nutrient depletion (Zhao et al., 2021; Li et al. 2023a, b). Additionally, studies show that historical soil erosion and redistribution can result in spatial variations of phosphorus stocks, with lower landscape positions accumulating more phosphorus due to sediment deposition (Vanden Bygaart et al., 2021). Furthermore, the use of soil amendments can mitigate phosphorus leaching by absorbing nutrients and reducing their mobility in leachate (Zhang et al., 2024). Overall, the interplay of erosion, runoff, and soil management practices plays a critical role in phosphorus loss from soils, highlighting the need for integrated management strategies to minimize nutrient loss (Chen et al., 2023; Zhang et al., 2024).

Excessive phosphorus input in agricultural systems can lead to environmental risks such as surface water deterioration due to phosphorus loss (Liu et al., 2015). Phosphorus loss from soil is a critical environmental issue that can lead to water pollution and eutrophication of water bodies (Redding et al., 2006; Liang et al., 2011). Various studies have highlighted different pathways and factors contributing to phosphorus loss from soil. Differences in phosphorus losses between soils can be attributed to variations in the age of the soils, with significant phosphorus loss occurring during the early stages of weathering (Ellis, 1969). Surface runoff and soil erosion are considered the main pathways for phosphorus loss from organic farming and forestry land. Phosphorus sorption capacity of soils and application of phosphatic materials can reduce P losses in runoff and leaching (Andersen et al., 2015; Cui et al., 2022; Xu et al., 2023). Phosphorus sorption by soil components like aluminum and iron plays a crucial role in phosphorus retention and loss dynamics, with leaching experiments confirming the kinetics of phosphorus sorption (Peng et al., 2020). This can be counteracted using sequential liming in clayey agricultural soil (Ulén & Etana, 2014).

Phosphate-solubilizing microorganisms have been found to enhance phosphorus availability in the soil by solubilizing complex phosphorus forms, making them more accessible to plants (Rawat et al., 2020). This mechanism is crucial for improving phosphorus uptake efficiency in plants, especially in phosphorus-deficient soils. In addition to root adaptations and microbial interactions, plants have developed various strategies to cope with phosphorus deficiency. These strategies include altering root conformation, exuding carboxylic acids, and forming symbiotic relationships with mycorrhizal fungi to enhance phosphorus acquisition (Yang et al., 2022). By modulating gene expression associated with different phosphorus acquisition strategies, plants can adapt to phosphorus-deficient soils and optimize their phosphorus uptake efficiency. The role of arbuscular mycorrhizal fungi (AMF) in enhancing plant nutrient uptake, including phosphorus, has been well-documented. AMF plays a crucial role in improving plant growth under diverse stress conditions by enhancing nutrient uptake, immobilization, and translocation of heavy metals (Chauhan et al., 2022). This symbiotic relationship between plants and AMF represents a promising avenue for boosting sustainable agriculture, particularly in phosphorus-deficient soils. The use of biofertilizers derived from bioelectrohydrogenesis residues has shown potential as a sustainable source of phosphorus for plants. These residues contain microorganisms that can provide plants with assimilable phosphorus, offering an eco-friendly alternative to traditional phosphorus fertilizers (Ndayisenga et al., 2022). By harnessing biofertilizers from innovative sources, farmers can improve phosphorus availability in soils while promoting sustainable agricultural practices. Seafatullah et al. (2015) found that biogas slurry and vermicompost can regulate phosphorus release in soil, potentially preventing phosphorus loss.

2.1.3 Potassium Acquisition, Assimilation, and Deficiency

Potassium plays a crucial role in the growth and development of plants, influencing various physiological processes essential for their overall health and productivity. The accumulation of potassium in plants has been shown to enhance the formation and translocation of carbohydrates, leading to improved water use efficiency and greater plant growth. Potassium transporters are essential for the acquisition, redistribution, and homeostasis of potassium in plants, highlighting the importance of potassium accumulation by roots and its distribution within plant cells. Potassium supply in the soil solution can stimulate the uptake and assimilation of other essential nutrients like nitrogen and phosphorus, further promoting plant growth (da Silva Lima et al., 2018). This highlights the interconnectedness of potassium with other vital elements in supporting plant functions. Furthermore, potassium application has been found to increase the rate of photosynthesis in plants, facilitating the translocation of photosynthates into different plant tissues (Sulistiani et al., 2020). This demonstrates the direct impact of potassium on enhancing the plant's ability to convert light energy into chemical energy, which is crucial for plant growth and development. Additionally, potassium deficiency can significantly impede plant growth, particularly affecting carbon assimilation processes and resulting in the accumulation of carbohydrates in source leaves (Chen et al., 2013). Moreover, potassium is closely linked to various aspects of plant physiology, including carbon assimilation, photosynthesis, protein synthesis, enzymatic activation, stomatal movement, and the translocation of nutrients from the soil to the plant (DeLima et al., 2020). These functions underscore the importance of potassium in supporting fundamental plant processes essential for growth, productivity, and overall plant health. Additionally, potassium has been identified as a vital regulator of plant responses and tolerance to abiotic stresses, highlighting its significance in enabling plants to cope with environmental challenges (Hasanuzzaman et al., 2018). Potassium not only influences plant growth but also enhances plant tolerance to various stressors such as frost, heat injuries, drought, diseases, and pest infestations (Shah et al., 2018). By maintaining anion balance in plants and supporting overall plant health, potassium improves plant resilience and ensures sustainable growth even under adverse conditions. Additionally, potassium has been shown to facilitate the transport of direct assimilates within plants, leading to higher assimilation rates and improved growth outcomes (Haeder et al., 1973). This further emphasizes the critical role of potassium in facilitating nutrient translocation and ensuring optimal plant physiological functions and plant vitality. Potassium transporters are essential for the acquisition, redistribution, and homeostasis of potassium in plants, highlighting the importance of potassium accumulation by roots and its distribution within plant cells (Gierth & Mäser, 2007). Factors such as organic compound exudation, root characteristics, and transpiration rates can influence potassium uptake in plants (White, 2013). Additionally, potassium regulates turgor pressure in plants, affecting processes like phloem translocation and ion distribution between roots and leaves (Lang, 1983). Potassium has been associated with improving enzyme activity involved in detoxification processes and enhancing membrane permeability and ATPase activity, particularly under low-temperature conditions (Jain et al., 2009).

These findings suggest that potassium plays a multifaceted role in plant metabolism, influencing biochemical pathways that are essential for plant survival and adaptation to varying environmental conditions. Furthermore, potassium has been identified as a key element required for plant growth and physiology, underscoring its significance in supporting fundamental plant functions (Pandey et al., 2018). This demonstrates how potassium not only influences aboveground growth but also belowground processes that are crucial for plant performance. In terms of nutrient management, potassium supplementation has been found to enhance the photosynthetic recovery of plants following rehydration, indicating its role in improving plant resilience to water stress (Catuchi et al., 2012). Moreover, potassium application has been shown to increase assimilate distribution to roots, particularly under water deficit conditions, highlighting its importance in optimizing resource allocation within plants (VanLaere et al., 2022).

Soil potassium loss is a significant concern that can have a substantial impact on agricultural productivity. Various factors contribute to potassium loss from soils, including leaching, erosion, and adsorption processes. Leaching is a major pathway through which potassium can be lost from soils, especially in regions with low potential charges in the soil and highly soluble potassium sources, leading to significant losses through leaching. Water movement in soils can result in potassium loss, reducing soil potassium content and ultimately affecting soil productivity and crop yields (Mendes et al., 2016; Liu et al., 2020; Zhao et al., 2021). Erosion poses a threat to soil resources, particularly in areas prone to water-induced soil erosion. The movement of soil due to erosion can result in the loss of potassium, among other nutrients, further exacerbating soil degradation. Potassium can be lost from ponds through seepage and soil adsorption processes, indicating that different environmental conditions can influence the mechanisms of potassium loss (Boyd et al., 2007; Belayneh et al., 2019). Adsorption processes, also play a role in potassium loss, where potassium is adsorbed by bottom soils in aquatic environments. This adsorption, whether exchangeable or non-exchangeable, contributes to the overall loss of potassium from the soil system. Understanding these various mechanisms of potassium loss is crucial for implementing effective soil management strategies to mitigate nutrient depletion and sustain agricultural productivity (Pine & Boyd, 2010). Interaction between soil properties and potassium dynamics is essential in determining the extent of potassium loss. Point out that soil texture and mineralogy influence potassium holding capacity, which, in turn, affects the susceptibility of soils to potassium losses (Rosolem & Steiner, 2017).

In agricultural settings, the application of potassium fertilizers is a common practice to replenish soil nutrients. According to Rosolem et al. (2018), potassium fertilization, particularly with certain compounds like KCl, can result in significant potassium losses through leaching, especially in tropical sandy soils. This highlights the importance of optimizing fertilizer application strategies to minimize nutrient losses and enhance nutrient use efficiency in agricultural systems. Environmental factors such as soiling in photovoltaic systems, can also influence nutrient dynamics in soils. Soiling on solar panels can lead to reduced energy generation efficiency, affecting the overall performance of photovoltaic systems. This demonstrates the interconnectedness of environmental factors and agricultural practices in influencing soil potassium levels and nutrient management strategies (Mejia & Kleissl, 2013). Soil potassium loss is a complex issue influenced by various processes such as leaching, erosion, and adsorption, as well as environmental factors and agricultural practices. Understanding the mechanisms underlying potassium loss and implementing sustainable soil management practices are crucial for maintaining soil fertility, enhancing crop productivity, and ensuring long-term agricultural sustainability.

2.1.4 Calcium Acquisition, Assimilation, and Deficiency

Calcium acquisition and assimilation in plants play crucial roles in various physiological processes. Calcium is essential for plant growth and development, influencing cell wall structure, enzyme activation, and signal transduction pathways (Paulsen & Harper, 1968). Studies have shown that increasing calcium concentrations can enhance nitrate assimilation and seedling growth, highlighting the importance of calcium in selective ion uptake by plants (Ward et al., 1986). Calcium has been implicated in nitrogen assimilation processes in wheat seedlings, further emphasizing its role in plant nutrition (Paulsen & Harper, 1968). Furthermore, the interaction between calcium and other elements like potassium is vital for plant health. For instance, the application of calcium oxide nanoparticles along with farmyard manure has been shown to induce cadmium tolerance in mung bean plants by influencing physiological and biochemical parameters, thereby reducing cadmium acquisition by roots and shoots (Mazhar et al., 2023). Ca plays a role in mitigating stress factors in plants. Exogenous Ca has been found to alleviate nitrogen and water deficits, as well as improve the growth of wheat seedlings exposed to high temperatures, indicating its involvement in osmoregulation and nitrogen assimilation under heat stress conditions (Wang et al., 2010). Additionally, calcium interacts with metabolites like oxalic acid to regulate acidity within plants, facilitating the normal conversion and transport of assimilated substances and preventing over-acidification and toxicity (Zhang, 2023). Calcium acquisition mechanisms in plants are vital for various physiological processes, and recent studies have elucidated the intricate role of calcium in plant biology. Mechanical stimulation in plant cells can immediately increase cytosolic calcium levels, indicating that calcium acts as a second messenger in cellular responses to external stimuli. This finding suggests that calcium plays a role in differentiating functions within the cytoplasm of plant cells (Knight et al., 1992). Additionally, Lapeyrie and Bruchet (1986) highlighted the potential use of active calcium efflux and insolubilization by mycorrhizal fungi to regulate excessive calcium absorption by calcifuge plants, indicating a complex interplay between fungi and plants in calcium acquisition processes. In suspension cell culture of tobacco, the involvement of endogenous hydrogen sulfide regulated by calcium in conferring thermotolerance was found, emphasizing the intricate relationship between calcium signaling and stress responses in plants (Li et al., 2015b). Rhizosheaths, known for their role in phosphorus acquisition, may also play a part in nitrogen acquisition in plants, showcasing the multifaceted strategies plants employ to acquire essential nutrients from the soil (Abrahão et al., 2018). The significance of calcium in plant growth and development is further underscored by Hepler (2005), who highlighted calcium as a central regulator of various physiological processes in plants. This regulatory role of calcium is supported by Trewavas and Knight (1994), who discussed calcium as a dynamic signaling molecule that transduces numerous signals in plant tissues, emphasizing its significance in plant form and function. Calcium signals are fundamental transducers and regulators in plant adaptation and developmental processes, further solidifying the pivotal role of calcium in plant biology (Kudla et al., 2010). Whalley and Knight (2012) elucidated how calcium signatures are decoded by plants to elicit specific gene responses, indicating the intricate mechanisms through which plants interpret calcium signals to regulate gene expression and physiological responses. This decoding mechanism of calcium signatures was also explored by Lenzoni et al. (2017), who delved into predicting plant immunity gene expression by understanding how calcium signals are decoded, highlighting the role of calcium in plant defense mechanisms.

Calcium loss from soil is a significant concern in agricultural and environmental contexts. Calcium losses from limed soils during fertilization can be 2–4 times higher than magnesium losses (Lapenis et al., 2008). This emphasizes the vulnerability of calcium in soil management practices. Additionally, it was found that for loamy soils, a substantial portion of annual calcium loss occurs between October and March, indicating a seasonal variation in calcium dynamics (Toky et al., 1989). Dale et al. (1990) further emphasized that despite variations in leaching due to different land uses, a consistent loss of calcium was observed across different land-use types in their study. Acid rain can lead to significant calcium loss in forest soils, especially in soils with poor buffer capacity like podzol soil (Wu et al., 2011). This underscores the role of external factors in influencing calcium dynamics in soil systems. Long-term effects of acid rain on forest ecosystems, showing substantial losses of calcium and magnesium from soil due to acid inputs (Likens et al., 1996). Calcium loss from the soil is not only influenced by external factors like acid deposition but also by agricultural practices, such as the cycling of nutrients in traditional agroforestry systems, and past land use can impact soil calcium levels (Adams, 1986; Samson, 2024). The growth and decay of plant species can affect soil calcium availability, emphasizing the intricate relationship between vegetation dynamics and soil nutrient cycling. Fertilizer application has been shown to influence an increase in soil nutrient loss, including calcium, due to changes in soil nutrient concentrations (Samson, 2024). Additionally, Adams (1986) explored the interaction between liming and nitrogen fertilizers, showing that liming can affect nitrogen recovery but not necessarily from calcium nitrate, indicating the complexity of nutrient interactions in soil systems. Soil erosion plays a role in the loss of essential nutrients like calcium from the soil (Breemen & Protz, 1988; Irawan & Indarto, 2011). Overall, the loss of calcium from soil is a multifaceted issue influenced by various factors such as agricultural practices, acid deposition, soil erosion, and land use changes. Understanding the dynamics of calcium in soil systems is crucial for sustainable soil management and agricultural productivity.

Calcium deficiency in plants can have significant effects on their growth and development, impacting various physiological processes. Plants that lack an adequate supply of calcium can influence their antioxidative defense system and rupturing of chloroplasts (Hall et al., 1972; Liu et al., 2009). Furthermore, calcium deficiency primarily affects the expanding pith cells below the apical shoot meristem, causing a collapse of the cell walls in these regions (Özgen et al., 2011). Calcium-deficient conditions can decrease flower production and reproductive losses (Harris & Brolmann, 1966). Additionally, symptoms of calcium deficiency often manifest as rot or necrosis at the extremities of the plant, such as leaf tips or fruits (Wang et al., 2020). In potato plants, calcium deficiency alters the translocation of carbohydrates, leading to their accumulation in leaves. This disruption in carbohydrate transport can have cascading effects on plant growth and development. Moreover, the deficiency of calcium can result in the reduced absorption of monovalent cations and an increase in the uptake of divalent cations (Singh & Sharma, 1972).

In the case of calcium-deficient plants, there is a decrease in tissue boron content, along with reduced calcium levels, sugar, and starch contents (Chatterjee et al., 1987). This interplay between calcium and other essential nutrients underscores the complex relationships within plant physiology and the cascading effects of deficiencies. Furthermore, the distribution of calcium in plant tissues plays a vital role in maintaining structural integrity and physiological functions. For example, the fractionation of calcium in tuber and leaf tissues can explain the symptoms of calcium deficiency in plants. Calcium deficiency can also impact the genetic and physiological characteristics of plants. Research has highlighted that plants exhibiting calcium deficiency symptoms tend to sequester calcium in vacuoles as calcium oxalate crystals (Gao et al., 2020a, b). This sequestration mechanism represents a plant's attempt to cope with the lack of available calcium, albeit with visible manifestations of deficiency. Moreover, the genetic characterization of calcium deficiency in maize has shed light on the underlying mechanisms that contribute to the manifestation of symptoms in plants (Wang et al., 2020). Calcium deficiency is more pronounced in meristematic regions due to calcium's role in cell wall structure and function. This specificity in symptom expression highlights the importance of calcium in supporting fundamental processes essential for plant growth (Santiago et al., 2018). Furthermore, the occurrence of blossom-end rot in pepper plants has been linked to factors such as prolonged high temperatures in the root zone and low root mass, emphasizing the multifactorial nature of calcium deficiency symptoms (Mándoki & Pénzes, 2012). Overall, calcium deficiency in plants results in a cascade of physiological, metabolic, and structural changes that impact their growth, development, and overall health. From alterations in cellular structures like chloroplasts to disruptions in metabolic processes such as carbohydrate translocation, the consequences of inadequate calcium levels are far-reaching. Understanding the intricate interplay between calcium and other nutrients, as well as the genetic and physiological responses to deficiency, is crucial for developing strategies to mitigate the detrimental effects of calcium deficiency in plants.

2.1.5 Magnesium Acquisition, Assimilation, and Deficiency

Magnesium is an essential element for plant growth and development, playing crucial roles in various physiological processes within plants. The acquisition and assimilation of magnesium in plants are intricately linked to their overall health and productivity. Magnesium's significance in plant biology extends to its role in photosynthesis. Magnesium is a constituent of chlorophyll, the pigment essential for capturing light energy during photosynthesis. Magnesium deficiency can negatively influence leaf expansion, starch, and sucrose accumulation, and net assimilation in *Phaseolus vulgaris*, emphasizing the importance of magnesium in maintaining optimal plant functions (Fischer & Bremer, 1993). Magnesium deficiency in sugar beet demonstrates how acclimation to low magnesium levels can differentially affect photosystems I and II, underscoring the impact of magnesium deficiency on plant productivity and quality (Hermans et al., 2004). This connection between magnesium and chlorophyll underscores the importance of magnesium in the photosynthetic process, which is fundamental for plant growth and biomass production (Rivelli et al., 2010). Additionally, magnesium is involved in the modulation of RuBP carboxylase in the stroma of chloroplasts, further emphasizing its role in photosynthetic reactions (Alaoui-Sossé et al., 2004). The study by Wang (2024) also highlighted the molecular mechanism of exogenous magnesium in regulating cation homeostasis in the roots of peanut seedlings under salt stress, showcasing the intricate interplay between magnesium and plant responses to environmental challenges.

Magnesium interacts with other nutrients and elements in the soil to influence plant growth. Egamberdieva et al. (2018) discussed the interactive effects of nutrients and *Bradyrhizobium japonicum* on soybean growth and root architecture, indicating the complex relationships between nutrient availability and plant development. Mycorrhizal fungi have been shown to enhance plant acquisition of not only nitrogen but also phosphorus, potassium, magnesium, and micronutrients, emphasizing the role of symbiotic relationships in improving plant nutrient uptake (Ruotsalainen & Kytöviita, 2004). Additionally, endophytic fungi contribute to the acquisition of essential soil nutrients, including macronutrients like phosphorus, nitrogen, potassium, and magnesium, as well as micronutrients such as zinc, iron, and copper, further highlighting the diverse mechanisms through which plants obtain magnesium and other vital elements (Diaz et al., 2021).

Magnesium depletion in soil is a critical issue affecting agricultural productivity and plant health. Intensive farming practices over the past decades have caused a significant decrease in soil mineral content, including up to a 30% reduction in magnesium levels (Pickering et al., 2020). The excessive use of chemical fertilizers, pollution, and climate change further depletes magnesium in the soil, impacting the food chain (Maier, 2024). One significant factor contributing to this depletion is the low levels of exchangeable magnesium (ex-Mg) in acidic soils (Zheng et al., 2020). Using chemical fertilizers and pesticides can exacerbate magnesium deficiency in soil by altering soil qualities and reducing essential mineral content (Uwitonze & Razzaque, 2018). Long-term application of fertilizers like ammonium nitrate can lead to the loss of the most mobile part of the soil's magnesium fund (Boiko, 2024). Soil erosion and heavy metal accumulation can also contribute to magnesium deficits in soil, affecting plant magnesium levels (Zhang et al., 2020a, b). Acid rain can induce soil acidification, disrupting metabolic balance and enzymatic activities and leading to the depletion of essential minerals like magnesium (Prakash et al., 2022). As soil substrates age and weatherable minerals decline, essential nutrients, including magnesium, are increasingly depleted through long-term leaching processes (Bauters et al., 2022). Competition for nutrients in the soil, particularly between magnesium and other cations, can also result in gradual magnesium depletion, leading to plant magnesium deficiency (Li et al., 2023a, b). Leaching of soil nutrients, such as calcium and magnesium, through nitrate transport in the soil profile can further deplete soil fertility and contribute to magnesium loss (Kelly, 2023). Overall, factors such as soil acidity, intensive farming practices, chemical inputs, and nutrient competition contribute to magnesium depletion in soil, impacting plant health and agricultural productivity.

The impact of magnesium deficiency on plant physiology and metabolism is a subject of extensive research. Studies have shown that magnesium deficiency can lead to reduced rates of leaf expansion, altered starch and sucrose accumulation, and decreased net assimilation in plants, highlighting the pivotal role of magnesium in sustaining plant growth and metabolic processes (Fischer & Bremer, 1993). Furthermore, magnesium deficiency can affect the regulation of photosynthesis and plant growth, underscoring the intricate relationship between magnesium availability and plant performance (Cao & Tibbitts, 1992). It appears that the co-application of effective microorganisms and nanomagnesium can enhance plant responses to salt stress, emphasizing the potential of magnesium supplementation in improving plant tolerance to environmental challenges (El-Mageed et al., 2022). Magnesium acquisition and assimilation in plants are critical processes that influence various aspects of plant growth, development, and metabolism. From its role in photosynthesis to its interactions with other nutrients and its impact on plant responses to environmental stressors, magnesium is a key element for plant health and productivity. Understanding the mechanisms underlying magnesium uptake and plant utilization is essential for optimizing agricultural practices and enhancing crop yields in diverse environmental conditions.

2.1.6 Sulfur Acquisition, Assimilation, and Deficiency

Sulfur (S) is an essential macronutrient that is pivotal in plant growth, metabolism, and stress tolerance. Its acquisition and assimilation are critical for the synthesis of vital biomolecules, including amino acids, proteins, and secondary metabolites. This article aims to provide a comprehensive overview of sulfur acquisition and assimilation in plants, highlighting the molecular mechanisms involved, the regulatory pathways, and the implications for plant health and productivity. The process of sulfur acquisition begins with the uptake of sulfate (SO_4^{2-}) from the soil through specialized transporters located in the root membranes. The SULTR (sulfate transporter) family of proteins is primarily responsible for this uptake, with different members exhibiting distinct expression patterns and substrate affinities. For instance, the SULTR4;1 transporter in *Arabidopsis thaliana* has been shown to play a significant role in sulfate uptake and compartmentalization within plant tissues, influencing overall sulfur metabolism and homeostasis (Zuber et al., 2010). Mutations in this transporter can lead to altered sulfate content and metabolic adaptations, underscoring the importance of efficient sulfur transport in plants (Zuber et al., 2010). Once sulfate is absorbed, it is transported to various tissues, where it undergoes assimilation into organic forms. The assimilation pathway involves several enzymatic steps, with the reduction of sulfate to sulfide being a critical regulatory point. The enzyme adenylyl sulfate reductase (APR) catalyzes the reduction of 5′-adenylylsulfate (APS) to sulfite, which is subsequently reduced to sulfide by sulfite reductase (Kettles et al., 2014). This pathway is tightly regulated by the availability of sulfur and the demand for sulfur-containing compounds, such as cysteine and glutathione, which serve as precursors for various metabolic processes (Kettles et al., 2014). Cysteine, a key amino acid in sulfur metabolism, is synthesized from sulfide through the action of O-acetyl serine (thiol) lyase (OAS-TL). The regulation of this enzyme is crucial for maintaining sulfur balance within the plant, particularly under conditions of sulfur deficiency. Studies have shown that mutations in OAS-TL can lead to altered sulfur metabolism and physiological responses, including early leaf death phenotypes in *Arabidopsis* (Shirzadian-Khorramabad et al., 2010). This highlights the intricate relationship between sulfur availability and plant development. In addition to its role in amino acid synthesis, sulfur is also involved in the production of secondary metabolites, such as glucosinolates, which are important for plant defense against herbivores and pathogens. The accumulation of these compounds is influenced by sulfur availability and can be modulated by environmental factors, including stress conditions (Koroleva et al., 2010). For example, the presence of sulfur-rich apoptotic cells in *Arabidopsis* has been linked to the regulation of glucosinolate biosynthesis, indicating a complex interplay between sulfur metabolism and plant defense mechanisms (Koroleva et al., 2010). The interaction between sulfur and other nutrients, particularly nitrogen, is another critical aspect of sulfur metabolism in plants. Sulfur deficiency can lead to an accumulation of nitrate in plant tissues, suggesting a direct impact on nitrogen metabolism (Alegria Terrazas et al., 2022). This relationship is particularly important in the context of nutrient management in agricultural systems, where the balance of sulfur and nitrogen can significantly affect crop yield and quality (Zenda et al., 2021). Recent advances in systems biology have provided new insights into the regulatory networks governing sulfur metabolism in plants. High-throughput omics approaches have revealed complex interactions between sulfur assimilation pathways and various signaling molecules, including hydrogen sulfide (H_2S), which acts as a signaling molecule in response to stress (Chen et al., 2015). The role of H_2S in modulating gene expression related to sulfur metabolism further emphasizes the dynamic nature of sulfur homeostasis in plants (Mondal et al., 2021). Moreover, the rhizosphere microbiota plays a crucial role in sulfur acquisition and transformation. Soil microorganisms can influence the availability of sulfur compounds and enhance plant uptake through various mechanisms, including the production of sulfur-containing metabolites (Zenda et al., 2021). Understanding these interactions is essential for developing sustainable agricultural practices that optimize sulfur use efficiency and improve crop resilience to environmental stressors.

The loss of sulfur from soil has emerged as a critical issue in agricultural practices worldwide, particularly in the context of increasing crop yields and the intensification of farming systems. The availability of sulfur in soils has been declining due to several factors, including changes in agricultural practices, environmental regulations, and the use of high-analysis fertilizers that are often deficient in sulfur. One of the primary reasons for the loss of sulfur from soils is the shift in agricultural practices towards high-yielding crop varieties that require greater nutrient inputs. The adoption of these varieties has led to an increased removal of sulfur from the soil, as crops take up more nutrients to support their growth and development. Studies have shown that the depletion of sulfur reserves in soils is exacerbated by the limited application of sulfur fertilizers, which has become a common practice in many regions (Jena et al., 2021; Raghunath et al., 2021; Singh et al., 2022; Ghosh et al., 2022). This trend is particularly evident in areas where intensive cropping systems are employed, leading to a significant reduction in the sulfur content of the soil (Yazhini, 2023). Moreover, the reduction in atmospheric sulfur deposition due to stricter environmental regulations aimed at controlling sulfur dioxide emissions has further contributed to the decline in soil sulfur levels. The decrease in industrial emissions has resulted in lower atmospheric inputs of sulfur, which historically contributed to soil fertility (Ghosh et al., 2022; Lornac et al., 2020; Yazhini, 2023). Consequently, the natural replenishment of sulfur in soils has been compromised, leading to widespread deficiencies that can adversely affect crop yields and quality (Erdem et al., 2016; Lee & Kim, 2012). The dynamics of sulfur in soil are complex and influenced by various factors, including soil type, pH, and organic matter content. For instance, the mineralization of organic sulfur compounds into plant-available forms is a slow process that depends on the microbial activity in the soil, which is influenced by environmental conditions (Singh et al., 2018; Staugaitis et al., 2014). The prevailing form of sulfur in soils is organic sulfur, and its conversion to inorganic forms that plants can utilize is critical for maintaining adequate sulfur levels (Staugaitis et al., 2014; Siwik-Ziomek et al., 2016). However, the decline in organic matter inputs, such as farmyard manure, has further exacerbated the sulfur deficiency problem (Siwik-Ziomek et al., 2016).

Research has indicated that sulfur deficiency can lead to a range of physiological disorders in plants, including reduced chlorophyll synthesis, impaired protein metabolism, and decreased crop resilience to environmental stresses (Godlewska, 2018; Lee & Kim, 2012). This deficiency is particularly concerning for crops such as oilseeds and pulses, which have shown significant yield reductions in sulfur-deficient soils (Zenda et al., 2021; Nepali & Bhandari, 2019). The implications of sulfur deficiency extend beyond individual crops, affecting overall soil health and ecosystem functioning. To address the issue of sulfur loss from soils, it is essential to implement effective management strategies that include the judicious use of sulfur fertilizers. The application of sulfur in various forms, such as elemental sulfur or sulfate fertilizers, can help replenish soil sulfur levels and improve crop yields (Erdem et al., 2016; Yadav, 2023). Additionally, integrating sulfur management into broader nutrient management practices can enhance the efficiency of fertilizer use and reduce the risk of nutrient imbalances in soils (Lee & Kim, 2012; Wang et al., 2014).

2.1.7 Iron acquisition, Assimilation, and Deficiency

Iron acquisition and assimilation in plants is a critical area of research due to the essential role of iron in various physiological processes, including photosynthesis, respiration, and overall plant growth. Iron is a micronutrient that is often limited in availability in soils, necessitating the development of sophisticated mechanisms by plants to acquire and utilize this vital element. This article aims to synthesize current knowledge on the mechanisms of iron acquisition and assimilation in plants, particularly focusing on the model organism *Arabidopsis thaliana*, while also considering the broader implications for agricultural practices and plant health. Iron uptake in plants primarily occurs through the roots, where it is absorbed from the soil solution. The process is highly regulated and involves several key genes and proteins. For instance, the genes FRO2, IRT1, and IRT2 have been identified as crucial for iron uptake in *Arabidopsis*, facilitating the reduction of Fe(III) to Fe(II) and its subsequent transport into root cells (Zhou, 2024). The expression of these genes is modulated by the plant's iron status, with increased expression observed under iron-deficient conditions, highlighting the adaptive responses of plants to fluctuating iron availability (Park et al., 2019). In addition to direct uptake, plants can also utilize microbial interactions to enhance iron acquisition. The presence of beneficial microorganisms, such as arbuscular mycorrhizal fungi (AMF), can significantly improve iron uptake efficiency. These fungi extend the root system's reach and enhance nutrient absorption, including iron, from the soil (Rajapitamahuni et al., 2023). Furthermore, certain plant growth-promoting rhizobacteria can enhance the efficiency of AMF symbiosis, leading to improved iron homeostasis (Lurthy et al., 2021). This symbiotic relationship underscores the importance of the rhizosphere microbiota in facilitating iron acquisition. The regulation of iron homeostasis is a complex process that involves various signaling pathways and transcription factors. For example, the bHLH transcription factors bHLH100 and bHLH101 have been shown to play significant roles in controlling iron homeostasis through a FIT-independent pathway, indicating that multiple regulatory mechanisms are at play (Sivitz et al., 2012). Additionally, the interaction between iron and other nutrients, such as phosphorus and zinc, further complicates the regulatory landscape, as these elements can influence each other's availability and uptake (Xie et al., 2019).

Iron deficiency in plants leads to a series of physiological responses aimed at mitigating the effects of limited iron availability, such as chlorosis (Schenkeveld et al., 2014). For instance, plants often decrease the expression of nonessential iron-containing proteins while maintaining the expression of critical proteins necessary for survival (Bakirbas & Walker, 2022). This selective expression is part of a broader adaptive mechanism that allows plants to prioritize essential functions under nutrient stress. Moreover, the role of phytohormones in mediating the response to iron deficiency has been highlighted, with salicylic acid and other hormones influencing both iron uptake and defense mechanisms against pathogens (Montejano-Ramírez & Valencia-Cantero, 2023). The interplay between iron acquisition and plant defense mechanisms is another critical aspect of iron homeostasis. Research has shown that plants can employ an iron-withholding defense strategy to limit pathogen access to iron, thereby enhancing their resistance to diseases (Hsiao et al., 2017). This strategy reflects a sophisticated balance between nutrient acquisition and defense, emphasizing the importance of iron in both growth and immunity. Recent studies have also explored the role of noncoding RNAs in regulating iron deficiency responses. For instance, the long noncoding RNA CAN OF SPINACH has been implicated in modulating the plant's response to iron deficiency, suggesting that posttranscriptional regulation plays a significant role in iron homeostasis (Bakirbas & Walker, 2022). This finding opens new avenues for understanding the molecular mechanisms underlying iron acquisition and assimilation.

The transport of iron within the plant is equally important as its uptake. Once absorbed, iron must be translocated to various tissues where it is utilized or stored. The vacuolar iron transporter proteins have been identified as key regulators of cellular iron accumulation, ensuring that iron levels remain within a safe range to prevent toxicity (Ram et al., 2021). This transport system is crucial for maintaining iron homeostasis, particularly in the context of varying environmental conditions. In addition to the intrinsic mechanisms of iron acquisition and transport, external factors such as soil composition and microbial interactions significantly influence iron availability. The competition for iron between plants and soil microorganisms, particularly fungi, can impact plant health and growth (López-Berges et al., 2013). Understanding these interactions is vital for developing strategies to enhance iron biofortification in crops, which is essential for addressing iron deficiency in human diets.

The implications of iron loss from soil are profound, particularly in agricultural contexts. The availability of iron in the soil can be significantly affected by its chemical form and the soil's redox potential. Under anaerobic conditions, such as those found in waterlogged soils, ferric iron (Fe^{3+}) can be reduced to ferrous iron (Fe^{2+}), which is more soluble and can leach from the soil profile, leading to iron loss (Nie, 2024). This process is particularly pronounced in flooded paddy fields, where the reduction of iron is facilitated by iron-reducing bacteria, which thrive in low-oxygen environments (Otoidobiga et al., 2015). The activation and loss of soil iron are also influenced by soil acidity. Acidic conditions promote the reductive dissolution of iron oxides, enhancing the mobility of iron in the soil (Nie, 2024). In regions with high rainfall and leaching, such as the red soil region of southern China, the combination of acidic pH and waterlogged conditions can lead to significant iron loss from the soil (Nie, 2024). This phenomenon is exacerbated by the presence of organic matter, which can chelate iron and facilitate its movement through the soil profile (Jiang et al., 2011). Moreover, the competition between iron and other metals for uptake by plants can further complicate the dynamics of iron in soil. For instance, in calcareous soils, high pH levels can lead to reduced iron availability, as iron becomes less soluble and is more likely to precipitate as ferric oxides (Kumar et al., 2020). This can result in iron deficiency in crops, particularly in environments where other metals, such as copper, compete for the same uptake pathways (Schenkeveld et al., 2014). The presence of calcium carbonate and bicarbonates can also impede iron absorption by plants, leading to further depletion of iron in the soil (Mahdii & Turki, 2020). In rice cultivation, for example, excessive iron availability due to waterlogging can lead to iron toxicity, which manifests as leaf bronzing and reduced photosynthetic efficiency (Rajonandraina et al., 2023). This highlights the delicate balance that must be maintained in soil management practices to ensure optimal iron availability for plant growth. In addition to agricultural impacts, iron loss from soil can have broader environmental consequences. The leaching of iron into waterways can contribute to the eutrophication of aquatic systems, as iron can interact with phosphorus and other nutrients, promoting algal blooms (Smith et al., 2023). Furthermore, the reduction of iron oxides in soil can influence the mobility of trace elements, such as arsenic and mercury, which can pose risks to human health and the environment (Hellal et al., 2017).

Iron loss from soil is a critical environmental issue that has significant implications for agricultural productivity, soil health, and ecosystem sustainability. To mitigate iron loss from soil, several management strategies can be employed. These include the application of iron fertilizers, which can help replenish iron levels in the soil, particularly in calcareous and acidic soils (Elham et al., 2014). Additionally, practices that enhance soil drainage and aeration can reduce the incidence of waterlogging and the associated reduction of iron (Otoidobiga et al., 2015). The incorporation of organic matter into the soil can also improve iron retention and availability, as organic matter can bind to iron and prevent its leaching (Jiang et al., 2011). Iron acquisition and assimilation in plants is a multifaceted process involving intricate regulatory mechanisms, adaptive responses to nutrient stress, and interactions with soil microorganisms. As research continues to uncover the complexities of iron homeostasis, it is imperative to consider the implications for agricultural practices and food security. Enhancing our understanding of these processes will be essential for developing sustainable strategies to improve iron nutrition in crops and, consequently, in human populations.

2.1.8 Zinc Acquisition, Assimilation, and Deficiency

Zinc is an essential micronutrient for plants, playing crucial roles in various physiological processes such as photosynthesis, respiration, and enzyme activation (Li et al., 2015a, b). The acquisition and assimilation of zinc in plants are tightly regulated processes that involve a complex network of mechanisms to ensure adequate zinc supply to all plant cells (Khan et al., 2021). One of the key players in zinc acquisition in plants is the Zinc/Iron-regulated Protein (ZIP) family, which includes transporters responsible for zinc uptake and transport within the plant (Jouvin et al., 2012; Basavarajeshwari et al., 2018). Additionally, phytosiderophores (PS) are known to be involved in zinc acquisition, aiding in the mobilization of unavailable zinc and increasing its assimilation by plants (Jouvin et al., 2012; Yadav et al., 2022). Plants have developed sophisticated strategies to enhance zinc uptake and assimilation. For instance, the amino acids present in controlled release matrices can complex zinc, making it more available for plant assimilation (Jiménez-Rosado et al., 2023). Zinc-solubilizing rhizobacteria have been shown to improve zinc biofertilization and biofortification in crops, leading to increased plant growth and yield (Yadav et al., 2022). In the context of zinc homeostasis, plants have evolved mechanisms to maintain intracellular zinc concentrations and coordinate processes like zinc sequestration, uptake from the soil, and transport within the plant (Khan et al., 2021). The expression profiling of zinc transporter genes in plants grown under different zinc concentrations has revealed the involvement of various ZIP family transporters in zinc uptake and transport mechanisms (Basavarajeshwari et al., 2018). Moreover, overexpression of specific zinc transporter genes like ZmIRT1 and ZmZIP3 has been shown to enhance iron and zinc accumulation in transgenic plants, emphasizing the role of these transporters in metal nutrient uptake (Li et al., 2015a, b). Low molecular weight organic acids such as malate and citrate in zinc tolerance mechanisms in plants have been studied (Harmens et al., 1994). While the exact mechanism of how these organic acids contribute to increased zinc tolerance is not fully understood, their presence in zinc-tolerant plants suggests a potential role in zinc uptake and detoxification processes.

Zinc loss from soil can occur through various mechanisms influenced by factors such as soil properties, agricultural practices, and plant interactions. Zinc deficiency in agricultural soils has been linked to limitations in crop production, with potential yield losses of up to 40% (Tessema et al., 2019). Studies have shown that the mobility of zinc in soil can be affected by the presence of organic matter, with higher organic matter content reducing the mobility and toxicity of zinc for plants. The presence of manure in fertilizers has been reported to increase the mobility and leaching of zinc in soil solutions (Rutkowska & Szulc, 2014; Kuziemska et al., 2022). Additionally, zinc deficiency is commonly encountered in plants growing in calcareous and sodic soils due to the formation of insoluble zinc carbonate, which reduces the availability of zinc for plant uptake (Sirohi et al., 2015). Furthermore, the rhizosphere activity plays a crucial role in the release of zinc in soil, indicating that plant roots can influence the mobilization of zinc in soil-plant systems (Houben et al., 2014). Factors like soil water content, pH, and root growth have also been identified that can impact zinc mobility, with slightly acidic soils exhibiting high mobility of zinc (Nambiar, 1976; Mihai et al., 2017). Losses of zinc from agroecosystems can occur through processes such as washing away of arable soils and alienation by the marketable parts of crops (Lukin & Zhulikov, 2022). Moreover, the depth-wise distribution and mobility of zinc in soil profiles can affect its phytoaccumulation by plants, highlighting the potential for phytoremediation of zinc-contaminated soils (Mani et al., 2012). In the microbial world, zinc acquisition is also a critical factor for pathogens like *Neisseria gonorrhoeae* and *Candida albicans*, which scavenge host zinc to support their growth and virulence (Citiulo et al., 2012). These pathogens employ high-affinity zinc uptake systems like ZnuABC to ensure sufficient zinc supply for their survival and proliferation within the host environment. Understanding the mechanisms by which these pathogens acquire zinc can provide insights into developing novel strategies to interfere with their metal acquisition systems and combat infections. In environmental bacteria such as *Francisella* species, distinct zinc acquisition mechanisms have been identified, highlighting the diversity of strategies employed by different bacterial species to acquire zinc based on their environmental niches. The presence of unique zinc uptake systems like ZupT in bacteria adapted to high-metal environments underscores the adaptability of microbial species to varying zinc concentrations in their habitats (Moreau et al., 2018). Zinc loss from soil is a complex process influenced by various factors such as soil properties, agricultural practices, and plant interactions. Understanding the mechanisms of zinc loss is essential for sustainable soil management practices and ensuring adequate zinc availability for plant growth and productivity.

2.2 Symptoms Representing Nutrient Stress

Nutrient stress in plants can manifest in a variety of ways, often visible through distinct symptoms. These symptoms can vary depending on the specific nutrient that is deficient and the stage of plant growth. Common symptoms of nutrient deficiency are described as follows.

(i) Chlorosis: yellowing of leaves due to a lack of chlorophyll. This is often associated with deficiencies in nitrogen, magnesium, or sulfur.
(ii) Necrosis: tissue death, typically characterized by brown or black spots on leaves or stems. This can be indicative of deficiencies in potassium, calcium, or phosphorus.
(iii) Stunted growth: reduced plant size and vigor, often resulting from deficiencies in nitrogen, phosphorus, or potassium.
(iv) Purpling: a purplish discoloration of leaves, particularly the undersides, which can be a sign of phosphorus deficiency.
(v) Interveinal chlorosis: yellowing between the veins of leaves, commonly associated with magnesium or iron deficiency.
(vi) Leaf curling: abnormal twisting or curling of leaves, often indicative of potassium or boron deficiency.
(vii) Fruit abnormalities: poor fruit development, small size, or premature ripening, which can be caused by deficiencies in calcium, boron, or zinc.
(viii) Root rot: decay or death of roots, often due to deficiencies in phosphorus or potassium, which can lead to poor water and nutrient uptake.

It is important to note that these symptoms can overlap and may not be specific to a single nutrient deficiency. Therefore, a combination of visual observations and soil testing is often necessary to accurately diagnose nutrient stress and determine the appropriate corrective measures. Some of common symptoms of nutrient deficiency in crops is summarized in Table 1.

3 Production of Biochar

In the modern usage of organic amendments, biochar happens to be a porous, carbonaceous solid substance that is derived through thermal decomposition of biomass (Narzari et al., 2015; Saravanan & Kumar, 2022; Abhishek et al., 2022; Amalina et al., 2022). Biochar possesses a variety of characteristics like high

Table 1 Some common nutrient deficiencies in plants and their symptoms

Nutrient	Deficiency symptoms	References
Nitrogen (N)	Stunted growth, yellowing of leaves (chlorosis), especially older leaves, reduced yield, delayed flowering	Mengel et al. (2001), Huang et al. (2020), Tian et al. (2020), Padilla et al. (2020), Mu and Chen (2021), Lu et al. (2021), Grzebisz et al. (2023), and Buchelt et al. (2020)
Phosphorus (P)	The purplish-red coloration of leaves, especially stems and lower leaves, stunted growth, root development issues, delayed flowering	Hammond et al. (2004), Goodman (2004), Ha and Tran (2014), Dixon et al. (2020), Reddy et al. (2020), de Bang et al. (2021), Chan et al. (2021), Fathi and Mehdiniya Afra (2023), and Khan et al. (2023)
Potassium (K)	Weak stems, yellowing of leaf margins, especially older leaves, wilting, reduced fruit quality	Gierth and Mäser (2007), White (2013), Zakari et al. (2020), Campos et al. (2021), Kusaka et al. (2021), Wakeel and Ishfaq (2022), Aleksandrov (2022), Huang et al. (2022), Attia et al. (2022), and Buchelt et al. (2020)
Calcium (Ca)	Blossom end rot in tomatoes, tip burn in lettuce, stunted growth, root tips die	Chatterjee et al. (1987), Gao et al. (2020a, b), Wang et al. (2020), Santiago et al. (2018), Mándoki and Pénzes (2012), Gao et al. (2020a, b), Wang et al. (2020), de Bang et al. (2021), Aras et al. (2021), and Duan et al. (2022)
Magnesium (Mg)	Interveinal chlorosis, yellowing between leaf veins, especially older leaves	Fischer and Bremer (1993), Cao and Tibbitts (1992), El-Mageed et al. (2022), Sun et al. (2017), Zhang et al. (2020a, b), He et al. (2020), Chaudhry et al. (2021), and Ishfaq et al. (2022)
Sulfur (S)	General chlorosis, stunted growth, similar symptoms to nitrogen deficiency	Godlewska (2018), Lee and Kim (2012), Zenda et al. (2021), Nepali and Bhandari (2019), Haneklaus et al. (2008), Santos et al. (2020), Li et al. (2020), Narayan et al. (2023), and Buchelt et al. (2020)
Iron (Fe)	Interveinal chlorosis, yellowing between leaf veins, especially young leaves, stunted growth	Schenkeveld et al. (2014), Bakirbas and Walker (2022), Montejano-Ramírez and Valencia-Cantero (2023), Tran and Camps (2021), Perkowska et al. (2021), Li et al. (2021), Merry et al. (2022), Mahawar et al. (2023), and Viets et al. (1954)
Zinc (Zn)	Rosette symptoms (small, clustered leaves), stunted growth, delayed flowering	Li et al. (2015a, b), Khan et al. (2021, 2022), Younas et al. (2023), Vadlamudi et al. (2020), and Viets et al. (1954)

surface area, high cation exchange capacity, high porosity high carbon content, etc. (Abhishek et al., 2022; Amalina et al., 2022). The enrichment of such physical–biological–chemical properties in biochar highly depends on the choice of feedstock before thermal decomposition (Yaashikaa et al., 2020; Amalina et al., 2022). Generally, agricultural wastes about lignocellulose, hemicellulose substances, algae biomass, energy crops, crop residues, etc. are found to be appropriate for the synthesis of biochar (Yakout, 2017; Fakayode et al., 2020; Yaashikaa et al., 2020). On the other hand, feedstocks from sewage wastes, animal wastes, fur, skeletal, and other non-lignocellulosic materials create significant difficulties in producing

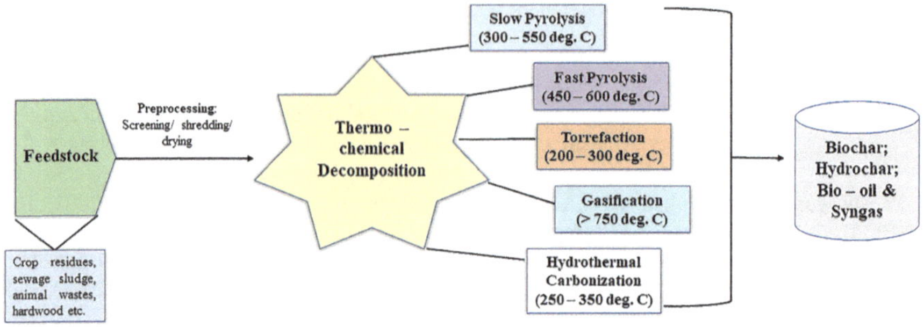

Fig. 1 Production of biochar by different types of thermochemical decompositions

high-grade biochar of such rich qualities (Hmid et al., 2014; Panahi et al., 2020). Non-lignocellulosic materials, such as biochar produced from heavy metal-containing sludge, may contain, along with the presence of heavy metals, elements like phosphorus, sulfur, etc. This can be harmful for biological systems as compared to biochar produced from lignocellulosic substances (Yaashikaa et al., 2020). The basic technology involved in the synthesis of biochar of different feedstock is thermochemical decomposition (Fig. 1) (Hmid et al., 2014; Yaashikaa et al. 2022; Amalina et al., 2022; Abhishek et al., 2022). However, drivers like decomposition temperature, heating rate, and residence time determine the rate and total yield of biochar from a particular feedstock. Conversion of biomass into different biofuels is not new, since biological decomposition like anaerobic and aerobic digestion with the help of bacteria or enzymes is prevalent for a long period in nature (Hossain, 2019). In contrast to biological decomposition techniques, thermal/thermochemical decompositions are fast in producing fuels from biomass (Panahi et al., 2020). Generally, thermal decomposition techniques like pyrolysis, torrefaction, hydrothermal carbonization, and gasification are used in producing biochar (Wang et al., 2018; Tang et al., 2019; Sakhiya et al., 2020; Yaashikaa et al., 2020; Amalina et al., 2022).

3.1 Pyrolysis

Pyrolysis is the most used thermal decomposition technique that converts biomass into biofuels especially biochar under an optimized condition of zero oxygen and high temperature (Amonette & Joseph, 2012; Brown et al., 2015; Leng & Huang, 2018; Senthil & Lee, 2021; Amalina et al., 2022). The other biofuels associated with this technique are the bio-oil (liquid) and syngas. Pyrolysis can be categorized into four methods based on the applied temperature and heating rate. The first method is known as the slow pyrolysis, where the temperature of thermal decomposition remains between 300 and 550 °C with an optimized heating rate of 0.1–0.8 °C/s and long residence time (5–30 min/25–35 h) (Yaashikaa et al., 2020). Followed by such a slow process, biochar when needed in huge amounts in less period, methods like fast pyrolysis, and flash carbonization are adopted. Fast pyrolysis is executed with a high temperature (450–600 °C) and a high heating rate (10–1000 °C/s) with a low residence time (0.5–2 s) (Santoso et al., 2020; Yaashikaa et al., 2020; Amalina et al., 2022). Like the previous, flash carbonization is exhibited at an extreme heating rate of >1000 °C/s and very little residence time (<0.5 s) (Brown et al., 2015). Unlike these two fast methods, a relatively modernized and innovative method that most of industries adopt is microwave pyrolysis, where the temperature is easily optimized with a quick heating rate, rapid on/off control, and selective heating. Conclusively, it is found that biochar produced slowly at a temperature between 250 and 700 °C with low heating rate and long residence period is better in quality with optimized content of carbon and other nutrients, high surface area, CEC, and porosity that suit as a better amendment for soil and plant growth (Venkatesh et al., 2022).

3.2 Torrefaction

The word "torrefaction" was derived from the French word "torréfaction" or "to torrefy" which has the literal meaning of "roasting" due to the practice of coffee bean roasting in France (Nordin et al., 2013). This thermal decomposition technique is executed at a low temperature between 200 and 300 °C with a heating rate of less than 50 °C/min (Chen et al., 2021; Amalina et al., 2022). Like slow pyrolysis, the torrefaction is performed for a long period where the biomass is charred at least for 30 min–1 h (Yaashikaa et al., 2020). Generally, this thermal decomposition technique can be categorized into four methods consisting of (a) oxidative torrefaction (assisted by oxidizing agents), (b) dry torrefaction, (c) wet torrefaction (carbonization of biomass is done under the contact of water), and (d) steam torrefaction (biomass treated with steam) (Acharya et al., 2015). A wide range of studies has found that dry torrefaction when accomplished with stages like heating, pre-heating (at 100 °C), drying (up to 200 °C), torrefaction (200–300 °C) for 30 min–1 h, and cooling produces not the best but a decent quality of biochar that has wide applications in environment management (Yaashikaa et al., 2020).

3.3 Gasification

Gasification is a combination of thermal and chemical decomposition where the carbonaceous biomass/matter is decomposed at a high temperature (>750 °C) with the assistance of gasification agents (oxygen, air, nitrogen, etc.) (Amalina et al., 2022). Charred matter along with syngas like H_2, CO, CO_2, and C_2H_2 is produced as final products (You et al., 2017). Biomass customized to a pellet size of 0.2–6 cm is exposed under complete aerobic conditions as to avoid devolatilization (Simone et al., 2012). Therefore, large amounts of greenhouse gases (GHGs) are released during gasification as by-products (You et al., 2017). Eventually, the charred product obtained after the gasification of biomass is much of a lower quality when compared to biochar obtained by pyrolysis.

3.4 Hydrothermal Carbonization

One of the most cost-effective thermochemical decomposition processes that can convert wet biomass into charred products is known as hydrothermal carbonization. Water being a major component of hydrothermal carbonization serves as a dual-purpose agent, viz., solvent and reactant (Sharma et al., 2020). Biomass based on a dynamic temperature profile of hydrothermal carbonization can be converted to charred products, bio-oil, and gases as desirable. Generally, "hydrochar" is a solid product that is obtained at a temperature of less than 250 °C and pressure between 4 and 6 MPa (Wang et al., 2016). With the gradual increase in temperature ranging between 250 and 350 °C, hydrothermal liquefaction takes place generating products like bio-oils (Amalina et al., 2022). Moreover, the temperature reaching over 374 °C and pressure close to 22.1 MPa, syngas, viz., CO_2, C_2H_2, CO, CH_4, etc., are produced in abundance (Wang et al., 2016; Yaashikaa et al., 2020). Similar to pyrolysis, the hydrochar produced by hydrothermal carbonization undergoes a series of processes, viz., condensation, dehydration, and polymerization, to generate a solid charred product similar to biochar.

4 Characteristics of Biochar (Physical, Chemical, and Biological)

Biochar is a carbon-rich material that is highly known for its porous structure, high water retention, high surface area, high CEC, and good nutrient profile (Sohi et al., 2009; Narzari et al., 2015; Abhishek et al., 2022). These properties of biochar vary at different levels, as these are inherited from the different feedstocks, governed by the adopted techniques, applied temperature, heating rate, residence time, and pressure at the time of thermochemical decomposition. Biomass provided for the synthesis of biochar generally consists of cellulose, hemicellulose, and lignin. Biomass when undergoing slow pyrolysis experiences de-polymerization; thus, disintegration of cellulose happens (Fakayode et al., 2020; Yaashikaa et al., 2020). Contrarily, during fast pyrolysis the high heating rate (10–1000 °C/s) helps in the development of levoglucosan (Yaashikaa et al., 2020). Following condensation, the levoglucosan forms hydroxymethyl furfural which is attributed to the property of aromaticity in biochar (Fakayode et al., 2020; Yaashikaa et al., 2020). Similarly, biomass containing hemicellulose gets depolymerized to form oligosaccharides followed by decarboxylation and aromatization to form biochar (Fakayode et al., 2020; Amalina et al., 2022). Unlike cellulose and hemicellulose, the decomposition of lignin takes place as a breakage of β-O-4 lignin to generate free radicals which in turn helps in generating a chain reaction between molecules to synthesize biochar (Amalina et al., 2022). The different stages of depolymerization, condensation, and re-polymerization help in producing biochar with different physical and chemical properties based on the feedstock provided. When examined under the higher resolution of scanning electron microscope, biochar was found to be coarse and porous in structure (Chia et al., 2015). Similarly, a group of researchers found that biochar experiences crackling and shrinkage as an effect of intensive thermal pressure (Brown et al., 2006; Chia et al., 2015). Unlike other C-based materials like coal, charcoal, coke, etc., the biochar under XRD examination was found to be amorphous rather than crystalline (Yakout, 2017). This amorphous nature of biochar is believed to be inherited due to cooking at a high temperature (>300 °C) and long periods (35 min–2 h). Simultaneously, under Brunauer-Emmett-Teller (BET) analysis, the surface area of biochar increased as the treatment temperature increased until it reached 800 °C, after which the surface area decreased with the increase in temperature. The presence of high surface area in biochar was believed to be found because of pores of different lengths, especially due to the presence of micropores (<2 nm) which had the greatest contribution in constructing high surface area in biochar (Chia et al., 2015). Adjacently, the studies have found that important chemical properties of biochar, viz., pH, and CEC that help in soil growth lie between 8.2–13.0 and 1.1–44.2 cmol/kg, respectively (Jha et al., 2010; Venkatesh et al., 2022). Noticeably, this wide range of pH and CEC was witnessed because of temperature change in pyrolysis. Biochar produced at high temperatures (>800 °C) reflected high pH, electrical conductivity, and CEC as compared to biochar produced at low temperatures (200–500 °C) (Venkatesh et al., 2022). According to the European Biochar Certificate, a good grade biochar for carbon sequestration must have an H/C ratio of less than 0.7 and O/C ratio of less than 0.5 correspondingly (Venkatesh et al., 2022). The H/C ratio can be defined as the degree of aromaticity, whereas the O/C ratio determines the stability of biochar in soil (Wiedemeier et al., 2015; Zhang et al. 2021a, b, c). Therefore, biochar with decent carbon content, high water holding capacity, CEC, better H/C and O/C profile, high surface area and porosity, etc. is completely dependent on the nature of feedstock, the treatment temperature, heating rate, and residence time during pyrolysis.

5 Utilization of Biochar in Reconciling Crop Nutrients

The growth and development of plants in any soil surface require balanced food, water, oxygen, carbon, and essential nutrients. Essential nutrients for plant tally over 13 where nitrogen, phosphorous, and potassium are considered as the primary nutrients while calcium, magnesium, sulfur, zinc, iron, chlorine copper, molybdenum, boron, magnesium, etc. are the secondary nutrients (Uchida, 2000; Abbas et al., 2021). Biochar is a nutrient-rich material with high content of carbon, a decent amount of phosphorous, sulfur, magnesium, wide C/N ratio, and low H/C, O/C ratio. These entities of biochar get enriched due to pyrolysis temperature, heating rate, residence time, and, most importantly, the nature of the feedstock that is pyrolyzed. The nutrient content of biochar is directly correlated to the amount of nutrients in feedstock. Therefore, biochar can be a great source of nutrients for crops to replenish their growth and productivity (Fig. 2).

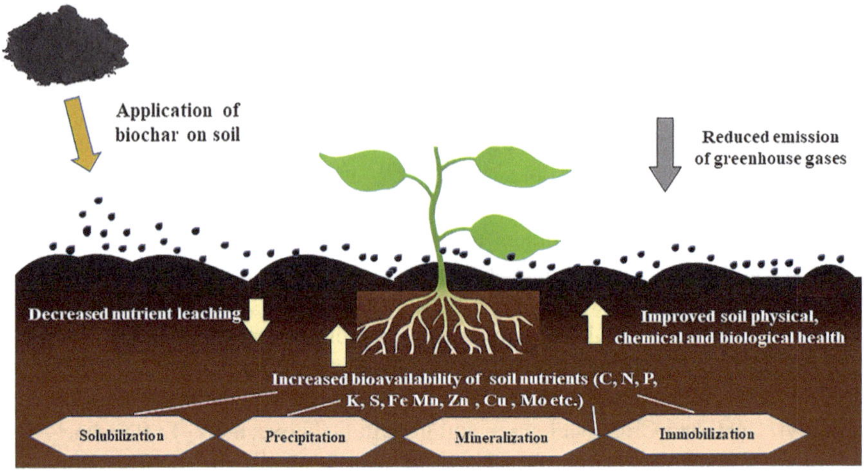

Fig. 2 Benefits of biochar application improving soil heath and better nutrient management

5.1 Essential Nutrients

5.1.1 Nitrogen

Nitrogen is one of the important nutrients for plant growth and productivity. In the modern era of conventional agriculture, the insertion of nitrogen as a nutrient in the soil is done by additives like synthetic fertilizer which on repetition reduces the nitrogen use efficiency in plants and increases the N_2O emission from soil. Biochar is a complete source of N; it contains both organic forms of N (hydrolyzable-N, water soluble-N, and non-hydrolyzable-N) and inorganic forms of N (NH_4^+-N, NO_3^--N, and N_2O-N) (Hossain et al., 2020). However, all forms of N are not present in biochar such as amino acids like arginine due to volatilization of some part of N during pyrolysis at different temperatures and heating rates (Hossain et al., 2020). A laboratory study of biochar produced from similar feedstock at different temperatures found that the suitable temperature for retention of all inorganic forms of nitrogen in biochar was around 600 °C (Zhu et al., 2016; Hossain et al., 2020). On the other hand, among the different feedstocks like chicken manure, crop residues, wood, and chlorella microalgae provided for biochar production, the microalgal-based biochar retained the highest N content (10.23–14.12%) when given a long residence time (20–60 min) at 500 °C (Chang et al., 2015; Hossain et al., 2020). With a wide range of pH, high surface area, and high C/N ratio, the biochar when applied on soil increases the soil pore structure and pH and provides aeration to increase the sorption of inorganic N forms. Such processes later help in mineralization–immobilization of N to increase the bioavailability of nitrogen as an essential nutrient to plants.

5.1.2 Phosphorus

Unlike nitrogen, the phosphorus content in biochar increased as the pyrolysis temperature increased (Hossain et al., 2020). Depending upon feedstocks, pyrolysis temperature, and residence time, phosphorus content in biochar can vary between 0.00% and 5.9% (DeLuca et al., 2015; Xiao et al., 2018; Hossain et al., 2020). Moreover, the type of feedstock provided during biochar production determines the P content in biochar where in the study it was found that in swine solid biochar the P content was the highest (>5%) and in rice husk biochar was the lowest (<0.15%) (Xiao et al., 2018). Simultaneously, for pyrolysis temperature increasing from 200 to 550 °C, the average P content in biochar increased by 1% overall (Hossain et al., 2020). Plants generally uptake phosphorus in the form of inorganic P such as HPO_4^{-2} and $H_2PO_4^-$ (DeLuca et al., 2015). Biochar itself being a P source cannot directly infuse P nutrient in plants due to less soluble P forms. Biochar when applied in soil alters the pH, enzyme efficiencies, and microbial community structure to form organomineral complexes for better P solubility (DeLuca et al., 2015). These cumulative actions assisted by biochar help in the retention of P in soil thus providing better bioavailability and solubility of nutrient P for plants.

5.1.3 Potassium

As with most of the other nutrients in biochar, the potassium (K) content varies with the type of feedstocks and pyrolysis temperature. For example, biochar derived from crop residues like wheat straw rice straw and bamboo biomass contains more K than rice husks, corn stalks, apple branches, and sewage sludge (Subedi et al., 2016; Xiao et al., 2018; Hossain et al., 2020; Karim et al., 2022). With a gradual increase of pyrolysis temperature from 250 to 550 °C, the K content in different biochars increased by an average percentage of 1.5 (Hossain et al., 2020). Several studies have shown revealed that, application of biochar has increased K concentration in plant tissue (Biederman & Harpole, 2013; Karim et al., 2022). Another aspect of K increase in soil is due to biochar's high surface area, large porous structure, and residence of negative charge on its surface area (Karim et al., 2022). These entities of biochar exert nutrient retention in soil, due to the increase of CEC which shows affinity toward K^+.

5.2 Secondary Nutrients and Trace Elements

The secondary nutrients and trace elements that provide essential nourishment to plants include elements like S, Ca, Mg, Fe, Cu, B, Zn, Mn, Mo, etc. (Hossain et al., 2020). Among all these secondary nutrients, sulfur (S) is one of the few important nutrients that help the plant in protein synthesis and transformation of energy in all living forms (DeLuca et al., 2015). Acting as the external source of energy, sulfur helps in oxidative decomposition under anaerobic conditions (DeLuca et al., 2015; Karim et al., 2022). Although sulfur is present in abundance from the Earth's geological bed, it needs mineralization like nitrogen for its assimilation by plants. Biochar-altering soil pH, aeration, and surface area in the soil provide a suitable substratum for S-solubilizing bacteria (Hossain et al., 2020; Karim et al., 2022). On the other hand, the content of S, Ca, and Mg in biochar from different feedstocks produced at a temperature between 250 and 750 °C varies between 0.001–0.63%, 0.40–6.15%, and 0.001–3.78%, respectively (Hossain et al., 2020). Feedstocks like wood, grass, rice husk, rice straw, and animal wastes were tested for estimating such nutrient content, where the biochar derived from animal wastes gave the highest amount of S, Ca, and Mg (Brantley et al., 2016; Xiao et al., 2018; Hossain et al., 2020). In the case of micronutrients or trace elements (Fe, Cu, B, Zn, Mn, and Mo, etc.), biochar helps in nutrient retention rather than adding such nutrients in soil and plants (Hossain et al., 2020; Mousavi et al., 2023). The content of such trace elements in biochar is less or highly negligible depending upon the feedstocks used. Moreover, a study regarding the estimation of trace elements in biochar is still under process where elements like Mn, Mo, and Cu still need proper evidence to be present in biochar. Fe content in biochar lies between 0.009 and 7480 mg/kg, with higher content in feedstock like animal manure (Chen et al., 2018; Hossain et al., 2020). Similarly, for Zn and Cu, the range lies between 131–4981 and 99–2446 mg/kg, respectively (Brantley et al., 2016; Hossain et al., 2020).

6 Efficacy of Biochar in Motility of Plant-Soil Nutrient Reaction

Since the emergence of "The Anthropocene," sustainable agriculture has been the determinant of food security for humans (Biswas et al., 2021). Deterioration of environmental health including rapid soil erosion, frequent floods and droughts, acidification of soil, emission of GHGs, and less productivity of crops has made sustainable agriculture a meticulous job to attain. As modern literatures suggest, practices involving precision farming using satellites, incorporation of fungi that coexist with different plants to benefit soil and plant growth, and introduction of soil bacteria degrading xenobiotics and promoting plant nutrition along with organic farming are quite a standout (Chia et al., 2015; Mousavi et al., 2023). Among these interventions, biochar has been identified as a multifaceted tool that is beneficial for soil health and plant growth. Biochar produced by slow pyrolysis at a temperature between (200 and 500 °C) has high surface area, porosity, CEC, and high carbon content (Amalina et al., 2022; Mousavi et al., 2023). Thus, when applied on soil–plant ecosystem, biochar exhibits dilution effect on soil to reduce bulk density while increasing surface area, porosity, and water retention capacity. On the other hand, biochar with high carbon content, high C/N ratio, good H/C and O/C ratio, etc. attributes to the properties like better stability, pertinent aromatics, and recalcitrance (Alkharabsheh et al., 2021). These properties of biochar get amplified when applied on soil, thus helping in retention of nutrients with zero nutrient leaching, nitrogen immobilization, and sequestration of carbon (Fig. 2). Thus, biochar helps in the motility of soil–plant nutrient reaction to promote better soil health, plant growth, and sustainable agriculture (Fig. 2). Some of the important contributions of biochar in soil–plant nutrient management are thus articulated in the following subcategories.

6.1 Nutrient Leaching

The phenomenon of nutrient leaching is highly dependent on the interaction between biochar and soil biota. Although biochar may be highly nutrient richby itself, when added to soil, it does not guarantee nutrient retention. Biochar highly influences the soil physical properties such as the bulk density, porosity, surface area, and water holding capacity. Biochar also alters the chemical attributes of soil, viz., pH, CEC, and anion exchange capacity (AEC) (Laird & Rogovska, 2015; Alkharabsheh et al., 2021; Mousavi et al., 2023). Simultaneously, biochar has a huge impact in improving the biological processes in soil as it promotes better substrata for integrated root–mycorrhizal system under perennial vegetation (Van Der Heijden et al., 2008; Krishnakumar et al., 2014; Mousavi et al., 2023). The complex interactions between biochar and soil's physical–chemical–biological attributes when run synergistically promote nutrient retention and recycling by minimizing nutrient leaching (Krishnakumar et al., 2014; Alkharabsheh et al., 2021). A study done by Laird et al. (2010) found that when hardwood biochar applied to a Midwestern agricultural soil, the total P-leaching drastically dropped from 29% to 5% in a duration of 45 week (Laird & Rogovska, 2015). Later in the same study, the NO_3^- N leaching was reduced from 60% to 40% with the assistance of swine manure (Laird et al., 2010; Laird & Rogovska, 2015). In a field study done by Rubin et al. (2020), it was found that biochar application with a rate of 10%(w/w) and the aggregate nutrient leaching of P and NH_4^+ were astonishingly reduced by a percentage of 37.7 and 50.2, respectively (Alkharabsheh et al., 2021). The affinity to nutrient retention in soil due to the application of biochar is completely dependent on the CEC and its interrelationship with organic functional groups. During oxidation, it is believed that the cation exchange capacity is much more expressive because of the formation of carboxylic functional group (Mousavi et al., 2023). Simultaneously, over a broad pH, the carboxylate anion gets delocalized by resonance to form carboxylic functional group. Such surface complexation helps the H_xPO_4 to undergo ligand formation with functional groups like carboxyl and hydroxide, thus retaining P nutrient in soil (Laird & Rogovska, 2015). On the other hand, the NO_3^-/NH_4^- gets retained by electrostatic forces to retain N nutrient (Laird & Rogovska, 2015; Alkharabsheh et al., 2021).

6.2 Nutrient Availability and Immobilization

The fate of plant productivity and soil health nourishment is highly dependent on the relationship between the availability of nutrients and a healthy microbial community. Application of biochar in soil has been witnessed to influence the nutrient profile of soil while providing a better substrate for microbial community survival (Hussain et al., 2017). Unweathered biochar is a source of nutrient salts, which is why when applied to soil nourishes the short-term source of nutrients for plants (Hossain et al., 2020). This mechanism on repeated terms can amplify nutrient cycling over a long period, thus increasing the turnover and pool size of labile organic nutrients (DeLuca et al., 2015). As mentioned earlier, temperature applied during slow pyrolysis designated a biochar with certain nutrient content. Biomass experiencing a temperature between 100 and 250 °C experiences volatilization of carbon (C) and nitrogen (N) followed by sulfur (S), potassium (K), and potassium (P) above 375, 700, and 800 °C, respectively (Fakayode et al., 2020; Yaashikaa et al., 2020; Amalina et al., 2022). Moreover, over 1000 °C elements like magnesium (Mg), calcium (Ca), and manganese (Mn) get volatilized. The application of good grade biochar with decent amount of C, N, P, Ca, Mg, K, and, S thus ameliorated CEC and surface oxidation of soil which in turn increase the bioavailability of plant nutrients (Alkharabsheh et al., 2021; Mousavi et al., 2023). A study done by Alkharabsheh et al. (2021) found that biochar application in calcareous soil at 1–2%(w/w) increased the content of N, P, K, Zn, Mn, etc. by 41%, 165%, 160%, 42%, and 21%, respectively. Usually, inorganic P precipitates with Ca^{2+} and aluminum (Al^{3+}) which makes the P insoluble in soil (Laird & Rogovska, 2015). Eventually, biochar when applied to soil transforms itself as a P source for plants, enhancing its solubility through alteration of soil pH and promotion of P-solubilizing bacteria. On the other hand, the higher C/N ratio of biochar influences the mineralization–immobilization of nitrogen in soil with better N bioavailability for plants (Krishnakumar et al., 2014; Purakayastha et al., 2019). Since biochar alters the soil pH, it helps in root colonization of arbuscular mycorrhizal fungi (AMF) in leguminous crops (Krishnakumar et al., 2014). The AMF enhances the biological nitrogen fixation and serves as a protector from soil predators. Thus, the enhanced symbiotic nature of AMF and associated plants, courtesy to biochar, helps in alteration of soil nutrient profile through better nutrient sorption, immobilization, transformation, and bioavailability.

6.3 GHG Emission

As articulated in this chapter, it is quite evident that biochar has the potential to enhance soil physicochemical and biological properties while retaining nutrients like C, N, P, K, Ca, etc. Biochar, therefore, is a low-cost tool that is highly efficient in sequestering carbon and reducing emission of GHGs majorly carbon dioxide (CO_2), nitrous oxide (N_2O), and methane (CH_4) especially from croplands and paddy fields (Purakayastha, 2012; Purakayastha et al., 2019). As Ibrahim et al. (2017) stated, the application of biochar in paddy fields derived from peanut shells and sewage sludge significantly decreased the emission of CH_4 and CO_2 (Mousavi et al., 2023). Although these studies have contradictories, it is seen that biochar application in upland soils increases CO_2 emission intensity by 9% ($P < 0.05$) (Lyu et al., 2022). Biochar produced at low temperatures alters the soil pH which provides a suitable condition for microbial communities; as a result, the production of organic matter gets increased so does the CO_2 emission intensity. On the other hand, biochar produced at high temperature (>500 °C) when applied on upland soil restricts such conditions due to having aromatic hydrocarbons which in return decreases CO_2 emission (4%, $P < 0.05$) to some extent (Lyu et al., 2022). In the scenario in the case of paddy fields and wetlands, biochar has prominent efficacy in the reduction of CO_2 as well as CH_4 emissions. Biochar when applied to paddy fields and wetland soil exhibits enzyme activities such as dehydrogenases protease, chitinase, cellulase, amylase, and xylanase which in return metabolize carbon and nitrogen, thus helping in the reduction of CO_2 emissions (Lyu et al., 2022). Simultaneously, biochar enlarges the pore structure of soil especially the clogged wetlands and paddy fields; as a result, it inhibits the anaerobic condition, which is not suitable for CH_4 emission (Purakayastha, 2012). Eventually, it was found that the application of biochar suppresses CH_4 emission from wetlands and paddy fields by a percentage of 7% ($P < 0.05$) (Guo et al., 2015; Pascual et al., 2020; Lyu et al., 2022). As a part of circular economy, crop residues like rice straw and wheat straw are often used for biochar production. This serves a dual purpose by restricting the emission of GHGs as a source by not burning crop residues under the open sky and transforming these highly nutritious materials into organic amendments. A study by Dong et al. (2013) showed that application of rice straw biochar on paddy field significantly reduced CH_4 emission (47.3–86.4%) while improving rice production by a maximum of 13.5% (Lyu et al., 2022). The emission of N_2O from any cropland desires a condition like alkalinity (Purakayastha, 2012; Purakayastha et al., 2019). Biochar when applied on soil decreases the soil pH and, as a result, restricts such condition of alkalinity which reduces the emission of N_2O from uplands, farmlands, and even paddy and wetland fields (Purakayastha et al., 2019; Alkharabsheh et al., 2021; Mousavi et al., 2023). A meta-analysis done by Lyu et al. (2022) showed that biochar application in upland soil and paddy soil decreased the emission of N_2O by 62% and 20%, respectively. The mechanism behind the suppression of N_2O from different croplands is due to the induction of aeration and oxygen content in soil by biochar. Such condition enhances soil porosity to adsorb NH_4^+ and NO_3^- (Laird & Rogovska, 2015; Karim et al., 2022; Mousavi et al., 2023). The decreased pH due to the presence of biochar in soil helps the AMF to minimize the concentration of chlorophyll and N in plants thereby restricting NH_4^+-N, NO_3^--N, and inorganic N (Krishnakumar et al., 2014; Purakayastha et al., 2019; Lyu et al., 2022), hence, reducing the emission of N_2O.

7 Conclusion

The exponential increase in human population and urbanization has ridiculously disturbed the Earth's equilibrium. The unprecedented rate and scale of pervasive pollution and drawdown of natural resources have questioned the human and ecological sustainability. To the blooming population of the world, one of the biggest challenges appears to be the food security for the present and future generations. Conventional agriculture which is chemical intensive is costly and unsustainable because it causes deterioration of soil health and plant productivity in the long run. A significant contribution of such agricultural practice is global warming-induced climate change through emission of GHGs. Despite having large production, the loss of soil fertility, and increased GHG emissions, the crops produced are found to be less nutritious and underdeveloped at times. The phenomena like soil erosion, nutrient leaching, and the exchange of nutrients as GHGs into the active pool of the atmosphere have led to a deficiency of nutrients in plants. Biochar among various organic amendments can be a low-cost, effective, and multifaceted tool that can replenish the nutrient content in soil as well as plants. Biochar as articulated in this chapter, when applied to soil, has the potential to ameliorate the soil's physical, chemical, and biological attributes. Increases in surface area, porosity, water retention capacity, and alteration in pH to support microbial community are the major contributions of biochar to soil. Soil properties like surface area and charge, along with the balance between mineralisation and immobilisation of nutrients such as nitrogen and phosphorus, collectively influence the bioavailability of nutrients for plants. For example, negatively charged soil surfaces attract positively charged cations like potassium (K+), while the mineralisation of organic matter releases plant-available forms of nitrogen and phosphorus. This chapter provides a complete insight into the mechanisms of action and contribution of biochar to nutrient retention and the regulation of nutrient availability. Furthermore, it warrants more conclusive experiments of biochar application in large agricultural fields accompanied by conservation agricultural practice to retain more soil nutrients with less emission of GHGs. It also suggests to blend biochar with other organic amendments including mustard cake, farmyard manure, vermicompost, etc. and apply that mixture in large agricultural fields to understand the nutrient dynamics in soil and plants.

References

Abbas, S., Javed, M. T., Ali, Q., Azeem, M., & Ali, S. (2021). Nutrient deficiency stress and relation with plant growth and development. In S. Fahad, O. Sönmez, S. Saud, D. Wang, C. Wu, M. Adnan, M. Arif, & Amanullah (Eds.), *Engineering tolerance in crop plants against abiotic stress* (pp. 239–262). CRC Press. https://doi.org/10.1201/9781003160717-12

Abhishek, K., Shrivastava, A., Vimal, V., Gupta, A. K., Bhujbal, S. K., Biswas, J. K., Singh, L., Ghosh, P., Pandey, A., Sharma, P., & Kumar, M. (2022). Biochar application for greenhouse gas mitigation, contaminants immobilization and soil fertility enhancement: A state-of-the-art review. *Science of the Total Environment, 853*, 158562. https://doi.org/10.1016/j.scitotenv.2022.158562

Abrahão, A., Costa, P., Lambers, H., Andrade, S., Sawaya, A., Ryan, M., & Oliveira, R. (2018). Soil types select for plants with matching nutrient-acquisition and -use traits in hyperdiverse and severely nutrient-impoverished campos rupestres and cerrado in central Brazil. *Journal of Ecology, 107*(3), 1302–1316. https://doi.org/10.1111/1365-2745.13111

Acharya, B., Dutta, A., & Minaret, J. (2015). Review on comparative study of dry and wet torrefaction. *Sustainable Energy Technologies and Assessments, 12*, 26–37. https://doi.org/10.1016/j.seta.2015.08.003

Adams, S. N. (1986). The interaction between liming and forms of nitrogen fertilizer on established grassland. *The Journal of Agricultural Science, 106*(3), 509–513. https://doi.org/10.1017/s0021859600063395

Alaoui-Sossé, B., Genet, P., Vinit-Dunand, F., Toussaint, M., Epron, D., & Badot, P. (2004). Effect of copper on growth in cucumber plants (*Cucumis sativus*) and its relationships with carbohydrate accumulation and changes in ion contents. *Plant Science, 166*(5), 1213–1218. https://doi.org/10.1016/j.plantsci.2003.12.032

Alegria Terrazas, R., Robertson-Albertyn, S., Corral, A. M., Escudero-Martinez, C., Kapadia, R., Balbirnie-Cumming, K., Morris, J., Hedley, P. E., Barret, M., Torres-Cortes, G., Paterson, E., Baggs, E. M., Abbott, J., & Bulgarelli, D. (2022). Defining composition and function of the rhizosphere microbiota of barley genotypes exposed to growth-limiting nitrogen supplies. *mSystems, 7*(6), e0093422. https://doi.org/10.1101/605204

Aleksandrov, V. (2022). Identification of nutrient deficiency in plants by artificial intelligence. *Acta Physiologiae Plantarum, 44*(3), 29. https://doi.org/10.1007/s11738-022-03363-0

Alkharabsheh, H. M., Seleiman, M. F., Battaglia, M. L., Shami, A., Jalal, R. S., Alhammad, B. A., Almutairi, K. F., & Al-Saif, A. M. (2021). Biochar and its broad impacts in soil quality and fertility nutrient leaching and crop productivity: A review. *Agronomy, 11*(5), 993. https://doi.org/10.3390/agronomy11050993

Amalina, F., Abd Razak, A. S., Krishnan, S., Sulaiman, H., Zularisam, A. W., & Nasrullah, M. (2022). Biochar production techniques utilizing biomass waste-derived materials and environmental applications–A review. *Journal of Hazardous Materials Advances, 7*, 100134. https://doi.org/10.1016/j.hazadv.2022.100134

Amonette, J. E., & Joseph, S. (2012). Characteristics of biochar: Microchemical properties. In J. Lehmann & S. Joseph (Eds.), *Biochar for environmental management* (pp. 65–84). Routledge. https://doi.org/10.4324/9781849770552

Andersen, D., Helmers, M., & Burns, R. (2015). Phosphorus sorption capacity of six Iowa soils before and after five years of use as vegetative treatment areas. *Applied Engineering in Agriculture, 31*(4), 611–620. https://doi.org/10.13031/aea.31.11128

Aras, S., Keles, H., & Bozkurt, E. (2021). Physiological and histological responses of peach plants grafted onto different rootstocks under calcium deficiency conditions. *Scientia Horticulturae, 281*, 109967. https://doi.org/10.1016/j.scienta.2021.109967

Attia, H., Rebah, F., Ouhibi, C., Saleh, M. A., Althobaiti, A. T., Alamer, K. H., Ben Nasri, M., & Lachaâl, M. (2022). Effect of potassium deficiency on physiological responses and anatomical structure of basil, *Ocimum basilicum* L. *Biology, 11*(11), 1557. https://doi.org/10.3390/biology11111557

Bakirbas, A., & Walker, E. (2022). Can of spinach, a novel long non-coding RNA, affects iron deficiency responses in *Arabidopsis thaliana*. *Frontiers in Plant Science, 13*. https://doi.org/10.3389/fpls.2022.1005020

Balashov, E., Buchkina, N., Šimanský, V., & Hořák, J. (2021). Effects of slow and fast pyrolysis biochar on n2o emissions and water availability of two soils with high water-filled pore space. *Journal of Hydrology and Hydromechanics, 69*(4), 467–474. https://doi.org/10.2478/johh-2021-0024

Basavarajeshwari, R., Yamunarani, R., Ramegowda, V., Geetha, K., & Shankar, A. (2018). Expression profiling of zinc transporter genes in tomato grown under different concentrations of zinc. *International Journal of Microbiology Research, 10*(6), 1252. https://doi.org/10.9735/0975-5276.10.6.1252-1255

Bauters, M., Grau, O., Doetterl, S., Heineman, K., Dalling, J., Prada, C., & Janssens, I. (2022). Tropical wood stores substantial amounts of nutrients, but we have limited understanding why. *Biotropica, 54*(3), 596–606. https://doi.org/10.1111/btp.13069

Belayneh, M., Yirgu, T., & Tsegaye, D. (2019). Potential soil erosion estimation and area prioritization for better conservation planning in Gumara watershed using RUSLE and GIS techniques'. *Environmental Systems Research, 8*(201). https://doi.org/10.1186/s40068-019-0149-x

Biederman, L. A., & Harpole, W. S. (2013). Biochar and its effects on plant productivity and nutrient cycling: A meta-analysis. *GCB Bioenergy, 5*(2), 202–214. https://doi.org/10.1111/gcbb.12037

Biswas, J. K., Mondal, B., Priyadarshini, P., Abhilash, P. C., Biswas, S., & Bhatnagar, A. (2021). Formulation of water sustainability index for India as a performance gauge for realizing the united nations sustainable development goal 6. *Ambio, 51*(6), 1569–1587. https://doi.org/10.1007/s13280-021-01680-1

Boiko, V. (2024). Magnesium in soils of agricultural landscapes of western Siberia. *Indian Journal of Agricultural Research*. https://doi.org/10.18805/ijare.af-854

Bonser, A., Lynch, J., & Snapp, S. (1996). Effect of phosphorus deficiency on growth angle of basal roots in *Phaseolus vulgaris*. *New Phytologist, 132*(2), 281–288. https://doi.org/10.1111/j.1469-8137.1996.tb01847.x

Boyd, C., Boyd, C., & Rouse, D. (2007). Potassium adsorption by bottom soils in ponds for inland culture of marine shrimp in Alabama. *Journal of the World Aquaculture Society, 38*(1), 85–91. https://doi.org/10.1111/j.1749-7345.2006.00076.x

Brantley, K. E., Savin, M. C., Brye, K. R., & Longer, D. E. (2016). Nutrient availability and corn growth in a poultry litter biochar-amended loam soil in a greenhouse experiment. *Soil Use and Management, 32*(3), 279–288. https://doi.org/10.1111/sum.12296

Breemen, N., & Protz, R. (1988). Rates of calcium carbonate removal from soils. *Canadian Journal of Soil Science, 68*(2), 449–454. https://doi.org/10.4141/cjss88-042

Brown, R. A., Kercher, A. K., Nguyen, T. H., Nagle, D. C., & Ball, W. P. (2006). Production and characterization of synthetic wood chars for use as surrogates for natural sorbents. *Organic Geochemistry, 37*(3), 321–333. https://doi.org/10.1016/j.orggeochem.2005.10.008

Brown, R., del Campo, B., Boateng, A. A., Garcia-Perez, M., & Mašek, O. (2015). Fundamentals of biochar production. In J. Lehmann & S. Joseph (Eds.), *Biochar for environmental management* (pp. 29–62). Routledge. https://doi.org/10.4324/9781849770552

Buchelt, A. C., Teixeira, G. C. M., Oliveira, K. S., Rocha, A. M. S., De Mello Prado, R., & Caione, G. (2020). Silicon contribution via nutrient solution in forage plants to mitigate nitrogen, potassium, calcium, magnesium, and sulfur deficiency. *Journal of Soil Science and Plant Nutrition, 20*(3), 1532–1548. https://doi.org/10.1007/s42729-020-00245-7

Butterbach-Bahl, K., Baggs, E., Dannenmann, M., Kiese, R., & Zechmeister-Boltenstern, S. (2013). Nitrous oxide emissions from soils: How well do we understand the processes and their controls? *Philosophical Transactions of the Royal Society B Biological Sciences, 368*(1621), 20130122. https://doi.org/10.1098/rstb.2013.0122

Campos, C. N. S., Teixeira, G. C. M., de Prado, R. M., Caione, G., da Silva Júnior, G. B., de David, C. H. O., Sales, A. C., Roque, C. G., & Teodoro, P. E. (2021). Macronutrient deficiency in cucumber plants: Impacts in nutrition, growth and symptoms. *Journal of Plant Nutrition, 44*(17), 2609–2626. https://doi.org/10.1080/01904167.2021.1921205

Cao, W., & Tibbitts, T. (1992). Growth, carbon dioxide exchange and mineral accumulation in potatoes grown at different magnesium concentrations. *Journal of Plant Nutrition, 15*(9), 1359–1371. https://doi.org/10.1080/01904169209364403

Cardoso, A., Longhini, V., Berça, A., Ongaratto, F., Siniscalchi, D., Reis, R., & Ruggieri, A. (2022). Pasture management and greenhouse gases emissions. *Bioscience Journal, 38*, e38099. https://doi.org/10.14393/bj-v38n0a2022-60614

Catuchi, T. A., Guidorizzi, F. V. C., Guidorizi, K. A., Barbosa, A. D. M., & Souza, G. M. (2012). Physiological responses of soybean cultivars to potassium fertilization under different water regimes. Pesquisa agropecuária brasileira, 47, 519-527. https://doi.org/10.1590/S0100-204X2012000400007

Chan, C., Liao, Y. Y., & Chiou, T. J. (2021). The impact of phosphorus on plant immunity. *Plant & Cell Physiology, 62*, 582–589. https://doi.org/10.1093/pcp/pcaa168

Chang, Y. M., Tsai, W. T., & Li, M. H. (2015). Chemical characterization of char derived from slow pyrolysis of microalgal residue. *Journal of Analytical and Applied Pyrolysis, 111*, 88–93. https://doi.org/10.1016/j.jaap.2014.12.004

Chatterjee, C., Sinha, P., Nautiyal, N., Agarwala, S., & Sharma, C. (1987). Metabolic changes associated with boron-calcium interaction in maize. *Soil Science and Plant Nutrition, 33*(4), 607–617. https://doi.org/10.1080/00380768.1987.10557609

Chaudhry, A. H., Nayab, S., Hussain, S. B., Ali, M., & Pan, Z. (2021). Current understandings on magnesium deficiency and future outlooks for sustainable agriculture. *International Journal of Molecular Sciences, 22*(4), 1819. https://doi.org/10.3390/ijms22041819

Chauhan, S., Mahawar, S., Jain, D., Upadhyay, S., Mohanty, S., Singh, A., & Maharjan, E. (2022). Boosting sustainable agriculture by arbuscular mycorrhiza under stress condition: Mechanism and future prospective. *BioMed Research International, 1*, 5275449. https://doi.org/10.1155/2022/5275449

Chen, J., Wu, F., Shang, Y., Wang, W., Hu, W., Simón, M., & Zheng, H. (2015). Hydrogen sulphide improves adaptation of *Zea mays* seedlings to iron deficiency. *Journal of Experimental Botany, 66*(21), 6605–6622. https://doi.org/10.1093/jxb/erv368

Chen Y, Yu M, Zhu Z, Zhang L, Guo Q (2013) Optimisation of potassium chloride nutrition for proper growth, physiological development and bioactive component production in Prunella vulgaris L. *PLoS One 8*(7). https://doi.org/10.1371/journal.pone.0066259

Chen, H., Ma, J., Wei, J., Gong, X., Yu, X., Guo, H., & Zhao, Y. (2018). Biochar increases plant growth and alters microbial communities via regulating the moisture and temperature of green roof substrates. *Science of the Total Environment, 635*, 333–342. https://doi.org/10.1016/j.scitotenv.2018.04.127

Chen, W. H., Lin, B. J., Lin, Y. Y., Chu, Y. S., Ubando, A. T., Show, P. L., Ong, H. C., Chang, J. S., Ho, S. H., Culaba, A. B., & Pétrissans, A. (2021). Progress in biomass torrefaction: Principles, applications and challenges. *Progress in Energy and Combustion Science, 82*, 100887. https://doi.org/10.1016/j.pecs.2020.100887

Chen, M., Li, Y., Wang, C., Walter, M. T., & Huang, L. (2023). Modelling rainfall-induced phosphorus loss with eroded clay and surface runoff. *Hydrological Processes, 37*(2), e14817. https://doi.org/10.1002/HYP.14817

Chia, C. H., Downie, A., & Munroe, P. (2015). Characteristics of biochar: Physical and structural properties. In J. Lehmann & S. Joseph (Eds.), *Biochar for environmental management* (pp. 89–109). Routledge. https://doi.org/10.4324/9781849770552

Citiulo, F., Jacobsen, I., Miramón, P., Schild, L., Brunke, S., Zipfel, P., & Wilson, D. (2012). *Candida albicans* scavenges host zinc via pra1 during endothelial invasion. *PLoS Pathogens, 8*(6), e1002777. https://doi.org/10.1371/journal.ppat.1002777

Craine, J. M., Morrow, C., & Fierer, N. (2007). Microbial nitrogen limitation increases decomposition. *Ecology, 88*(8), 2105–2113. https://doi.org/10.1890/06-1847.1

Cui, G., Liu, Y., Wang, P., Bai, X., Wang, H., Xu, Y., & Dong, L. (2022). Distribution characteristics and risk assessment of agricultural land use non-point source pollution in typical biofuel ethanol planting areas. *International Journal of Environmental Research and Public Health, 19*(3), 1394. https://doi.org/10.3390/ijerph19031394

Dale, V., Mann, L., Olson, R., Johnson, D., & Dearstone, K. (1990). The long-term influence of past land use on the walker branch forest. *Landscape Ecology, 4*(4), 211–224. https://doi.org/10.1007/bf00129829

de Bang, T. C., Husted, S., Laursen, K. H., Persson, D. P., & Schjoerring, J. K. (2021). The molecular–physiological functions of mineral macronutrients and their consequences for deficiency symptoms in plants. *New Phytologist, 229*, 2446–2469. https://doi.org/10.1111/nph.17074

De Lima GS, Da Silva JB, Pinheiro FWA, Soares LADA, Gheyi HR (2020) Potassium does not attenuate salt stress in yellow passion fruit under irrigation management strategies. Revista Caatinga 33(4):1082–1091. https://doi.org/10.1590/1983-21252020v33n423rc

De Vries, F. T., & Bardgett, R. D. (2012). Plant–microbial linkages and ecosystem nitrogen retention: Lessons for sustainable agriculture. *Frontiers in Ecology and the Environment, 10*(8), 425–432. https://doi.org/10.1890/110162

DeLuca, T. H., Gundale, M. J., MacKenzie, M. D., & Jones, D. L. (2015). Biochar effects on soil nutrient transformations. In S. Joseph (Ed.), *Lehmann J* (pp. 421–454). Routledge. https://doi.org/10.4324/9781849770552

Diaz, P., Santos, R., Baron, N., Aguirre-Gil, O., & Rigobelo, E. (2021). Effect of aspergillus and bacillus concentration on cotton growth promotion. *Frontiers in Microbiology, 12*, 737385. https://doi.org/10.3389/fmicb.2021.737385

Dixon, M., Simonne, E., Obreza, T., & Liu, G. (2020). Crop response to low phosphorus bioavailability with a focus on tomato. *Agronomy, 10*(5), 617. https://doi.org/10.3390/agronomy10050617

Dong, D., Yang, M., Wang, C., Wang, H., Li, Y., Luo, J., & Wu, W. (2013). Responses of methane emissions and rice yield to applications of biochar and straw in a paddy field. *Journal of Soils and Sediments, 13*, 1450–1460. https://doi.org/10.1007/s11368-013-0732-0

Duan, S., Zhang, C., Song, S., Ma, C., Zhang, C., Xu, W., Bondada, B., Wang, L., & Wang, S. (2022). Understanding calcium functionality by examining growth characteristics and structural aspects in calcium-deficient grapevine. *Scientific Reports, 12*, 3233. https://doi.org/10.1038/s41598-022-06867-4

Egamberdieva, D., Jabborova, D., Wirth, S., Alam, P., Alyemeni, M., & Ahmad, P. (2018). Interactive effects of nutrients and bradyrhizobium japonicum on the growth and root architecture of soybean (*Glycine max* L.). *Frontiers in Microbiology, 9*, 1000. https://doi.org/10.3389/fmicb.2018.01000

Eickhout, B., Bouwman, A. F., & van Zeijts, H. (2006). The role of nitrogen in world food production and environmental sustainability. *Agriculture, Ecosystems and Environment, 116*(1–2), 4–14. https://doi.org/10.1016/j.agee.2006.03.009

Elham, J., Golchin, A., & Ghassem, H. (2014). The effect of soil and foliar fertilization with iron on yield and leaf chemical composition of four spotted bean cultivars in a calcareous soil. *Greener Journal of Biological Sciences, 4*(4), 116–127. https://doi.org/10.15580/gjbs.2014.4.012714069

Ellis, R. (1969). Mineral losses during soil formation in granitic material. *Soil Research, 7*(3), 317. https://doi.org/10.1071/sr9690317

El-Mageed, T., Gyushi, M., Hemida, K., El-Saadony, M., El-Mageed, S., Abdalla, H., & Abdelkhalik, A. (2022). Coapplication of effective microorganisms and nanomagnesium boosts the agronomic, physio-biochemical, osmolytes, and antioxidants defenses against salt stress in *Ipomoea batatas*. *Frontiers in Plant Science, 13*, 883274. https://doi.org/10.3389/fpls.2022.883274

Erdem, H., Torun, M., Erdem, N., Yazici, A., Tolay, I., Günal, E., & Özkutlu, F. (2016). Effects of different forms and doses of sulphur application on wheat. *Turkish Journal of Agriculture – Food Science and Technology, 4*(11), 957. https://doi.org/10.24925/turjaf.v4i11.957-961.837

Fakayode, O. A., Aboagarib, E. A. A., Zhou, C., & Ma, H. (2020). Co-pyrolysis of lignocellulosic and macroalgae biomasses for the production of biochar–a review. *Bioresource Technology, 297*, 122408. https://doi.org/10.1016/j.biortech.2019.122408

Fan, M., Zhu, J., Richards, C., Brown, K., & Lynch, J. (2003). Physiological roles for aerenchyma in phosphorus-stressed roots. *Functional Plant Biology, 30*(5), 493. https://doi.org/10.1071/fp03046

Fathi, A., & Mehdiniya Afra, J. (2023). Plant growth and development in relation to phosphorus: A review. *Bulletin of University of Agricultural Sciences and Veterinary Medicine Cluj-Napoca Agriculture, 80*(1), 1–7. https://doi.org/10.15835/buasvmcn-agr:2022.0012

Fidel, R. B., Laird, D. A., & Parkin, T. B. (2018). Effect of biochar on soil greenhouse gas emissions at the laboratory and field scales. *Soil Systems, 3*(1), 8. https://doi.org/10.3390/soilsystems3010008; https://doi.org/10.20944/preprints201810.0315.v1

Firestone, M. K. (1982). Biological denitrification. *Nitrogen in Agricultural Soils, 22*, 289–326. https://doi.org/10.2134/agronmonogr22.c8

Fischer, E., & Bremer, E. (1993). Influence of magnesium deficiency on rates of leaf expansion, starch and sucrose accumulation, and net assimilation in *Phaseolus vulgaris*. *Physiologia Plantarum, 89*(2), 271–276. https://doi.org/10.1111/j.1399-3054.1993.tb00153.x

Frink, C. R., Waggoner, P. E., & Ausubel, J. H. (1999). Nitrogen fertilizer: Retrospect and prospect. *Proceedings of the National Academy of Sciences, 96*(4), 1175–1180. https://doi.org/10.1073/pnas.96.4.1175

Gao, H., Wu, X., Zorrilla, C., Vega, S., & Palta, J. (2020a). Fractionating of calcium in tuber and leaf tissues explains the calcium deficiency symptoms in potato plant overexpressing cax1. *Frontiers in Plant Science, 10*, 1793. https://doi.org/10.3389/fpls.2019.01793

Gao, W., Gao, D., Song, L., Sheng, H., Cai, T., & Liang, H. (2020b). Contribution of the non-growing season to annual N2O emissions from the continuous permafrost region in Northeast China. *Biogeosciences Discuss*, 1–43. https://doi.org/10.5194/bg-2020-305

Ghosh, D., Murmu, S., Adhikary, S., Mukherjee, A., Mal, S., Khanam, R., & Mandal, B. (2022). Sulphur flow in soil–plant system with long-term agricultural practices on a vertisol. *European Journal of Soil Science, 73*(6), e13318. https://doi.org/10.1111/ejss.13318

Gierth, M., & Mäser, P. (2007). Potassium transporters in plants – Involvement in K+ acquisition, redistribution and homeostasis. *FEBS Letters, 581*, 2348–2356. https://doi.org/10.1016/j.febslet.2007.03.035

Godlewska, A. (2018). Sulphur content in test plants and arylsulfatase activity in soil after application of waste materials. *Applied Ecology and Environmental Research, 16*(1), 145–152. https://doi.org/10.15666/aeer/1601_145152

Goodman, R. M. (2004). *Encyclopedia of plant and crop science*. Routledge.

Grzebisz, W., Zielewicz, W., & Przygocka-Cyna, K. (2023). Deficiencies of secondary nutrients in crop plants—A real challenge to improve nitrogen management. *Agronomy, 13*(1), 66. https://doi.org/10.3390/agronomy13010066

Guo, W., Zhang, Z., Liu, J., Zhu, P., & Yang, J. (2014). Studies on changes of soil carbon and nitrogen dynamics under long-term fertilizer in the black soil based on DSSAT model. *Advanced Materials Research, 1073–1076*, 1743–1750. https://doi.org/10.4028/www.scientific.net/amr.1073-1076.1743

Guo, Y. L., Wang, D. D., Zheng, J. Y., Zhao, S. W., & Zhang, X. C. (2015). Effect of biochar on soil greenhouse gas emissions in semi-arid region. *Huan Jing ke Xue = Huanjing Kexue, 36*(9), 3393–3400.

Hasanuzzaman M, Bhuyan MHMB, Nahar K, Hossain MS, Al Mahmud J, Hossen MS, Masud AAC, Moumita, Fujita M (2018) Potassium: A vital regulator of plant responses and tolerance to abiotic stresses. Agronomy 8(3), 31. https://doi.org/10.3390/agronomy8030031

Haeder HE, Mengel K, Forster H (1973) The effect of potassium on translocation of photosynthates and yield pattern of potato plants. J Sci Food Agric. 24(12):1479–1487. https://doi.org/10.1002/jsfa.2740241203

Ha, S., & Tran, L. S. (2014). Understanding plant responses to phosphorus starvation for improvement of plant tolerance to phosphorus deficiency by biotechnological approaches. *Critical Reviews in Biotechnology, 34*, 16–30. https://doi.org/10.3109/07388551.2013.783549

Hall, J., Barr, R., Al-Abbas, A., & Crane, F. (1972). The ultrastructure of chloroplasts in mineral-deficient maize leaves. *Plant Physiology, 50*(3), 404–409. https://doi.org/10.1104/pp.50.3.404

Hammond, J. P., Broadley, M. R., & White, P. J. (2004). Genetic responses to phosphorus deficiency. *Annals of Botany, 94*(3), 323–332. https://doi.org/10.1093/aob/mch156

Haneklaus, S., Bloem, E., & Schnug, E. (2008). History of sulfur deficiency in crops sulfur-yellow poison or jilted plant nutrient? In J. Jez (Ed.), *Sulfur: A missing link between soils, crops, and nutrition* (Vol. 50, pp. 45–58). American Society of Agronomy. https://doi.org/10.2134/agronmonogr50.c4

Harmens, H., Koevoets, P., Verkleij, J., & Ernst, W. (1994). The role of low molecular weight organic acids in the mechanism of increased zinc tolerance in *Silene vulgaris* (Moench) Garcke. *New Phytologist, 126*(4), 615–621. https://doi.org/10.1111/j.1469-8137.1994.tb02956.x

Harris, H., & Brolmann, J. (1966). Comparison of calcium and boron deficiencies of the peanut I. Physiological and yield differences1. *Agronomy Journal, 58*(6), 575–578. https://doi.org/10.2134/agronj1966.00021962005800060007x

He, H., Jin, X., Ma, H., Deng, Y., Huang, J., & Yin, L. (2020). Changes of plant biomass partitioning, tissue nutrients and carbohydrates status in magnesium-deficient banana seedlings and remedy potential by foliar application of magnesium. *Scientia Horticulturae, 268*, 109377. https://doi.org/10.1016/j.scienta.2020.109377

Hellal, J., Burnol, A., Locatelli, A., & Battaglia-Brunet, F. (2017). Experimental column setup for studying anaerobic biogeochemical interactions between iron (oxy)hydroxides, trace elements, and bacteria. *Journal of Visualized Experiments, 130*, 56240. https://doi.org/10.3791/56240

Hepler, P. (2005). Calcium: A central regulator of plant growth and development. *The Plant Cell, 17*(8), 2142–2155. https://doi.org/10.1105/tpc.105.032508

Hermans, C., Johnson, G., Strasser, R., & Verbruggen, N. (2004). Physiological characterisation of magnesium deficiency in sugar beet: Acclimation to low magnesium differentially affects photosystems I and II. *Planta, 220*(2), 344–355. https://doi.org/10.1007/s00425-004-1340-4

Hina, N. (2024). Global meta-analysis of nitrate leaching vulnerability in synthetic and organic fertilizers over the past four decades. *Water, 16*(3), 457. https://doi.org/10.3390/w16030457

Hmid, A., Mondelli, D., Fiore, S., Fanizzi, F. P., Al Chami, Z., & Dumontet, S. (2014). Production and characterization of biochar from three-phase olive mill waste through slow pyrolysis. *Biomass and Bioenergy, 71*, 330–339. https://doi.org/10.1016/j.biombioe.2014.09.024

Hossain, S. Z. (2019). Biochemical conversion of microalgae biomass into biofuel. *Chemical Engineering and Technology, 42*(12), 2594–2607. https://doi.org/10.1002/ceat.201800605

Hossain, M. Z., Bahar, M. M., Sarkar, B., Donne, S. W., Ok, Y. S., Palansooriya, K. N., Kirkham, M. B., Chowdhury, S., & Bolan, N. (2020). Biochar and its importance on nutrient dynamics in soil and plant. *Biochar, 2*, 379–420. https://doi.org/10.1007/s42773-020-00065-z

Houben, D., Sonnet, P., Tricot, G., Mattielli, N., Couder, E., & Opfergelt, S. (2014). Impact of root-induced mobilization of zinc on stable Zn isotope variation in the soil–plant system. *Environmental Science and Technology, 48*(14), 7866–7873. https://doi.org/10.1021/es5002874

Hsiao, P., Cheng, C., Koh, K., & Chan, M. (2017). The *Arabidopsis* defensin gene, AatPDF1.1, mediates defence against *Pectobacterium carotovorum* subsp. carotovorum via an iron-withholding defence system. *Scientific Reports, 7*(1), 9175. https://doi.org/10.1038/s41598-017-08497-7

Huang, J., Chen, Y., Sui, P., Nie, S., & Gao, W. (2014). Soil nitrous oxide emissions under maize-legume intercropping system in the north China plain. *Journal of Integrative Agriculture, 13*(6), 1363–1372. https://doi.org/10.1016/s2095-3119(13)60509-2

Huang, C. H., Singh, G. P., Park, S. H., Chua, N. H., Ram, R. J., & Park, B. S. (2020). Early diagnosis and management of nitrogen deficiency in plants utilizing Raman spectroscopy. *Frontiers in Plant Science, 11*, 663. https://doi.org/10.3389/fpls.2020.00663

Huang, W., Lin, M., Liao, J., Li, A., Tsewang, W., Chen, X., Sun, B., Liu, S., & Zheng, P. (2022). Effects of potassium deficiency on the growth of tea (Camelia sinensis) and strategies for optimizing potassium levels in soil: A critical review. *Horticulturae, 8*(7), 660. https://doi.org/10.3390/horticulturae8070660

Hussain, M., Farooq, M., Nawaz, A., Al-Sadi, A. M., Solaiman, Z. M., Alghamdi, S. S., Ammara, U., Ok, Y. S., & Siddique, K. H. (2017). Biochar for crop production: Potential benefits and risks. *Journal of Soils and Sediments, 17*, 685–716. https://doi.org/10.1007/s11368-016-1360-2

Ibrahim, A., Usman, A. R. A., Al-Wabel, M. I., Nadeem, M., Ok, Y. S., & Al-Omran, A. (2017). Effects of conocarpus biochar on hydraulic properties of calcareous sandy soil: Influence of particle size and application depth. *Archives of Agronomy and Soil Science, 63*(2), 185–197. https://doi.org/10.1080/03650340.2016.1193785

Irawan, J. F., & Indarto, I. (2011). Application of a multispectral SPOT image for land use classification in Sampean watershed. *Journal of Tropical Soils, 16*(2), 175–181. https://doi.org/10.5400/jts.2011.16.2.175

Ishfaq, M., Wang, Y., Yan, M., Wang, Z., Wu, L., Li, C., & Li, X. (2022). Physiological essence of magnesium in plants and its widespread deficiency in the farming system of China. *Frontiers in Plant Science, 13*, 802274. https://doi.org/10.3389/fpls.2022.802274

Jain, R., Solomon, S., Shrivastava, Shrivastava A.K., & Lal, P. (2009). Nutrient application improves stubble bud sprouting under low temperature conditions in sugarcane. Sugar Tech 11, 83–85. https://doi.org/10.1007/s12355-009-0016-6

Jena, R., Yakadri, M., Bhatt, S., & Reddy, K. (2021). Effect of various sources and levels of sulphur on growth, dry matter production and nutrient uptake of Indian mustard (*Brassica juncea* L.). *International Journal of Environment and Climate Change, 11*(12), 417–423. https://doi.org/10.9734/ijecc/2021/v11i1230592

Jha, P., Biswas, A. K., Lakaria, B. L., & Rao, A. S. (2010). Biochar in agriculture–prospects and related implications. *Current Science, 99*(9), 1218–1225.

Jiang, M., Xia, L., Wang, H., Zou, Y., & Wu, H. (2011). Transfer and transformation of soil iron and implications for hydrogeomorphological changes in Naoli river catchment, Sanjiang Psanjiang plain, northeast China. *Chinese Geographical Science, 21*(2), 149–158. https://doi.org/10.1007/s11769-011-0454-4

Jiménez-Rosado, M., Foggia, M., Rosignoli, S., Guerrero, A., Rombolà, A., & Romero, A. (2023). Effect of zinc and protein content in different barley cultivars: Use of controlled release matrices. *Renewable Agriculture and Food Systems, 38*, e34. https://doi.org/10.1017/s1742170523000297

Jouvin, D., Weiss, D., Mason, T., Bravin, M., Louvat, P., Zhao, F., & Benedetti, M. (2012). Stable isotopes of Cu and Zn in higher plants: Evidence for Cu reduction at the root surface and two conceptual models for isotopic fractionation processes. *Environmental Science and Technology, 46*(5), 2652–2660. https://doi.org/10.1021/es202587m

Ju, O. (2023). Assessing cumulative nitrous oxide emissions during both rice cultivation and fallow phases: A three-year case study in South Korea. *Preprints 2023*, 2023091823. https://doi.org/10.20944/preprints202309.1823.v1

Júnior, V., Comin, J., Ferreira, G., Tavares, J., Couto, R., & Filho, P. (2021). Nitrous oxide emissions in soils fertilized with pig manure: Soil processes and strategies of control and mitigation. *Research Society and Development, 10*(2), e23910212427. https://doi.org/10.33448/rsd-v10i2.12427

Karim, A. A., Kumar, M., Singh, E., Kumar, A., Kumar, S., Ray, A., & Dhal, N. K. (2022). Enrichment of primary macronutrients in biochar for sustainable agriculture: A review. *Critical Reviews in Environmental Science and Technology, 52*(9), 1449–1490. https://doi.org/10.1080/10643389.2020.1859271

Kelly, C. (2023). Tree species influence nitrate and nitrous oxide production in forested riparian soils. *Nitrogen, 4*(4), 311–330. https://doi.org/10.3390/nitrogen4040023

Kettles, N., Kopriva, S., & Malin, G. (2014). Insights into the regulation of DMSP synthesis in the diatom *Thalassiosira pseudonana* through APR activity, proteomics and gene expression analyses on cells acclimating to changes in salinity, light and nitrogen. *PLoS One, 9*(4), e94795. https://doi.org/10.1371/journal.pone.0094795

Khan, S., Malik, A., & Ahmad, F. (2021). Role of zinc homeostasis in plant growth. In S. T. Khan & A. Malik (Eds.), *Microbial biofertilizers and micronutrient availability* (pp. 179–195). Springer. https://doi.org/10.1007/978-3-030-76609-2_9

Khan, S. T., Malik, A., Alwarthan, A., & Shaik, M. R. (2022). The enormity of the zinc deficiency problem and available solutions; an overview. *Arabian Journal of Chemistry, 15*(3), 103668. https://doi.org/10.1016/j.arabjc.2021.103668

Khan, F., Siddique, A. B., Shabala, S., Zhou, M., & Zhao, C. (2023). Phosphorus plays key roles in regulating plants' physiological responses to abiotic stresses. *Plants, 12*(15), 2861. https://doi.org/10.3390/plants12152861

Knight, M., Smith, S., & Trewavas, A. (1992). Wind-induced plant motion immediately increases cytosolic calcium. *Proceedings of the National Academy of Sciences, 89*(11), 4967–4971. https://doi.org/10.1073/pnas.89.11.4967

Koroleva, O., Gibson, T., Cramer, R., & Stain, C. (2010). Glucosinolate-accumulating s-cells in *Arabidopsis* leaves and flower stalks undergo programmed cell death at early stages of differentiation. *The Plant Journal, 64*(3), 456–469. https://doi.org/10.1111/j.1365-313x.2010.04339.x

Krishnakumar, S., Rajalakshmi, A. G., Balaganesh, B., Manikandan, P., Vinoth, C., & Rajendran, V. (2014). Impact of biochar on soil health. *International Journal of Advanced Research, 2*(4), 933–950.

Kudla, J., Batistič, O., & Hashimoto, K. (2010). Calcium signals: The lead currency of plant information processing. *The Plant Cell, 22*(3), 541–563. https://doi.org/10.1105/tpc.109.072686

Kumar, E., Surekha, K., Rekha, K., & Sharma, S. (2020). Effect of various sources of zinc and iron on dry matter yield, nutrient content and nutrient uptake of finger millet (*Eleusine coracana* L.). *Current Journal of Applied Science and Technology, 39*(4), 102–112. https://doi.org/10.9734/cjast/2020/v39i430536

Kusaka, M., Kalaji, H. M., Mastalerczuk, G., Dąbrowski, P., & Kowalczyk, K. (2021). Potassium deficiency impact on the photosynthetic apparatus efficiency of radish. *Photosynthetica, 59*(1), 127–136. https://doi.org/10.32615/ps.2020.077

Kuziemska, B., Klej, P., Wysokiński, A., Jaremko, D., & Pakuła, K. (2022). Yielding and bioaccumulation of zinc by cocksfoot under conditions of different doses of this metal and organic fertilization. *Agronomy, 12*(3), 686. https://doi.org/10.3390/agronomy12030686

Laird, D., & Rogovska, N. (2015). Biochar effects on nutrient leaching. In J. Lehmann & S. Joseph (Eds.), *Biochar for environmental management* (pp. 521–542). Routledge. https://doi.org/10.4324/9781849770552

Laird, D., Fleming, P., Wang, B., Horton, R., & Karlen, D. (2010). Biochar impact on nutrient leaching from a Midwestern agricultural soil. *Geoderma, 158*(3–4), 436–442. https://doi.org/10.1016/j.geoderma.2010.05.012

Lang, A. (1983). Turgor-regulated translocation. *Plant, Cell & Environment, 6*, 683–689. https://doi.org/10.1111/1365-3040.ep11589312

Lapenis, A., Lawrence, G., Bailey, S., Aparin, B. F., Shiklomanov, A., Speranskaya, N., & Calef, M. (2008). Climatically driven loss of calcium in steppe soil as a sink for atmospheric carbon. *Global Biogeochemical Cycles, 22*(2), GB2010. https://doi.org/10.1029/2007gb003077

Lapeyrie, F., & Bruchet, G. (1986). Calcium accumulation by two strains, calcicole and calcifuge, of the mycorrhizal fungus *Paxillus involutus*. *New Phytologist, 103*(1), 133–141. https://doi.org/10.1111/j.1469-8137.1986.tb00602.x

Lee, B., & Kim, T. (2012). Sulphur supply level effects on the assimilation of nitrate and sulphate into amino acids and protein in forage rape (*Brassica napus* L.). *Journal of the Korean Society of Grassland and Forage Science, 32*(4), 343–352. https://doi.org/10.5333/kgfs.2012.32.4.343

Leng, L., & Huang, H. (2018). An overview of the effect of pyrolysis process parameters on biochar stability. *Bioresource Technology, 270*, 627–642. https://doi.org/10.1016/j.biortech.2018.09.030

Lenzoni, G., Liu, J., & Knight, M. (2017). Predicting plant immunity gene expression by identifying the decoding mechanism of calcium signatures. *New Phytologist, 217*(4), 1598–1609. https://doi.org/10.1111/nph.14924

Li, S., Zhou, X., Li, H., Liu, Y., Zhu, L., Guo, J., & Chen, R. (2015a). Overexpression of ZmIRT1 and ZmZIP3 enhances iron and zinc accumulation in transgenic *Arabidopsis*. *PLoS One, 10*(8), e0136647. https://doi.org/10.1371/journal.pone.0136647

Li, Z., Long, W., Yang, S., Wang, Y., Tang, J., Wen, L., & Xiong, M. (2015b). Endogenous hydrogen sulfide regulated by calcium is involved in thermotolerance in tobacco *Nicotiana tabacum* L. suspension cell cultures. *Acta Physiologiae Plantarum, 37*(10), 219. https://doi.org/10.1007/s11738-015-1971-z

Li, Q., Gao, Y., & Yang, A. (2020). Sulfur homeostasis in plants. *International Journal of Molecular Sciences, 21*(23), 8926. https://doi.org/10.3390/ijms21238926

Li, J., Cao, X., Jia, X., Liu, L., Cao, H., Qin, W., & Li, M. (2021). Iron deficiency leads to chlorosis through impacting chlorophyll synthesis and nitrogen metabolism in Areca catechu L. *Frontiers in Plant Science, 12*, 710093. https://doi.org/10.3389/fpls.2021.710093

Li, T., Ma, F., Wang, J., Qiu, P., Zhang, N., Guo, W., Xu, J., & Dai, T. (2023a). Study on the mechanism of rainfall-runoff induced nitrogen and phosphorus loss in hilly slopes of black soil area, China. *Water, 15*(17), 3148. https://doi.org/10.3390/w15173148

Li, J., Yang, X., Zhang, M., Li, D., Jiang, Y., Yao, W., & Zhang, Z. (2023b). Yield, quality, and water and fertilizer partial productivity of cucumber were promoted by the interaction of water, nitrogen, and magnesium. *Agronomy*. https://doi.org/10.21203/rs.3.rs-2449234/v1

Liang, Z., Peng, X., Luan, Z., Li, W., & Zhao, Y. (2011). Reduction of phosphorus release from high phosphorus soil by red mud. *Environmental Earth Sciences, 65*(3), 581–588. https://doi.org/10.1007/s12665-011-1105-x

Likens, G., Driscoll, C., & Buso, D. (1996). Long-term effects of acid rain: Response and recovery of a forest ecosystem. *Science, 272*(5259), 244–246. https://doi.org/10.1126/science.272.5259.244

Lin, F., Wang, H., Shaghaleh, H., Ali Adam Hamad, A., Zhang, Y., Yang, B., & Alhaj Hamoud, Y. (2024). Effects of biochar amendment on N2O emissions from soils with different pH levels. *Atmosphere, 15*(1), 68. https://doi.org/10.3390/atmos15010068

Liu, C., Weiqian, C., Lü, Y., Huang, H., Chen, L., Liu, X., & Hong, F. (2009). Cerium under calcium deficiency—Influence on the antioxidative defense system in spinach plants. *Plant and Soil, 323*(1–2), 285–294. https://doi.org/10.1007/s11104-009-9937-9

Liu, W., Kerr, P., Wu, Y., & Fang, Y. (2015). The application of soil amendments benefits to the reduction of phosphorus depletion and the growth of cabbage and corn. *Environmental Science and Pollution Research, 22*(21), 16772–16780. https://doi.org/10.1007/s11356-015-4870-3

Liu, W., Li, Y., Huang, G., Wu, J., Duan, H., Liao, Y., Huang, R., Xu, Z., & Fan, H. (2019). Nutrient leaching of Chinese fir (*Cunninghamia lanceolata*) seedlings under simulated nitrogen deposition. *Pakistan Journal of Botany, 51*(5). https://doi.org/10.30848/pjb2019-5(35)

Liu, R., Pan, Y., Bao, H., Shen, L., Jiang, Y., Tu, H., & Huang, W. (2020). Variations in soil physicochemical properties along slope position gradient in secondary vegetation of the hilly region, Guilin, southwest China. *Sustainability, 12*(4), 1303. https://doi.org/10.3390/su12041303

Liu, S., Qin, T., Dong, B., Shi, X., Lv, Z., & Zhang, G. (2021). The influence of climate, soil properties and vegetation on soil nitrogen in sloping farmland. *Sustainability, 13*(3), 1480. https://doi.org/10.3390/su13031480

López-Arredondo, D., Leyva-González, M., González-Morales, S., López-Bucio, J., & Herrera-Estrella, L. (2014). Phosphate nutrition: Improving low-phosphate tolerance in crops. *Annual Review of Plant Biology, 65*(1), 95–123. https://doi.org/10.1146/annurev-arplant-050213-035949

López-Berges, M., Turrà, D., Capilla, J., Schafferer, L., Matthijs, S., Jöchl, C., & Pietro, A. (2013). Iron competition in fungus-plant interactions. *Plant Signaling and Behavior, 8*(2), e23012. https://doi.org/10.4161/psb.23012

Lornac, A., Havé, M., Chardon, F., Soulay, F., Clément, G., Avice, J., & Masclaux-Daubresse, C. (2020). Autophagy controls sulphur metabolism in the rosette leaves of Arabidopsis and facilitates S remobilization to the seeds. *Cells, 9*(2), 332. https://doi.org/10.3390/cells9020332

Lu, L., Zhang, Y., Li, L., Yi, N., Liu, Y., Qaseem, M. F., Li, H., & Wu, A. M. (2021). Physiological and transcriptomic responses to nitrogen deficiency in *Neolamarckia cadamba. Frontiers in Plant Science, 12*. https://doi.org/10.3389/fpls.2021.747121

Lukin, S., & Zhulikov, D. V. (2022). Content and balance of trace elements (Co, Mn, Zn) in agroecosystems of the central chernozemic region of Russia. *Agriculture, 12*(2), 154. https://doi.org/10.3390/agriculture12020154

Lurthy, T., Pivato, B., Lemanceau, P., & Mazurier, S. (2021). Importance of the rhizosphere microbiota in iron biofortification of plants. *Frontiers in Plant Science, 12*. https://doi.org/10.3389/fpls.2021.744445

Lyu, H., Zhang, H., Chu, M., Zhang, C., Tang, J., Chang, S. X., Masek, O., & Ok, Y. S. (2022). Biochar affects greenhouse gas emissions in various environments: A critical review. *Land Degradation and Development, 33*(17), 3327–3342. https://doi.org/10.1002/ldr.4405

Mahawar, L., Ramasamy, K. P., Pandey, A., & Prasad, S. M. (2023). Iron deficiency in plants: an update on homeostasis and its regulation by nitric oxide and phytohormones. *Plant Growth Regulation, 100*, 283–299. https://doi.org/10.1007/s10725-022-00853-6

Mahdii, B. A., & Turki, A. M. (2020). Estimating soil pollution range with heavy metals in some areas of Baghdad city. *Medico-Legal Update, 20*(2), 747. https://doi.org/10.37506/mlu.v20i2.1204

Maier, J. (2024). Novel insights into an old story: Magnesium and vascular tone. *Acta Physiologica, 240*(3), e14110. https://doi.org/10.1111/apha.14110

Mándoki, Z., & Pénzes, B. (2012). Effects of using chemical-free root-knot nematode (Meloidogyne incognita) control methods on the occurrence of blossom-end rot in pepper. *Journal of Plant Protection Research, 52*(3), 337–341. https://doi.org/10.2478/v10045-012-0055-y

Mani, D., Sharma, B., Kumar, C., & Balak, S. (2012). Depth-wise distribution, mobility and naturally occurring glutathione based phytoaccumulation of cadmium and zinc in sewage-irrigated soil profiles. *International journal of Environmental Science and Technology, 10*(6), 1167–1180. https://doi.org/10.1007/s13762-012-0121-z

Martínez-Espinosa, C., Sauvage, S., Al Bitar, A., Green, P. A., Vörösmarty, C. J., & Miguel Sánchez-Pérez, J. (2020). Denitrification in wetlands: A review towards a quantification at global scale. *Science of the Total Environment, 754*, 142398. https://doi.org/10.1016/j.scitotenv.2020.142398

Mazhar, M., Ishtiaq, M., Maqbool, M., Ajaib, M., Hussain, I., Hussain, T., & Gul, A. (2023). Synergistic application of calcium oxide nanoparticles and farmyard manure induces cadmium tolerance in mung bean (Vigna radiata L.) by influencing physiological and biochemical parameters. *PLoS One, 18*(3), e0282531. https://doi.org/10.1371/journal.pone.0282531

Mejia, F., & Kleissl, J. (2013). Soiling losses for solar photovoltaic systems in California. *Solar Energy, 95*, 357–363. https://doi.org/10.1016/j.solener.2013.06.028

Mendes, W., Júnior, J., Cunha, P., Silva, A., Evangelista, A., & Casaroli, D. (2016). Potassium leaching in different soils as a function of irrigation depths. *Revista Brasileira De Engenharia Agrícola E Ambiental, 20*(11), 972–977. https://doi.org/10.1590/1807-1929/agriambi.v20n11p972-977

Mengel, K., Kirby, E. A., Kosegarten, H., & Appel, T. (2001). *Principles of plants nutrition* (p. 849). Kluwer Academic Publishers.
Merry, R., Dobbels, A. A., Sadok, W., Naeve, S., Stupar, R. M., & Lorenz, A. J. (2022). Iron deficiency in soybean. *Crop Science, 62*(1), 36–52. https://doi.org/10.1002/csc2.20661
Mihai, O., Pantea, O., Popovici, D., & Gheorghe, C. (2017). Evaluation of metal contents in correlation with phytosanitary treatments at vineyard. *Revista De Chimie, 68*(6), 1289–1293. https://doi.org/10.37358/rc.17.6.5659
Mondal, R., Madhurya, K., Saha, P., Chattopadhyay, S., Antony, S., Kumar, A., & Roy, D. (2021). Expression profile, transcriptional and post-transcriptional regulation of genes involved in hydrogen sulphide metabolism connecting the balance between development and stress adaptation in plants: A data-mining bioinformatics approach. *Plant Biology, 24*(4), 602–617. https://doi.org/10.1111/plb.13378
Montejano-Ramírez, V., & Valencia-Cantero, E. (2023). Cross-talk between iron deficiency response and defense establishment in plants. *International Journal of Molecular Sciences, 24*(7), 6236. https://doi.org/10.3390/ijms24076236
Moreau, G., Qin, A., & Mann, B. (2018). Zinc acquisition mechanisms differ between environmental and virulent Francisella species. *Journal of Bacteriology, 200*(4), e00587–e00517. https://doi.org/10.1128/jb.00587-17
Mousavi, S. M., Srivastava, A. K., & Cheraghi, M. (2023). Soil health and crop response of biochar: An updated analysis. *Archives of Agronomy and Soil Science, 69*(7), 1085–1110. https://doi.org/10.1080/03650340.2022.2054998
Mu, X., & Chen, Y. (2021). The physiological response of photosynthesis to nitrogen deficiency. *Plant Physiology and Biochemistry, 158*, 76–82. https://doi.org/10.1016/j.plaphy.2020.11.019
Nambiar, E. (1976). The uptake of zinc-65 by oats in relation to soil water content and root growth. *Soil Research, 14*(1), 67–74. https://doi.org/10.1071/sr9760067
Narayan, O. P., Kumar, P., Yadav, B., Dua, M., & Johri, A. K. (2023). Sulfur nutrition and its role in plant growth and development. *Plant Signaling & Behavior, 18*, 2030082. https://doi.org/10.1080/15592324.2022.2030082
Nardi, N., Anwar, S., Yani, M., & Hendrizal, M. (2021). Nitrous oxide emission from conservation forest of Kampar Peninsula peatland ecosystem. *Jurnal Pengelolaan Sumberdaya Alam Dan Lingkungan (Journal of Natural Resources and Environmental Management), 11*(3), 442–452. https://doi.org/10.29244/jpsl.11.3.442-452
Narzari, R., Bordoloi, N., Chutia, R. S., Borkotoki, B., Gogoi, N., Bora, A., & Kataki, R. (2015). Biochar: An overview on its production, properties and potential benefits. *Biology, Biotechnology and Sustainable Development, 1*(1), 13–40. https://doi.org/10.13140/RG.2.1.3966.2560
Nath, M., & Tuteja, N. (2016). NPKS uptake, sensing, and signaling and miRNAs in plant nutrient stress. *Protoplasma, 253*, 767–786. https://doi.org/10.1007/s00709-015-0845-y
Ndayisenga, F., Yu, Z., Wang, B., Yang, J., Wu, G., & Zhang, H. (2022). Using bioelectrohydrogenesis left-over residues as a future potential fertilizer for soil amendment. *Scientific Reports, 12*(1), 17779. https://doi.org/10.1038/s41598-022-22715-x
Nepali, B., & Bhandari, D. (2019). Enhancing the yield and quality of oilseed crops in Nepal through application of sulphur fertilizers. *Big Data in Agriculture., 1*(2), 10–11. https://doi.org/10.26480/bda.02.2019.10.11
Nie, S. (2024). Rice planting acceleration of the activation and loss of soil iron in the red soil region of southern China. *Soil Ecology Letters, 6*(2), 230198. https://doi.org/10.1007/s42832-023-0198-y
Nordin, A., Pommer, L., Nordwaeger, M., & Olofsson, I. (2013). Biomass conversion through torrefaction. In E. Dahlquist (Ed.), *Technologies for converting biomass to useful energy: Combustion, gasification, pyrolysis, torrefaction and fermentation* (pp. 217–240). Routledge. https://doi.org/10.1201/b14561
Otoidobiga, C., Keïta, A., Yacouba, H., Traoré, A., & Dianou, D. (2015). Dynamics and activity of iron-reducing bacterial populations in a west African rice paddy soil under subsurface drainage: Case study of Kamboinse in Burkina Faso. *Agricultural Sciences, 6*(8), 860–869. https://doi.org/10.4236/as.2015.68083

Özgen, Ş., Busse, J., & Palta, J. (2011). Influence of root zone calcium on shoot tip necrosis and apical dominance of potato shoot: Simulation of this disorder by ethylene glycol tetra acetic acid and prevention by strontium. *Hortscience, 46*(10), 1358–1362. https://doi.org/10.21273/hortsci.46.10.1358

Padilla, F. M., Farneselli, M., Gianquinto, G., Tei, F., & Thompson, R. B. (2020). Monitoring nitrogen status of vegetable crops and soils for optimal nitrogen management. *Agricultural Water Management, 241*, 106356. https://doi.org/10.1016/j.agwat.2020.106356

Panahi, H. K. S., Dehhaghi, M., Ok, Y. S., Nizami, A. S., Khoshnevisan, B., Mussatto, S. I., Aghbashlo, M., Tabatabaei, M., & Lam, S. S. (2020). A comprehensive review of engineered biochar: Production, characteristics, and environmental applications. *Journal of Cleaner Production, 270*, 122462. https://doi.org/10.1016/j.jclepro.2020.122462

Pandey, N. (2018). Role of Plant Nutrients in Plant Growth and Physiology. In: Hasanuzzaman, M., Fujita, M., Oku, H., Nahar, K., Hawrylak-Nowak, B. (eds) Plant Nutrients and Abiotic Stress Tolerance. Springer, Singapore. https://doi.org/10.1007/978-981-10-9044-8_2

Park, E., Tsuyuki, K., Hu, F., Lee, J., & Jeong, J. (2019). Prc2-mediated h3k27me3 contributes to transcriptional regulation of fit-dependent iron deficiency response. *Frontiers in Plant Science, 10*, 627. https://doi.org/10.3389/fpls.2019.00627

Pascual, M. B., Sanchez-Monedero, M. A., Chacon, F. J., Sanchez-Garcia, M., & Cayuela, M. L. (2020). Linking biochars properties to their capacity to modify aerobic CH_4 oxidation in an upland agricultural soil. *Geoderma, 363*, 114179. https://doi.org/10.1016/j.geoderma.2020.114179

Paulsen, G., & Harper, J. (1968). Evidence for a role of calcium in nitrate assimilation in wheat seedlings. *Plant Physiology, 43*(5), 775–780. https://doi.org/10.1104/pp.43.5.775

Peng, L., Xue, X., Tang, Q., Zhu, Y., Xiao, L., Yang, Y., & Lin, Q. (2020). Phosphorus retention and loss in three types of soils with implications for geographical pattern of eutrophication in China. *Water and Environment Journal, 34*(S1), 9–18. https://doi.org/10.1111/wej.12490

Perkowska, I., Potrykus, M., Siwinska, J., Siudem, D., Lojkowska, E., & Ihnatowicz, A. (2021). Interplay between coumarin accumulation, iron deficiency and plant resistance to Dickeya spp. *International Journal of Molecular Sciences, 22*(12), 6449. https://doi.org/10.3390/ijms22126449

Pickering, G., Mazur, A., Trousselard, M., Bieńkowski, P., Yaltsewa, N., Amessou, M., & Pouteau, E. (2020). Magnesium status and stress: The vicious circle concept revisited. *Nutrients, 12*(12), 3672. https://doi.org/10.3390/nu12123672

Pine, H., & Boyd, C. (2010). Adsorption of magnesium by bottom soils in inland brackish water shrimp ponds in Alabama. *Journal of the World Aquaculture Society, 41*(4), 603–609. https://doi.org/10.1111/j.1749-7345.2010.00400.x

Prakash, J., Agrawal, S., & Agrawal, M. (2022). Global trends of acidity in rainfall and its impact on plants and soil. *Journal of Soil Science and Plant Nutrition, 23*(1), 398–419. https://doi.org/10.1007/s42729-022-01051-z

Purakayastha, T. J. (2012). Preparation and utilization of biochar for soil amendment. In H. Pathak, P. K. Agarwal, & S. D. Singhj (Eds.), *Climate change impact, adaptation and mitigation in agriculture: Methodology for assessment and applications* (pp. 280–294). IRRI.

Purakayastha, T. J., Bera, T., Bhaduri, D., Sarkar, B., Mandal, S., Wade, P., Kumari, S., Biswas, S., Menon, M., Pathak, H., & Tsang, D. C. (2019). A review on biochar modulated soil condition improvements and nutrient dynamics concerning crop yields: Pathways to climate change mitigation and global food security. *Chemosphere, 227*, 345–365. https://doi.org/10.1016/j.chemosphere.2019.03.170

Raghunath, R., Pandian, P., Mahendran, P., Ragavan, T., & Geetha, R. (2021). Influence of sulphur fertilization on yield, nutrient content and uptake by sugarcane in sulphur deficient soil of Thiruppuvanam block of Sivagangai district. *Current Journal of Applied Science and Technology, 40*(16), 38–45. https://doi.org/10.9734/cjast/2021/v40i1631422

Rajapitamahuni, S., Kang, B. R., & Lee, T. K. (2023). Exploring the roles of arbuscular mycorrhizal fungi in plant–iron homeostasis. *Agriculture, 13*(10), 1918. https://doi.org/10.3390/agriculture13101918

Rajonandraina, T., Ueda, Y., Wissuwa, M., Kirk, G., Rakotoson, T., Manwaring, H., Andriamananjara, A., & Razafimbelo, T. (2023). Magnesium supply alleviates iron toxicity-induced leaf bronzing in rice through exclusion and tissue-tolerance mechanisms. *Frontiers in Plant Science, 14*, 1213456. https://doi.org/10.3389/fpls.2023.1213456

Ram, H., Sardar, S., & Gandass, N. (2021). Vacuolar iron transporter (like) proteins: Regulators of cellular iron accumulation in plants. *Physiologia Plantarum, 171*(4), 823–832. https://doi.org/10.1111/ppl.13363

Rashti, M., Wang, W., Harper, S., Moody, P., Chen, C., Ghadiri, H., & Reeves, S. H. (2015). Strategies to mitigate greenhouse gas emissions in intensively managed vegetable cropping systems in subtropical Australia. *Soil Research, 53*(5), 475. https://doi.org/10.1071/sr14355

Rawat, P., Das, S., Shankhdhar, D., & Shankhdhar, S. (2020). Phosphate-solubilizing microorganisms: Mechanism and their role in phosphate solubilization and uptake. *Journal of Soil Science and Plant Nutrition, 21*(1), 49–68. https://doi.org/10.1007/s42729-020-00342-7

Redding, M., Ghani, A., Kear, M., O'Connor, M., & Catto, W. (2006). Phosphorus leaching from pastures can be an environmental risk and even a significant fertiliser expense. *Proceedings of the New Zealand Grassland Association*, 293–296. https://doi.org/10.33584/jnzg.2006.68.2603

Reddy, V. R. P., Aski, M. S., Mishra, G. P., Dikshit, H. K., Singh, A., Pandey, R., Singh, M. P., Gayacharan, Ramtekey, V., Priti, R. N., & Nair, R. M. (2020). Genetic variation for root architectural traits in response to phosphorus deficiency in mungbean at the seedling stage. *PLoS One, 15*(6), e0221008. https://doi.org/10.1371/journal.pone.0221008

Rivelli, A., Maria, S., Pizza, S., & Gherbin, P. (2010). Growth and physiological response of hydroponically-grown sunflower as affected by salinity and magnesium levels. *Journal of Plant Nutrition, 33*(9), 1307–1323. https://doi.org/10.1080/01904167.2010.484092

Rosolem, C., & Steiner, F. (2017). Effects of soil texture and rates of k input on potassium balance in tropical soil. *European Journal of Soil Science, 68*(5), 658–666. https://doi.org/10.1111/ejss.12460

Rosolem, C., Almeida, D., Rocha, K., & Bacco, G. (2018). Potassium fertilisation with humic acid coated KCl in a sandy clay loam tropical soil. *Soil Research, 56*(3), 244–251. https://doi.org/10.1071/sr17214

Rubin, R. L., Anderson, T. R., & Ballantine, K. A. (2020). Biochar simultaneously reduces nutrient leaching and greenhouse gas emissions in restored wetland soils. *Wetlands, 40*(6), 1981–1991. https://doi.org/10.1007/s13157-020-01380-8

Ruotsalainen, A., & Kytöviita, M. (2004). Mycorrhiza does not alter low temperature impact on *Gnaphalium norvegicum*. *Oecologia, 140*(2), 226–233. https://doi.org/10.1007/s00442-004-1586-3

Rutkowska, B., & Szulc, W. (2014). Speciation of cu and Zn in soil solution in a long-term fertilization experiment. *Soil Science Annual, 65*(1), 25–28. https://doi.org/10.2478/ssa-2014-0004

Sohi, S., Krull, E., Lopez-Capel, E., & Bol, R. (2009). A Review of Biochar and Its Use and Function in Soil. Advances in Agronomy, 105, 47–82. https://doi.org/10.1016/S0065-2113(10)05002-9

Sakhiya, A. K., Anand, A., & Kaushal, P. (2020). Production, activation, and applications of biochar in recent times. *Biochar, 2*, 253–285. https://doi.org/10.1007/s42773-020-00047-1

Samson, V. (2024). Influence of fertilizer application on soil and nutrient loss during carrot harvesting. *Egyptian Journal of Agricultural Research, 102*(1), 42–54. https://doi.org/10.21608/ejar.2024.243106.1449

Shah T, Khan H, Noor MA, Ghoneim A, Wang X, Sher A, Nasir M, Basahi MA (2018) Effects of potassium on phenological, physiological, and agronomic traits of maize (Zea mays L.) under high nitrogen nutrition with optimum and reduced irrigation. Appl Ecol Environ Res 16(5):7079–7097. https://doi.org/10.15666/aeer/1605_70797097

Santiago, F., Santiago, F., Filho, J., & Ratke, R. (2018). Plant growth and symptomatology of macronutrient deficiencies in cowpea plants. *Comunicata Scientiae, 9*(3), 503–508. https://doi.org/10.14295/cs.v9i3.2593

Santos, E. F., Mateus, N. S., Rabêlo, F. H. S., Macedo, F. G., & Lavres, J. (2020). Diagnosing early disorders in Jatropha curcas to calcium, magnesium and sulfur deficiency. *Journal of Plant Nutrition, 43*(11), 1604–1616. https://doi.org/10.1080/01904167.2020.1730899

Santoso, E., Ediati, R., Kusumawati, Y., Bahruji, H., Sulistiono, D. O., & Prase-tyoko, D. (2020). Review on recent advances of carbon based adsorbent for methylene blue removal from waste water. *Materials Today Chemistry, 16*, 100233. https://doi.org/10.1016/j.mtchem.2019.100233

Saravanan, A., & Kumar, P. S. (2022). Biochar derived carbonaceous material for various environmental applications: Systematic review. *Environmental Research, 214*, 113857. https://doi.org/10.1016/j.envres.2022.113857

Schenkeveld, W., Schindlegger, Y., Oburger, E., Puschenreiter, M., & Kraemer, S. (2014). Geochemical processes constraining iron uptake in strategy II Fe acquisition. *Environmental Science and Technology, 48*(21), 12662–12670. https://doi.org/10.1021/es5031728

Seafatullah, M., Hoque, M., Islam, M., Islam, M., & Islam, M. (2015). Effect of cow dung, biogas slurry and vermicompost on phosphorus adsorption behavior of soil. *Journal of Scientific Research, 7*(3), 167–175. https://doi.org/10.3329/jsr.v7i3.23756

Senthil, C., & Lee, C. W. (2021). Biomass-derived biochar materials as sustainable energy sources for electrochemical energy storage devices. *Renewable and Sustainable Energy Reviews, 137*, 110464. https://doi.org/10.1016/j.rser.2020.110464

Silva Lima AF da, de Sousa Almeida M, Gomes KJS, Gomes de Sousa G, Freire MH da C, Marques VB (2022) Potassium fertilization and salt stress on beet crop. J Plant Nutr 45(7):984–991. https://doi.org/10.1080/01904167.2021.1994597

Sulistiani R, Rosmayati, Siregar LAM, Harahap F (2020) The Effects of Temperature and Potassium Fertilizer on the Growth, Yield, and Biochemical Parameters of Ipomoea batatas var. Antin-1. Acta Agrobot 73(3). https://doi.org/10.5586/aa.7337

Shang, Z., Abdalla, M., Albanito, F., Zhou, F., Xia, L., & Smith, P. (2020). Measurement of n2o emissions over the whole year is necessary for estimating reliable emission factors. *Environmental Pollution, 259*, 113864. https://doi.org/10.1016/j.envpol.2019.113864

Sharma, R., Jasrotia, K., Singh, N., Ghosh, P., Srivastava, S., Sharma, N. R., Singh, J., Kanwar, R., & Kumar, A. (2020). A comprehensive review on hydrothermal carbonization of biomass and its applications. *Chemistry Africa, 3*, 1–19. https://doi.org/10.1007/s42250-019-00098-3

Shirzadian-Khorramabad, R., Jing, H., Everts, G., Schippers, J., Hille, J., & Dijkwel, P. (2010). A mutation in the cytosolic o-acetylserine (thiol) lyase induces a genome-dependent early leaf death phenotype in Arabidopsis. *BMC Plant Biology, 10*(80), 1–12. https://doi.org/10.1186/1471-2229-10-80

Simone, M., Barontini, F., Nicolella, C., & Tognotti, L. (2012). Gasification of pelletized biomass in a pilot scale downdraft gasifier. *Bioresource Technology, 116*, 403–412. https://doi.org/10.1016/j.biortech.2012.03.119

Singh, O., & Sharma, V. (1972). Alterations in growth and metabolism of potato plants by calcium deficiency. *Plant and Soil, 36*(1–3), 363–369. https://doi.org/10.1007/bf01373490

Singh, S., Sarkar, D., Mehjabeen, Bhudevi, M., Rakesh, S., Singh, R., Kar, S., & Rakshit, A. (2018). Advanced forms of sulphur formulations for improving use efficiency in crop species. *Annual Research and Review in Biology, 27*(1), 1–14. https://doi.org/10.9734/arrb/2018/39763

Singh, A., Singh, A., & Singh, S. (2022). Sulphur nutrition and its response on the seed, oil and protein yield of Indian mustard (*Brassica juncea* L.) genotypes. *Plant Archives, 22*(2), 339–344. https://doi.org/10.51470/plantarchives.2022.v22.no2.059

Sirohi, G., Upadhyay, A., Srivastava, P., & Srivastava, S. (2015). PGPR mediated zinc biofertilization of soil and its impact on growth and productivity of wheat. *Journal of Soil Science and Plant Nutrition, 15*(1), 202–216. https://doi.org/10.4067/s0718-95162015005000017

Sivitz, A., Hermand, V., Curie, C., & Vert, G. (2012). *Arabidopsis* bHLH100 and bHLH101 control iron homeostasis via a fit-independent pathway. *PLoS One, 7*(9), e44843. https://doi.org/10.1371/journal.pone.0044843

Siwik-Ziomek, A., Lemanowicz, J., & Koper, J. (2016). Arylsulphatase activity and sulphate content in relation to crop rotation and fertilization of soil. *International Agrophysics, 30*(3), 359–367. https://doi.org/10.1515/intag-2015-0098

Smith, G. J., McDowell, R. W., Daly, K., hUallacháin, D. O., Condron, L., & Fenton, O. (2023). Factors controlling shallow subsurface dissolved reactive phosphorus concentration and loss kinetics from poorly drained saturated grassland soils. *Journal of Environmental Quality, 52*(2), 355–366. https://doi.org/10.1002/jeq2.20442

Staugaitis, G., Brazienė, Z., Marcinkevičius, A., Mažeika, R., Antanaitis, Š., & Staugaitienė, R. (2014). Spring barley as affected by nitrogen and sulphur fertiliser rates calculated using different diagnostic methods. *Zemdirbyste-Agriculture, 101*(4), 373–380. https://doi.org/10.13080/z-a.2014.101.047

Subedi, R., Taupe, N., Pelissetti, S., Petruzzelli, L., Bertora, C., Leahy, J. J., & Grignani, C. (2016). Greenhouse gas emissions and soil properties following amendment with manure-derived biochars: Influence of pyrolysis temperature and feedstock type. *Journal of Environmental Management, 166*, 73–83. https://doi.org/10.1016/j.jenvman.2015.10.007

Sun, Y., Yang, R., Li, L., & Huang, J. (2017). The magnesium transporter MGT10 is essential for chloroplast development and photosynthesis in Arabidopsis thaliana. *Molecular Plant, 10*, 1584–1587. https://doi.org/10.1016/j.molp.2017.09.017

Tang, S., Shao, N., Zheng, C., Yan, F., & Zhang, Z. (2019). Amino-functionalized sewage sludge-derived biochar as sustainable efficient adsorbent for Cu (II) removal. *Waste Management, 90*, 17–28. https://doi.org/10.1016/j.wasman.2019.04.042

Tessema, M., Brouwer, I., Feskens, E., Belachew, T., Zerfu, D., Belay, A., Demelash, Y., & Gunaratna, N. (2019). Soil zinc is associated with serum zinc but not with linear growth of children in Ethiopia. *Nutrients, 11*(2), 221. https://doi.org/10.3390/nu11020221

Tian, Z., Liu, X., Yu, J., Gu, S., Zhang, L., Jiang, D., Cao, W., & Dai, T. (2020). Early nitrogen deficiency favors high nitrogen recovery efficiency by improving deeper soil root growth and reducing nitrogen loss in wheat. *Archives of Agronomy and Soil Science, 66*(10), 1384–1398. https://doi.org/10.1080/03650340.2019.1671972

Toky, O., Kumar, P., & Khosla, P. (1989). Structure and function of traditional agroforestry systems in the western Himalaya. II. Nutrient cycling. *Agroforestry Systems, 9*(1), 71–89. https://doi.org/10.1007/bf00120156

Tran, D., & Camps, C. (2021). Early diagnosis of iron deficiency in commercial tomato crop using electrical signals. *Frontiers in Sustainable Food Systems, 5*, 631529. https://doi.org/10.3389/fsufs.2021.631529

Trewavas, A., & Knight, M. (1994). Mechanical signalling, calcium and plant form. *Plant Molecular Biology, 26*(5), 1329–1341. https://doi.org/10.1007/bf00016478

Tripolskaja, L., & Verbylienė, I. (2014). The effect of different forms of nitrogen fertilizers on nitrogen leaching. *Zemdirbyste-Agriculture, 101*(3), 243–248. https://doi.org/10.13080/z-a.2014.101.031

Trozzo, L., Francioni, M., Kishimoto-Mo, A. W., Foresi, L., Bianchelli, M., Baldoni, N., D'Ottavio, P., & Toderi, M. (2020). Soil N_2O emissions after perennial legume termination in an alfalfa-wheat crop rotation system under Mediterranean conditions. *Italian Journal of Agronomy, 15*(3), 229–238. https://doi.org/10.4081/ija.2020.1613

Uchida, R. (2000). Essential nutrients for plant growth: Nutrient functions and deficiency symptoms. In J. A. Silva & R. Uchida (Eds.), *Plant nutrient management in Hawaii's soils, approaches for tropical and subtropical agriculture* (pp. 31–55). College of Tropical Agriculture and Human Resources, University of Hawaii at Manoa.

Udverdi, M., Brodie, E. L., Riley, W., Kaeppler, S., & Jonathan, L. (2015). Impacts of agricultural nitrogen on the environment and strategies to reduce these impacts. *Procedia Environmental Sciences, 29*, 303. https://doi.org/10.1016/j.proenv.2015.07.275

Ulén, B., & Etana, A. (2014). Phosphorus leaching from clay soils can be counteracted by structure liming. *Acta Agriculturae Scandinavica Section B – Soil and Plant Science, 64*(5), 425–433. https://doi.org/10.1080/09064710.2014.920043

Uwitonze, A., & Razzaque, M. (2018). Role of magnesium in vitamin D activation and function. *The Journal of the American Osteopathic Association, 118*(3), 181–189. https://doi.org/10.7556/jaoa.2018.037

Vadlamudi, K., Upadhyay, H., Singh, A., & Reddy, M. (2020). Influence of zinc application in plant growth: An overview. *European Journal of Molecular & Clinical Medicine, 7*(7), 2321–2327.

Van Der Heijden, M. G., Bardgett, R. D., & Van Straalen, N. M. (2008). The unseen majority: Soil microbes as drivers of plant diversity and productivity in terrestrial ecosystems. *Ecology Letters, 11*(3), 296–310. https://doi.org/10.1111/j.1461-0248.2007.01139.x

Vanden Bygaart, A. J., Geng, X., & He, J. (2021). Spatial decoupling of legacy phosphorus in cropland: Soil erosion and deposition as a mechanism for storage. *Soil and Tillage Research, 211*, 105050. https://doi.org/10.1016/j.still.2021.105050

Venkatesh, G., Gopinath, K. A., Reddy, K. S., Reddy, B. S., Prabhakar, M., Srinivasarao, C., Visha Kumari, V., & Singh, V. K. (2022). Characterization of biochar derived from crop residues for soil amendment, carbon sequestration and energy use. *Sustainability, 14*(4), 2295. https://doi.org/10.3390/su14042295

Viets, F. G., Boawn, L. C., & Crawford, C. L. (1954). Zinc content of bean plants in relation to deficiency symptoms and yield. *Plant Physiology, 29*(1), 76–79. https://doi.org/10.1104/2Fpp.29.1.76

Van Laere, J., Willemen, A., Ding, Y., Said, H., Resch, C., Hood-Nowotny, R., Merckx, R., & Dercon, G. (2020). Potassium application to alleviate drought stress in cassava production: A growth chamber based carbon-13 pulse labelling experiment. In EGU General Assembly Conference Abstracts (p. 4839). https://ui.adsabs.harvard.edu/link_gateway/2020EGUGA..22.4839V/doi:10.5194/egusphere-egu2020-4839

Wakeel, A., & Ishfaq, M. (2022). Potassium in plants. In *Potash use and dynamics in agriculture* (pp. 19–27). Springer Singapore. https://doi.org/10.1007/978-981-16-6883-8

Wang, R. (2024). Molecular mechanism of exogenous magnesium in regulating cation homeostasis in roots of peanut seedlings under salt stress. *Agronomy, 14*(4), 724. https://doi.org/10.3390/agronomy14040724

Wang, Z., Ma, Z., Wang, C., & Lin, T. (2010). Exogenous Ca^{2+} alleviates nitrogen and water deficit, and improves growth of wheat (*Triticum aestivum*) seedlings exposed to high temperature. *Plant Growth Regulation, 61*(3), 223–229. https://doi.org/10.1007/s10725-010-9466-9

Wang, X., Wang, X., Zhang, H., & Jiang, W. (2014). Effects of sulphur application times on the sulphur accumulation and distribution for summer maize in fluvo-aquic soil. *Advance Journal of Food Science and Technology, 6*(6), 797–801. https://doi.org/10.19026/ajfst.6.113

Wang, L., Li, A., & Chang, Y. (2016). Hydrothermal treatment coupled with mechanical expression at increased temperature for excess sludge dewatering: Heavy metals, volatile organic compounds and combustion characteristics of hydrochar. *Chemical Engineering Journal, 297*, 1–10. https://doi.org/10.1016/j.cej.2016.03.131

Wang, Y., Zeng, Z., Tian, X., Dai, L., Jiang, L., Zhang, S., & Ruan, R. (2018). Production of bio-oil from agricultural waste by using a continuous fast microwave pyrolysis system. *Bioresource Technology, 269*, 162–168. https://doi.org/10.1016/j.biortech.2018.08.067

Wang, Y., Martins, L., Sermons, S., & Balint-Kurti, P. (2020). Genetic and physiological characterization of a calcium deficiency phenotype in maize. *G3 Genes|Genome|Genetics, 10*(6), 1963–1970. https://doi.org/10.1534/g3.120.401069

Ward, M., Aslam, M., & Huffaker, R. (1986). Enhancement of nitrate uptake and growth of barley seedlings by calcium under saline conditions. *Plant Physiology, 80*(2), 520–524. https://doi.org/10.1104/pp.80.2.520

Whalley, H., & Knight, M. (2012). Calcium signatures are decoded by plants to give specific gene responses. *New Phytologist, 197*(3), 690–693. https://doi.org/10.1111/nph.12087

White, P. J. (2013). Improving potassium acquisition and utilisation by crop plants. *Journal of Plant Nutrition and Soil Science, 176*, 305–316. https://doi.org/10.1002/jpln.201200121

Wiedemeier, D. B., Abiven, S., Hockaday, W. C., Keiluweit, M., Kleber, M., Masiello, C. A., McBeath, A. V., Nico, P. S., Pyle, L. A., Schneider, M. P., & Smernik, R. J. (2015). Aromaticity and degree of aromatic condensation of char. *Organic Geochemistry, 78*, 135–143. https://doi.org/10.1016/j.orggeochem.2014.10.002

Wu, J., Pan, Y., Tan, P., & Liu, T. (2011). Effect of acid deposition on soil Ca^{+2} contents of two forest ecosystems in Dinghushan biosphere reserve. *Advanced Materials Research, 356–360*, 2601–2605. https://doi.org/10.4028/www.scientific.net/amr.356-360.2601

Xiao, R., Wang, J. J., Gaston, L. A., Zhou, B., Park, J. H., Li, R., Dodla, S. K., & Zhang, Z. (2018). Biochar produced from mineral salt-impregnated chicken manure: Fertility properties and potential for carbon sequestration. *Waste Management., 78*, 802–810. https://doi.org/10.1016/j.wasman.2018.06.047

Xie, X., Hu, W., Fan, X., Chen, H., & Tang, M. (2019). Interactions between phosphorus, zinc, and iron homeostasis in nonmycorrhizal and mycorrhizal plants. *Frontiers in Plant Science, 10*, 1172. https://doi.org/10.3389/fpls.2019.01172

Xu, Z., Jiang, X., Xia, T., Wei, X., Wang, L., Zhu, L., & Gao, Z. (2023). Characteristics of phosphorus loss from organic farming and forestry land under simulated rainfall in the mountainous areas of western Anhui. *Polish Journal of Environmental Studies, 32*(5), 4907–4918. https://doi.org/10.15244/pjoes/168719

Yaashikaa, P. R., Kumar, P. S., Varjani, S., & Saravanan, A. J. B. R. (2020). A critical review on the biochar production techniques, characterization, stability and applications for circular bioeconomy. *Biotechnology Reports (Amsterdam, Netherlands), 21*(28), e00570. https://doi.org/10.1016/j.btre.2020.e00570

Yadav, S. (2023). Impact of sulphur nanoparticles on growth and biological yield of groundnut-mustard crop sequence under sandy loam soils of central Gujarat. *International Journal of Plant and Soil Science, 35*(20), 1105–1112. https://doi.org/10.9734/ijpss/2023/v35i203907

Yadav, R., Sharma, S., Varma, A., Rajawat, M., Khan, M., Sharma, P. K., Malviya, D., Singh, U. B., Rai, J. P., & Saxena, A. K. (2022). Modulation in biofertilization and biofortification of wheat crop by inoculation of zinc-solubilizing rhizobacteria. *Frontiers in Plant Science, 13*, 777771. https://doi.org/10.3389/fpls.2022.777771

Yakout, S. M. (2017). Physicochemical characteristics of biochar produced from rice straw at different pyrolysis temperature for soil amendment and removal of organics. *Proceedings of the National Academy of Sciences, India Section A: Physical Sciences, 87*, 207–214. https://doi.org/10.1007/s40010-017-0343-z

Yang, T., Yang, S., Chen, Z., Tan, Y., Bol, R., Duan, H., & He, J. (2022). Global transcriptomic analysis reveals candidate genes associated with different phosphorus acquisition strategies among soybean varieties. *Frontiers in Plant Science, 13*. https://doi.org/10.3389/fpls.2022.1080014

Yao, X., Wang, J., He, M., Liu, Z., Zhao, Y., Li, Y., Chi, T., Zhu, L., Zheng, P., Jetten, M. S. M., & Hu, B. (2024). Methane-dependent complete denitrification by a single Methylomirabilis bacterium. *Nature Microbiology, 9*(2), 464–476. https://doi.org/10.1038/s41564-023-01578-6

Yazhini, R. (2023). Synthesis and characterization of nano sulphur: Exploring its potential as slow release fertilizer. *Journal of Applied and Natural Science, 15*(3), 937–944. https://doi.org/10.31018/jans.v15i3.4665

Yläranta, T., Uusi-Kämppä, J., & Jaakkola, A. (1993). Leaching of nitrogen in barley, grass ley and fallow lysimeters. *Agricultural and Food Science, 2*(4), 281–291. https://doi.org/10.23986/afsci.72651

You, S., Ok, Y. S., Chen, S. S., Tsang, D. C. W., Kwon, E. E., Lee, J., & Wang, C. H. (2017). A critical review on sustainable biochar system through gasification: Energy and environmental applications. *Bioresource Technology, 246*, 242–253. https://doi.org/10.1016/j.biortech.2017.06.177

Younas, N., Fatima, I., Ahmad, I. A., & Ayyaz, M. K. (2023). Alleviation of zinc deficiency in plants and humans through an effective technique; biofortification: A detailed review. *Acta Ecologica Sinica, 43*(3), 419–425. https://doi.org/10.1016/j.chnaes.2022.07.008

Zakari, S. A., Asad, M. A. U., Han, Z., Zhao, Q., & Cheng, F. (2020). Relationship of nitrogen deficiency-induced leaf senescence with ROS generation and ABA concentration in rice flag leaves. *Journal of Plant Growth Regulation, 39*(4), 1503–1517. https://doi.org/10.1007/s00344-020-10128-x

Zenda, T., Liu, S., Dong, A., & Duan, H. (2021). Revisiting sulphur—The once neglected nutrient: It's roles in plant growth, metabolism, stress tolerance and crop production. *Agriculture, 11*(7), 626. https://doi.org/10.3390/agriculture11070626

Zhang, H. (2023). Aluminum supplementation mediates the changes in tea plant growth and metabolism in response to calcium stress. *International Journal of Molecular Sciences, 25*(1), 530. https://doi.org/10.3390/ijms25010530

Zhang, Y., Xun, P., Chen, C., Lu, L., Shechter, M., Rosanoff, A., & He, K. (2020a). Magnesium levels in relation to rates of preterm birth: A systematic review and meta-analysis of ecological, observational, and interventional studies. *Nutrition Reviews, 79*(2), 188–199. https://doi.org/10.1093/nutrit/nuaa028

Zhang, B., Cakmak, I., Feng, J., Yu, C., Chen, X., Xie, D., Wu, L., Song, Z., Cao, J., & He, Y. (2020b). Magnesium deficiency reduced the yield and seed germination in wax gourd by affecting the carbohydrate translocation. *Frontiers in Plant Science, 11*, 797. https://doi.org/10.3389/fpls.2020.00797

Zhang, T., Xi, Y., & Ma, X. (2021a). Finding optimal nutrients with farming techniques to influence maize productivity, and mitigate global warming potential under water-saving agriculture in semi-arid regions. *Research Square*. https://doi.org/10.21203/rs.3.rs-899404/v1

Zhang, T., Hao, P., Yang, B., Cao, H., Liao, B., & Zheng, X. (2021b). Effects of partial Blackwater substitution on soil potential nitrogen leaching in a summer maize field on the North China plain. *Life, 12*(1), 53. https://doi.org/10.3390/life12010053

Zhang, X., Wang, Y., Wen, J., Zhang, Y., Su, S., Wen, Y., Yan, M., Bai, L., Wu, C., & Zeng, X. (2021c). The C/N ratio and phenolic groups of exogenous dissolved organic matter together as an indicator for evaluating the stability of mineral-organic associations in red soil. *Journal of Soils and Sediments, 21*, 821–831. https://doi.org/10.1007/s11368-020-02874-y

Zhang, B., Zhi-jian, G., Zhi, J., Bai, X., Yang, L., & Xia, W. (2023). Effects of irrigation and nitrogen fertilizer on soil carbon leaching in cotton fields in arid areas. *Sustainability, 15*(14), 11356. https://doi.org/10.3390/su151411356

Zhang, Y., Li, L., Zhang, F., Wu, X., Liu, J., Wei, J., & Zheng, J. (2024). Effect of soil amendment of greening waste on leaching loss of soil nitrogen and phosphorus. *Journal of Physics Conference Series, 2679*(1), 012043. https://doi.org/10.1088/1742-6596/2679/1/012043

Zhao, Y., Chen, X., Yi, S., Liu, C., Huang, B., & Zhao, M. (2012). Impact of fertilization and soil phosphorus status on phosphorus leaching from soil in vegetable greenhouse. *Advanced Materials Research, 610–613*, 2968–2973. https://doi.org/10.4028/www.scientific.net/amr.610-613.2968

Zhao, X., Gao, S., Lu, D., Chen, X., Wang, H., & Zhang, J. (2021). Evaluation of soil potassium-holding capacity based on waterlogging-simulation experiments. *Agronomy Journal, 113*(3), 2872–2883. https://doi.org/10.1002/agj2.20697

Zheng, W., Hassan, M., Nadeem, F., Wu, L., Zhang, F., & Li, X. (2020). Magnesium fertilization improves crop yield in most production systems: A meta-analysis. *Frontiers in Plant Science, 10*. https://doi.org/10.3389/fpls.2019.01727

Zhou, X. (2024). Comparative transcriptomic analysis revealed important processes underlying the static magnetic field effects on *Arabidopsis*. *Frontiers in Plant Science, 15*. https://doi.org/10.3389/fpls.2024.1390031

Zhu, X., Yang, S., Wang, L., Liu, Y., Qian, F., Yao, W., Zhang, S., & Chen, J. (2016). Tracking the conversion of nitrogen during pyrolysis of antibiotic mycelial fermentation residues using XPS and TG-FTIR-MS technology. *Environmental Pollution, 211*, 20–27. https://doi.org/10.1016/j.envpol.2015.12.032

Zuber, H., Davidian, J., Wirtz, M., Hell, R., Belghazi, M., Thompson, R., & Gallardo, K. (2010). *Sultr4;1* mutant seeds of *Arabidopsis* have an enhanced sulphate content and modified proteome suggesting metabolic adaptations to altered sulphate compartmentalization. *BMC Plant Biology, 10*(1), 78. https://doi.org/10.1186/1471-2229-10-78

Zumft, W. G. (1997). Cell biology and molecular basis of denitrification. *Microbiology and Molecular Biology Reviews, 61*(4), 533–616. https://doi.org/10.1128/mmbr.61.4.533-616.1997

Biochar-Mediated Restoration: Transforming Invasive Plants and Crop Residues into Ecological Assets

S. Saraniya, Rupali Nayal, Manu Thomas, Anuj Tripathi Kumar, and Shalini Dhyani

Abstract

Land degradation is a major concern affecting biodiversity, ecosystem services, livelihoods, and access to food. This issue is exacerbated by the proliferation of Invasive Alien Species (IAPs), which destroy natural habitats and stifle native species. Biochar, a carbon-rich material generated during the pyrolysis of organic wastes, promises a viable solution to mitigate the negative impacts of IAPs, while also enhancing soil health. This study attempts to investigate the possibilities of biochar production from crop waste and invasive plant materials and understand its potential role in restoring damaged ecosystems. The paper examines initiatives made for the production of biochar, evaluates the related policy and governance frameworks, and extracts lessons from a number of case studies. Additionally, the challenges and impediments to the adoption of biochar and possible solutions were explored. Enhancing soil fertility and sequestering carbon, biochar helps to support sustainable agriculture practices by converting ecological risks into valuable resources. The incorporation of biochar into circular economy models presents economic potential, decreases environmental degradation, and further encourages waste valuation in the context of agricultural and land restoration. The study intends to demonstrate the value of biochar in ecological restoration, sustainable agriculture, and climate change mitigation.

Keywords

Invasive Alien Species · Carbon sequestration · Land restoration · Soil amendment · Sustainability

S. Saraniya · R. Nayal · M. Thomas · A. T. Kumar · S. Dhyani (✉)
CSIR-National Environmental Engineering Research Institute, Nehru Marg, Nagpur, Maharashtra, India

S. Singh, S. Srivastava (eds.), *Biochar: A Precious Resource from Biological Waste*,
https://doi.org/10.1007/978-981-95-0425-1_11

1 Introduction

Land degradation is defined as an alteration in soil health that reduces the ecosystem's ability to produce commodities and services for its users (FAO, 2014). Land degradation is a combination of numerous variables, such as unsustainable farming methods, the introduction of alien species, excessive grazing, deforestation, soil erosion, climate fluctuations, etc. Over 30% of forests, 20% of cultivated areas, and 10% of grasslands globally are degraded to varied degrees, which negatively impact the lives of approximately 1.5 billion people (Bai et al., 2008). It is estimated that invasive species cause economic losses exceeding $33 billion annually across Southeast Asia alone (Liu et al., 2021). A 2019 report by the Asian Development Bank (ADB) highlights that invasive species can significantly reduce agricultural productivity in Asia, with losses ranging from 10% to 90% depending on the crop and invasive species.

According to an extensive study conducted by the Space Application Centre (SAC) of the Indian Space Research Organisation (ISRO), 96.40 million hectares, or 29.32% of India's total geographic area (TGA), were classified as degraded during the 2011–2013 period, compared to 94.53 million hectares (28.76% of TGA) in 2003–2005. This reflects an increase of 1.87 million hectares in degraded land over less than a decade (Gupta, 2019).

India, one of 12 global centers of origin and diversity for diverse floral species, is home to 8% of the world's biodiversity (El-Barougy et al., 2021). Over the past 50 years, biological invasions have surged, contributing to environmental degradation (Jackson et al., 2002). Invasive alien plant species (IAPs) now dominate native ecosystems, disrupting biodiversity and ecosystem processes. This has led to local extinctions and negatively impacted livelihoods by suppressing native species (El-Barougy et al., 2021).

United Nation's (UN) Intergovernmental Platform for Biodiversity and Ecosystem Services (IPBES) estimated that biotic invaders pose a threat to around 25% of the Earth's terrestrial surface, making them a primary cause of biodiversity loss (IPBES, 2019). The fast-paced spread of invasive alien plants (IAPs) in the Anthropocene has significantly impacted both living and non-living elements of ecosystems (Rathee et al., 2021). IAPs have contributed to land degradation by changing nutrient and pollutant cycles, microbial diversity, habitat structures, hydro-morphology, and soil chemistry (Qu et al., 2021). Moreover, decreases in agricultural output and significant financial losses are being brought on by land degradation, endangering farmers' livelihoods and access to food (Bhattacharyya et al., 2015). As a result, global food security is thus at risk due to the deterioration of soil productivity and nutrients (Ortiz-Bobea et al., 2021). Soil amendments are added to the soil to restore deteriorated areas, and recently, there is growing interest in biochar utilization as one of the innovative amendments.

Biochar is a material that has been pyrolyzed and is produced by burning organic matter (waste from animals and plants) at high temperatures without oxygen (Lehmann, 2007). The most common method for biochar production for agricultural and environmental uses involves slowly pyrolyzing organic matter at

350–700 °C with limited oxygen. Changes in temperature, source material, moisture content, and pyrolysis duration are important variables to take into account while producing biochar. For biochar production, a variety of organic raw materials can be used, including those from agroforestry, invasive plants, weeds (wood, wood shavings, tree bark, roots, and shells), animal-related sources (beddings and animal waste products), agricultural residues (corn stalks, root residues, rice husks and forage crops), industrial waste (carbon black and bagasse), agro-urban biosolids, and other carbon-rich materials (Das et al., 2021). In recent years, the application of various biochar sources has gained a lot of attention because of its ability to be added to soil as a soil amendment to reduce compaction, increase soil porosity, and improve pore characteristics, high adsorption capacity, surface activity, water holding, and nutrients (Liu et al., 2018).

As an emerging environmental protection material, biochar is recognized for being cost-effective and efficient (Beesley et al., 2011). Producing biochar from invasive plants is a viable strategy to mitigate the ecological damage caused by these species (Feng et al., 2021). The process of producing biochar from invasive plants enables the efficient use of waste and, to some extent, restricts the spread of invasive species. Thus, the problem of invasive plants can be minimized. Invasive plants are so abundant and diverse, utilizing them as the raw material for biochar will be less expensive and more convenient than using other biomass wastes (Wang et al., 2021). Biochar from invasive plants has been mainly used in agricultural soil enhancement and pollutant adsorption (Beckinghausen et al., 2020).

According to reports, the hydrophilic surface bond (OH) of fine-grained water hyacinth biochar increased water retention capacity by decreasing soil macropores and facilitating capillary action (Bordoloi et al., 2019). Through field trials, it was reported that *Lantana camara* biochar greatly increased soil total porosity, pH, organic carbon, total nitrogen, and accessible potassium and phosphorus while significantly reducing exchangeable acidity and bulk density (Berihun et al., 2017). Furthermore, maize treated with *Eupatorium adenophorum* biochar showed a considerable improvement in biomass and grain production. This was partially caused by modifications in the physical characteristics of the soil, such as enhanced air permeability and drainage of heavy clay following the addition of biochar (Obia et al., 2018). Since biochar is typically alkaline, it can be added to soil as a soil modification to increase pH and neutralize acidity (Yuan et al., 2011). Biochar is an excellent soil conditioner, hence effective to improve the quality of soil. By modifying the physical, chemical, and microbiological characteristics of the soil, it improves the soil environment (Yang et al., 2022).

Present chapter focuses on the production of biochar from invasive plants to address ecological issues brought about by these unwanted species while, offering financial benefits, maximizing waste resource utilization, as an effective climate change mitigation strategy by satisfying sustainable use criteria.

2 Biochar and Circular Economy

In recent years, the concept of the circular economy has gained prominence as a response to the challenges of traditional linear economic models, which often lead to resource depletion, environmental degradation, and waste generation. The circular economy focuses on keeping resources in use for as long as possible through practices like reuse, repair, and recycling (Temmes & Peck, 2020). One area where the principles of the circular economy can be particularly impactful is the bioeconomy, which focuses on the production and utilization of renewable, biobased resources. Within the bioeconomy, the concept of the circular bioeconomy has emerged, which combines the principles of the circular economy with the production and use of biobased materials and products (Bhat & Bhat, 2021). Biochar is one example of a biobased product that can be integrated into a circular bioeconomy. Biochar has a wide range of potential applications, including as a soil amendment, a carbon sequestration agent, and a feedstock for the production of biobased chemicals and fuels (Singh et al., 2022).

The production of biochar from waste streams, such as sewage sludge and microalgae, can be particularly beneficial in the context of the circular economy, as it allows for the valorization of these waste materials by converting them into valuable products that can be reintegrated into various processes, such as agriculture, energy production, etc. (Konwar et al., 2018; Bolognesi et al., 2021). By incorporating biochar into a circular bioeconomy framework, it is possible to create closed-loop systems that minimize waste, maximize resource utilization, and contribute to the overall sustainability of the bioeconomy (Holden et al., 2023).

The circular economy also emphasizes the reuse of by-products, and the biochar ecosystem offers numerous opportunities in this regard. For instance, the ash generated during the combustion of biochar can be repurposed as a source of nutrients for plant growth, thereby closing the nutrient loop and reducing the reliance on external inputs (Rafiullah et al., 2017; Kurniawan et al., 2023).

Additionally, the syngas and bio-oil produced during the pyrolysis process can be utilized as energy sources, either for on-site operations or for broader energy generation or as feedstocks for other industrial processes (Digman et al., 2009; Srinivasan et al., 2015). This integration of various by-products within the biochar system exemplifies the principles of the circular economy (Fig. 1).

Biochar Production from Agricultural Residues and Invasive Species

Each year, approximately 500 million tons of agricultural waste are recorded globally. The harvesting and processing of crops like oil palm, sorghum, peanuts, corn, sugarcane, and wheat generate substantial agricultural residues (Pehlivan et al., 2013). Converting these wastes into biochar not only reduces disposal costs but also provides an opportunity to utilize biochar in various sectors. Furthermore, other types of waste can also be used as feedstocks, though this might incur additional costs (Amalina et al., 2022).

Biochar production from agricultural residues aligns with circular economy principles by reducing waste, helping to divert them from landfills or open burning and

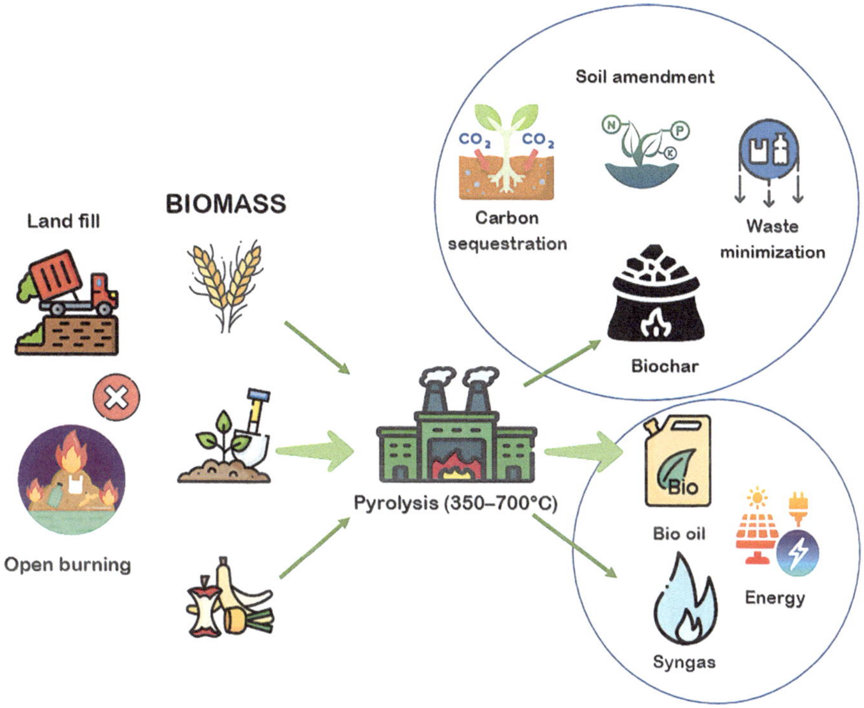

Fig. 1 Integrating biochar production into the circular economy

closing nutrient loops in agriculture (Feng et al., 2021; Patel & Panwar, 2023; Zhu et al., 2022). Biochar can be used contributing to sustainability across multiple sectors (Singh et al., 2022). Production methods, such as microwave-assisted pyrolysis, can be optimized using machine learning techniques to improve efficiency and reduce costs (Rex et al., 2023). The development of a biochar economy offers opportunities for job creation and increased GDP, particularly through small-scale enterprises (Jindo et al., 2020; Rex et al., 2023).

Interestingly, biochar production from invasive plant species also offers a promising solution for waste management and circular economy, particularly in agriculture sector. Invasive plants, which pose threats to ecosystems, can be converted into biochar, providing multiple benefits (Feng et al., 2021). This process manages invasive species and creates a valuable resource for soil amendment, carbon sequestration, remediating heavy metal pollution, etc. (Singh et al., 2022; Ceriani et al., 2024). The characteristics of biochar vary depending on the species and production conditions, with woody and herbaceous invasive species showing different properties suitable for various applications (Ahmed et al., 2020).

Global Relevance of Using Invasive Plants and Crop Residues in Biochar Preparation
In various regions, biochar production and application have demonstrated significant potential for improving soil health, enhancing crop productivity, and contributing to environmental sustainability. For instance, in India, biochar derived from invasive weeds like *Lantana camara* and *Parthenium hysterophorus* has been shown to improve fodder yield, quality, and soil microbial activity when used as a soil amendment along with inorganic fertilizers. However, ensuring consistent effectiveness across different weed sources and gaining farmer acceptance pose challenges. The demonstrated benefits in both agriculture and sustainable weed management could drive adoption and policy support (Choudhary et al., 2023).

Similarly, in coal mine-degraded lands, biochar from these invasive weeds has proven effective in improving soil properties, increasing enzyme activity, and enhancing carbon sequestration potential. Yet, variability in biochar performance across different soil types presents a challenge to the consistency of land restoration outcomes (Ghosh & Maiti, 2023).

In another Indian context, converting agricultural residues, such as rice husks and corn stover, into biochar addresses the issue of crop residue burning by enhancing soil fertility, sequestering carbon, and boosting crop production. However, the adoption of this practice at the farm level faces challenges related to cost and scalability (Mohan et al., 2018). Nonetheless, biochar presents a viable and sustainable solution to the problem of biomass burning, making it particularly relevant to Indian agriculture (Table 1).

Expanding to a global perspective, in Vietnam, the potential for producing biochar from rice straw at the household scale has been explored, with findings indicating reductions in agricultural waste, improved soil health, and the provision of useful thermal energy for household activities like cooking. Despite these benefits, the feasibility of large-scale adoption is constrained by resource availability and technical knowledge at the household level (Hung & Ky, 2018).

In China, biochar production from variety of invasive plants has shown promise in decreasing the bioavailability of cadmium in contaminated soils, providing a sustainable method for environmental management. Yet, the variability in biochar properties depending on the source species may affect the consistency of its remediation effects (Wang et al., 2024).

In southern Ethiopia, biochar applications have effectively mitigated soil acidity and enhanced other physicochemical soil properties, supporting sustainable agriculture in the region. However, the variability in outcomes depending on soil type suggests that broader applications require careful consideration (Berihun et al., 2017). Thailand's research has demonstrated that biochar produced from agricultural and wood residues possesses properties conducive to soil amendment and carbon sequestration. Nevertheless, variability in biochar quality can pose challenges for its application across different environments (Wijitkosum, 2022). Moreover, in China, the combination of biochar with metal oxides (Fe) from water hyacinth has been effective in degrading pollutants such as 2,4,6-trichlorophenol and coal gasification

Table 1 Case studies of biochar production from invasive species

Country	Initiative	Benefits	Challenges	Enabling conditions	References
India	Biochar production from invasive weed (*Parthenium hysterophorus* L. *and Lantana camara* L.)	Improve fodder yield, quality, plant growth, and soil microbial activity when used as a soil amendment with inorganic fertilizers	Ensuring consistent biochar effectiveness from different invasive weeds and convincing farmers to adopt this novel practice pose significant challenges	Demonstrated benefits in fodder yield, soil health, and sustainable weed management can drive farmer adoption and policy support	Choudhary et al. (2023)
India	Application of biochar made from invasive weeds to coal mine degraded land	Improved the soil properties, increased enzyme activity, and carbon sequestration potential of the soil	Variability in biochar effectiveness across different soil types may affect the consistency of land restoration outcomes	The study shows that invasive weed biochar can enhance soil enzyme activity and carbon sequestration, supporting its use in land restoration efforts	Ghosh and Maiti (2023)
India	Solution to burning of biomass by converting agricultural waste (rice husks and corn stover) into biochar	Improve soil fertility, increase crop production, and sequester carbon	Adoption of biochar production from crop residues faces challenges due to the cost and scalability at the farm level	Biochar offers a sustainable solution to crop-residue burning, improving soil fertility and aiding in carbon sequestration, making it a viable option for Indian agriculture	Mohan et al. (2018)

(continued)

Table 1 (continued)

Country	Initiative	Benefits	Challenges	Enabling conditions	References
India	Biochar produced from *Pontederia crassipes*, an invasive species	Effectively remove cadmium from aqueous solutions, with a high Cd sorption capacity of 70.3 mg/g	The effectiveness of biochar in removing heavy metals like Cd may vary depending on the biochar's source and production method	Water hyacinth-derived biochar demonstrates high efficiency in Cd removal, offering a sustainable solution for water contamination management	Zhang et al. (2015)
Vietnam	Potential of using rice straw residue to produce biochar at the household scale	Protects soil, reduces agricultural waste, and creates useful thermal energy for household activities (e.g. cooking)	Household-scale biochar production from agricultural residues may be limited by resource availability and technical knowledge	The study highlights the potential for sustainable biochar production at the household level, offering an accessible method for improving soil health in rural Vietnam	Hung and Ky (2018)
China	Biochar production from six invasive plant species	Biochar application significantly decreased Cd bioavailability in the soils	The variability in biochar properties from different invasive species may impact its consistency in remediating cadmium-contaminated soils	Biochar produced from diverse invasive species has shown promise in improving the remediation of contaminated soils, providing a sustainable method for environmental management	Wang et al. (2024)
Southern Ethiopia	To examine the impact of biochar addition on soil acidity and various other physicochemical properties of the soil.	Increase in soil organic carbon, reduction of soil acidity	The effectiveness of biochar in addressing soil acidity and improving other soil properties may vary across different soil types in Ethiopia	The use of biochar has been shown to mitigate soil acidity and enhance soil physicochemical properties, supporting its use in sustainable agriculture in southern Ethiopia	Berihun et al. (2017)

Thailand	Biochar production from agricultural and wood residues	All five types of biochar were found to have suitable properties for use as soil amendments and for carbon sequestration	The variability in biochar quality from different agricultural and wood residues can pose challenges in its application across various environments	Biochar derived from agricultural and wood wastes supports sustainable agricultural practices, offering environmental benefits and improving soil health	Wijitkosum (2022)
China	*Pontederia crassipes* [modification with metal oxides (Fe)]	Degradation of 2,4,6-trichlorophenol and coal gasification wastewater	The complexity of integrating biochar with Fe3O4 for environmental applications may limit its practical adoption at a larger scale	The study demonstrates enhanced degradation of pollutants using biochar-supported Fe3O4, providing an innovative approach for wastewater treatment and soil remediation	Zhuang et al. (2020)
Indonesia	To convert durian seed waste into biochar	Act as soil amendment fertilizer containing excellent carbon content and elemental components such as potassium, magnesium, phosphor, and sulfur	The effectiveness of biochar as a soil amendment varies with pyrolysis temperature, complicating its standardized application	Biochar produced from agricultural waste at optimal pyrolysis temperatures has shown potential as a soil amendment, offering a sustainable solution for soil management	Mutolib et al. (2023)

wastewater. However, the complexity of integrating this technology into broader environmental applications may limit its scalability (Zhuang et al., 2020).

In Indonesia, converting agricultural waste like durian seed into biochar provides a nutrient-rich soil amendment. However, the effectiveness of this biochar can vary significantly based on pyrolysis temperatures, highlighting the need for standardization in its production. These examples illustrate the diverse applications and benefits of biochar across different regions and contexts, while also emphasizing the challenges that must be addressed to realize its full potential in sustainable agriculture and environmental management (Mutolib et al., 2023).

3 Community-Led Biochar Initiatives

Community-led biochar initiatives involving local communities in the production and use of biochar promote sustainable agricultural practices and waste management. These initiatives can be organized through local cooperatives, farmer groups, or community-based organizations.

One of the key benefits of community-led biochar initiatives is their ability to tackle waste management challenges. The community-led biochar initiatives by production and sale of biochar can provide a source of income for communities, while the use of biochar in agriculture can lead to increased crop yields and improved livelihoods for smallholder farmers (Marmiroli et al., 2018; Bolognesi et al., 2021). Ultimately, community-led biochar initiatives offer a holistic approach to addressing environmental, agricultural, and socioeconomic challenges. By empowering local communities to take an active role in the production and utilization of biochar, these initiatives can foster sustainable practices, improve food security, and contribute to the overall resilience of communities (Boyette et al., 2012). There have been initiatives all around the globe to increase biochar production and use by involving local cooperatives or smallholder farmer groups (Table 2).

In Jeli, Malaysia, the biochar initiative helped to address the issues of open dumping and burning of agricultural wastes. In spite of the enthusiasm for biochar initiatives among small groups in the region, challenges arise due to the dominance of major industry players. Biochar initiatives have improved soil fertility, produced a sustainable energy source, and improved water filtering in Nepal (Amaltari). However, the scheduling of service learning projects and climatic restrictions pose a threat to the ongoing community engagement. Obstacles like air pollution, need for village mobilization, and training hindered the local communities in pursuing biochar initiatives. In the Slateng village in East Java, biochar is primarily used as a fertilizer; however, locals' views on access to livelihood assets and dependence on government agencies were the major challenges. Finally, in India, biochar production has the potential to contribute to future technical improvement in several areas. However, it is hindered by various barriers like resource access, labor demand, and market limitations.

Table 2 Initiatives taken for biochar production

Place	Country	Benefits	Challenges	References
Jeli, Kelantan	Malaysia	Addresses the issues of open dumping and burning of agricultural waste, resulting in enhanced income sustainability	Market entry barriers due to dominance of established industry players	Abdullah et al. (2024)
Amaltari	Nepal	Improved soil fertility, sustainable energy source, water filtration, community engagement, and educational opportunities	Challenges with the timing of service learning projects due to climate constraints and ensuring continuity of engagement	Henderson et al. (2018)
Thailand	Thailand	Affordable for farmers to convert crop waste into biochar, does not require external intervention, and is viable for community-led or small NGO-driven initiatives	Air pollution concerns and the need for training and social mobilization to encourage biochar production in villages	Shafer (2020)
Slateng Village	East Java, Indonesia	Use of biochar as a fertilizer	Perceived challenges in livelihood asset access, dependence, and trust in governmental bodies	Sari et al. (2021)
Chandagal & Manoor, Karnataka	India	Potential for biochar systems to set the foundation for future novel technologies, adaptable beyond tropical regions	Limitations in resource access, labor demand, and market scalability	Müller et al. (2019)

These initiatives illustrate the diverse benefits and challenges faced by community-led biochar projects. By fostering local involvement, these projects can address both environmental and socioeconomic issues, leading to more resilient and sustainable communities. To enable meaningful comparison and wider adoption of biochar, the establishment of internationally agreed standards for its production and use is crucial (Hu et al., 2021). Currently, most studies on biochar application are conducted in small-scale laboratory settings. In this context, innovative commercial models should be integrated into the biochar-based circular economy. There is a need for increased research, innovation, and development within the biochar industry to adapt to evolving markets. Hence, there should be greater collaboration among researchers, biochar producers, consumers, and government entities to advance research and promote biochar production and use.

4 Socioeconomic Benefits

Biochar is considered a promising sustainable and eco-friendly tool for remediation because it can reduce environmental contamination, enhance soil quality, and capture carbon (Masud et al., 2023b; Li et al., 2023; Abhishek et al., 2022). It has proven to be highly effective in numerous green remediation practices, such as groundwater and soil remediation, wastewater treatment (WWT), waste management, carbon sequestration, and capturing greenhouse gases (Abhishek et al., 2022; Khalid et al., 2020; Masud et al., 2023a, b). Biochar has been employed as an adsorbent and filtration medium in water treatment, proficiently removing contaminants such as organic compounds, emerging pollutants, and heavy metals through effective adsorption and ion exchange techniques (Abhishek et al., 2023; Masud et al., 2023a, b).

The economic viability of biochar production varies greatly by region. In countries with lower labor costs, such as Ethiopia, Kenya, and Vietnam, simple kiln technologies have proven to be profitable. In underdeveloped rural regions of low-income countries, the addition of biochar to poor, often unproductive soils can have a significant positive impact on agriculture (Owsianiak et al., 2021; Robb et al., 2020; Martinsen et al., 2014). However, in middle-income countries where labor and other costs are higher, alternative technical innovations may achieve increased crop yields more effectively. Therefore, it is crucial to assess the trade-offs between different biochar production methods and alternative biowaste uses. Both environmental and economic considerations should be carefully weighed before advocating for biochar use in agriculture, especially as the economic conditions of a country improve (Owsianiak et al., 2021; Roberts et al., 2010).

Biochar produced from woody residues offers extensive benefits for forest and other ecosystem services. When applied to forest soils, biochar can significantly enhance forest health by reducing wildfire risks, improving water quality, retaining essential nutrients, and decreasing soil compaction. Producing biochar on-site is an effective way to utilize low-cost, low-quality raw materials, transforming them into valuable resources. This process not only creates economic opportunities by generating jobs and promoting rural development, but it also boosts forest productivity and supports carbon sequestration efforts. By integrating biochar into the soil, carbon is effectively sequestered, which helps mitigate climate change while simultaneously improving the health and resilience of forest ecosystems (Rodriguez Franco et al., 2024).

Biochar holds potential benefits across numerous sectors, including water treatment, construction, food processing, reducing fertilizers, cosmetics, metallurgy, wastewater management, and various chemical processes (Fig. 2). However, in India, its application is predominantly limited to small towns and rural areas, despite its broad capabilities. Expanding and promoting the use of biochar across the country could substantially enhance its advantages and increase its value in diverse industries. By broadening its application, India could unlock significant benefits and fully capitalize on biochar's potential (Bhuvaneshwari et al., 2019).

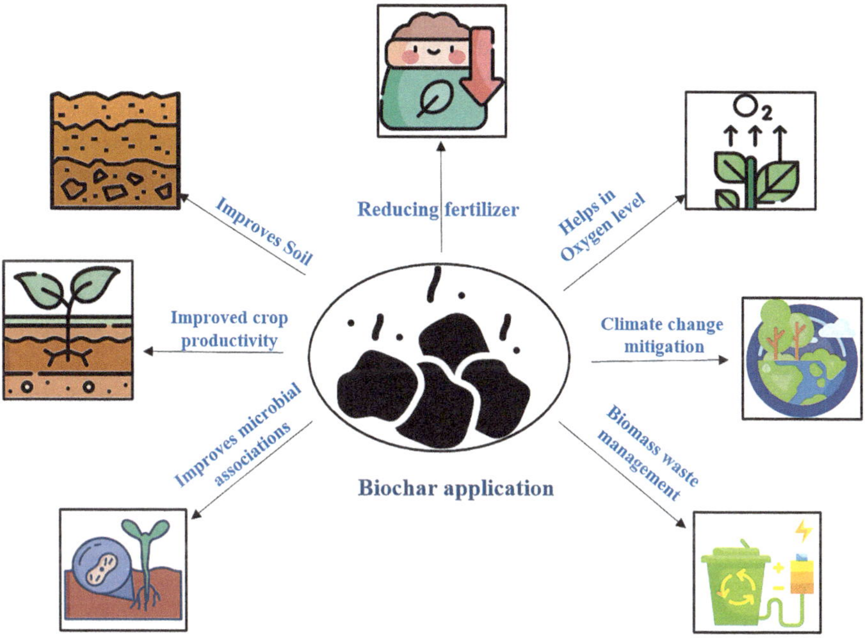

Fig. 2 Application of biochar products

Biochar has the potential to significantly enhance soil organic carbon (SOC) sequestration, modify crop nutrient uptake, and boost ecosystem productivity. These improvements can lead to increased agricultural yields and long-term soil health, thereby enhancing the socioeconomic value for farmers and communities (Bo et al., 2023; Inyang et al., 2016).

Biochar has demonstrated significant potential in reducing the need for fertilizers in crop production, contributing to more sustainable agricultural practices. Several studies have highlighted the benefits of biochar application, particularly in enhancing soil fertility and crop yields while reducing chemical inputs. Research indicates that biochar can substantially lower fertilizer requirements in agriculture. For instance, De Gryze et al. (2010) reported a reduction in fertilizer needs by approximately 7% with the application of 5 tha^{-1} per hectare of biochar. Biochar application can lead to a reduction in nitrogen fertilizer use ranging from 10% to 30% (Yeboah et al., 2022).

Biochar-producing cookstoves provide additional environmental and health benefits (Sundberg et al., 2020). By participating in carbon markets, biochar projects can generate revenue through carbon credits, incentivizing broader adoption and supporting sustainable practices (Thengane et al., 2021).

Socioeconomic factors like education, income, and age play a significant role in the adoption of biochar. Higher education levels can lead to a better understanding and acceptance of biochar technology, while higher income can provide the necessary financial resources for investment. Additionally, younger farmers may be more open to adopting new technologies. Key motivators for adopting biochar include increased crop yields and enhanced food security, which are especially critical in developing regions (Rogers et al., 2022).

Biochar production and application present significant opportunities for both farmers and unemployed youth in India. Farmers can utilize crop residues and agricultural waste materials to produce biochar, generating additional income streams while managing agricultural by-products more sustainably. By converting these residues into biochar, farmers not only enhance their soil fertility and productivity but also create valuable products that can be sold in local markets. Furthermore, biochar projects offer promising prospects for unemployed youth. Training programs can equip young individuals with the necessary skills in biochar production techniques, empowering them to establish their biochar production units. This entrepreneurial approach not only addresses the pressing issue of unemployment but also stimulates local economies and fosters innovation in rural areas.

These findings emphasize the importance of considering both environmental and economic impacts when implementing biochar technology in developing and middle-income countries. To ensure successful adoption, it is crucial to address the economic viability of biochar production and recognize the socioeconomic factors that influence its acceptance.

The outlook for the biochar market is highly promising, fueled by its potential role in several key areas. Carbon offset programs increasingly recognize biochar's ability to sequester carbon, making it an attractive tool for climate change mitigation. Circular economy initiatives are also embracing biochar for its capacity to recycle organic waste into valuable soil amendments, thereby reducing waste and enhancing resource efficiency. Furthermore, sustainable agricultural practices are integrating biochar to improve soil health, increase crop yields, and reduce dependency on chemical fertilizers. Together, these trends position biochar as a crucial element in achieving both environmental sustainability and agricultural productivity goals.

5 Policy and Governance

Biochar has the potential to reduce global carbon emissions by up to 6% annually (Lefebvre et al., 2023). However, its effectiveness is highly dependent on the quality of the biochar and the methods used in its production. Optimal conditions and high-quality feedstocks can maximize biochar's environmental benefits, while poor-quality production can limit its positive impact and introduce potential risks.

Biochar policy governance encompasses the creation and enforcement of regulations and incentives aimed at fostering the production and application of biochar to achieve environmental and agricultural benefits. Effective policy frameworks are crucial in accelerating the adoption of biochar on a commercial scale (Pourhashem et al., 2019).

Regulatory authorities in the region are actively promoting the adoption of biochar. In December 2023, the Indian government established a task force to investigate the use of biochar in the steel industry as a means to reduce carbon emissions. Additionally, the India Biochar and Bioresources Network (IBBN) launched a project in Byrkuru, Kolar, aimed at promoting the use of bioresources and biochar across India.

The Government of India acknowledges the significant potential of biochar and has implemented various policies and incentives to encourage its use (Table 3). These programs promote the use of biochar, aligning with their overarching goals of enhancing soil fertility, reducing chemical fertilizer dependency, and supporting environmental sustainability.

Supportive government policies, subsidies, and incentives designed to promote the use of biochar in environmental conservation and sustainable agriculture are driving the growth of the biochar market. Table 4 provides the names of the organizations which is working on biochar.

Clean Development Mechanism (CDM)

The Clean Development Mechanism (CDM), established under the United Nations Framework Convention on Climate Change (UNFCCC) and the Kyoto Protocol, is the largest and most comprehensive regulatory framework for project-based emission reduction initiatives. This mechanism allows public and private sectors in developed countries to purchase carbon credits from emission reduction projects in low- and middle-income countries, thus promoting investment in sustainable development and addressing global climate change (https://cdm.unfccc.int/).

Table 3 Various policies and incentives for biochar use and production

Sr. no.	Policy	Year	Objectives
1	Soil Health Card Scheme	2015	The Soil Health Card provides comprehensive information to farmers regarding the nutrient composition of their soil. This includes detailed insights into the levels of essential nutrients present and offers specific, tailored recommendations on how to improve soil health and fertility through precise and targeted nutrient applications (Soil Health Card, 2015)
2	National Mission for Sustainable Agriculture (NMSA)	2014	Promotes the adoption of sustainable agricultural practices that align seamlessly with the use of biochar. By incorporating biochar into farming, it supports soil health, enhances nutrient retention, reduces dependency on chemical fertilizers, and contributes to carbon sequestration, thus fostering long-term environmental sustainability and resilience in agricultural systems (NMSA, 2014)
3	National Policy for Management of Crop Residue	2014	Promoting biochar production as an alternative to agricultural residue burning provides a comprehensive solution that benefits the environment, agriculture, economy, and society. This strategy not only prevents the negative impacts of residue burning but also fosters sustainable development and resilience in agricultural communities

Table 4 Various organizations working on biochar

Geographical area	Regional group
Eastern Asia	Japan Biochar Association
	Korea Biochar Association
	Korea Biochar Research Center (KBRC), South Korea
	Mindanao Terra Preta – Biochar Initiative
South Eastern Asia	Biochar Malaysia Association
	Philippine Biochar Association
	South East Asia Biochar Interest Group
Southern Asia	Bangladesh Biochar Initiative
	India Biochar and Bioresources Network (IBBN)
Western Asia	Jordan Biochar Research Initiative (JBRI)
	North Sinai School Biochar
	Saudi Biochar Research Group

Biochar projects can qualify for benefits under the Clean Development Mechanism (CDM), which offers financial incentives for initiatives that reduce emissions and sequester carbon. These projects are vital for environmental sustainability, as biochar effectively captures and stores carbon in the soil, significantly contributing to climate change mitigation.

In summary, the CDM provides a robust and effective framework for advancing climate-friendly initiatives like biochar projects. By facilitating collaboration between developed and developing countries, the CDM helps achieve substantial environmental and socioeconomic benefits, contributing to a more sustainable and resilient global future.

6 Barriers and Solutions

Despite the potential benefits of biochar, several barriers hinder its widespread adoption. These barriers can be categorized into technical, economic, and social challenges. Addressing these barriers requires a comprehensive approach that includes education, policy support, and financial incentives.

1. *Policy, Legal, and Institutional*

Regulations governing biochar production and use vary widely between countries and, in certain cases, from state to state within the same country. This lack of regulatory clarity can create challenges for companies, making it harder for them to confidently enter the market. There is a strong need for regulations that specifically address biochar as a final product. Like many emerging industries, the biochar sector depends on clear guidance from regulators to support product development and marketing strategies (Jirka & Tomlinson, 2015). Policy-related challenges can stymie the production of biochar (Pierson et al., 2024). Currently, a broad variety of carbonized products are referred to as biochar, which is the term for all black solid residues left behind after thermal processing of materials produced from biomass. Protocols should be created to direct and control the processes of producing biochar because the kind of feedstock, the carbonization conditions, and the posttreatment procedures all affect the properties of biochar (Guo et al., 2016).

2. *Financial*

   Low-cost, sustainable biochar production systems are still under development; biochar production must be affordable for promotion and adoption at the local level (Omulo & Omulo, 2020). All biochar systems must generate money to remain financially sustainable, especially those in developing nations with limited access to finance and other resources. The amount of money allocated to research and development has generally fallen short of what is required to comfortably scale up particular biochar systems.

3. *Technical*

   The majority of documented biochar research was conducted on small sizes in labs or using unfeasibly high rates of amendment (Guo et al., 2016). A key factor for the growth of the industry is having access to reliable, ready-to-use technologies that can consistently produce quality biochar and generate positive returns within a reasonable period (Jirka & Tomlinson, 2015). There is also uncertainty in the durability and stability of carbon from biochar technology; the variation of biochar is a major obstacle in measuring and quantifying the influence of carbon content (Ganesapillai et al., 2020).

4. *Social*

   Biochar has to be more widely accepted and in demand in the area by showcasing its special benefits in extensive commercial plantings. The unfavorable attitudes and perceptions around biochar technology, particularly among the bulk of risk-averse smallholder farmers, are another obstacle to its upscaling. In some communities, traditional farming practices may be deeply ingrained, making it difficult to introduce new technologies like biochar. There is a lack of knowledge among the general public regarding biochar and its applications; there is also a common misconception that biochar might be applied once to permanently substitute for fertilizers that contain chemicals (Guo et al., 2016). The widely used traditional charring method poses risks of pollution (Tang et al., 2016).

Way Forward

Increasing education and outreach efforts, along with strategic marketing, will raise awareness and understanding of biochar's benefits among diverse stakeholders. Addressing regulatory barriers is crucial; this involves creating opportunities to educate policymakers and ensuring that regulations support biochar adoption. Enhancing dialogue between research and commercial communities, both within and outside the biochar sector, will foster innovation and practical applications.

Emphasizing long-term field trials with well-characterized biochar and standardized reporting will provide robust data to support biochar's efficacy. Additionally, investing in technology research and development, alongside exploring alternative financing mechanisms, will drive advancements and broaden access to biochar solutions, ultimately supporting sustainable agricultural and environmental practices.

7 Conclusions and Recommendations

The rapid spread of invasive species poses a major threat to Asia's rich biodiversity. Invasive species degrade land by creating extensive, unproductive monocultures. These invasions negatively impact biodiversity, water supplies, crop and pasture yields, as well as human and animal health, hindering the country's progress toward its Sustainable Development Goals. Furthermore, landscapes degraded by unsustainable land-use practices are more prone to invasion by non-native plant species, which significantly complicates restoration efforts. This ecological disruption, combined with the challenge of managing massive volumes of crop residues, strains the region's land and water resources.

Biochar, a carbon-rich material created through the pyrolysis of biomass, offers a unique solution to these problems. By transforming invasive plants and crop residues into biochar, this approach not only tackles waste disposal issues but also has the potential to sequester carbon, enhance nutrient cycling, improve soil health, and support sustainable agricultural practices. However, obstacles to widespread usage include uneven biochar efficiency, legal restrictions, and low farmer acceptability.

Awareness and outreach programs need to be conducted on biochar's benefits, concentrating on its impact on improving agricultural productivity and ecosystem restoration. Policies should focus on improving cooperation between communities, legislators, and researchers. Long-term biochar field experiments should be done to ensure consistent performance while fostering community-led efforts to offer economic possibilities and enhance sustainable agriculture practices.

References

Abdullah, P. S., Azmin, S. N. H. M., Latiff, Z. A. A., Nawi, A. S. M., Abd Kharim, M. N., & Idris, A. A. (2024). Socio-economic empowerment of poor community through biochar-themed social enterprise initiative in Jeli, Kelantan(Chapter 6). Universiti Malaysia Kelantan.

Abhishek, K., Shrivastava, A., Vimal, V., Gupta, A. K., Bhujbal, S. K., Biswas, J. K., Singh, L., Ghosh, P., Pandey, A., Sharma, P., & Kumar, M. (2022). Biochar application for greenhouse gas mitigation, contaminants immobilization and soil fertility enhancement: A state-of-the-art review. *Science of the Total Environment, 853*, 158562. https://doi.org/10.1016/j.scitotenv.2022.158562

Ahmed, A., Bakar, M. S. A., Hamdani, R., Park, Y. K., Lam, S. S., Sukri, R. S., et al. (2020). Valorization of underutilized waste biomass from invasive species to produce biochar for energy and other value-added applications. *Environmental Research, 186*, 109596.

Amalina, F., Abd Razak, A. S., Krishnan, S., Sulaiman, H., Zularisam, A. W., & Nasrullah, M. (2022). Biochar production techniques utilizing biomass waste-derived materials and environmental applications–A review. *Journal of Hazardous Materials Advances, 7*, 100134.

Bai, Z. G., Dent, D. L., Olsson, L., & Schaepman, M. E. (2008). *Global assessment of land degradation and improvement 1.* Identification by remote sensing. Report, 2008.

Beckinghausen, A., Reynders, J., Merckel, R., Wu, Y. W., Marais, H., & Schwede, S. (2020). Post-pyrolysis treatments of biochars from sewage sludge and A. mearnsii for ammonia (NH4-n) recovery. *Applied Energy, 271*, 115212.

Beesley, L., Moreno-Jiménez, E., Gomez-Eyles, J. L., Harris, E., Robinson, B., & Sizmur, T. (2011). A review of biochars' potential role in the remediation, revegetation and restoration of contaminated soils. *Environmental Pollution, 159*(12), 3269–3282.

Berihun, T., Tadele, M., & Kebede, F. (2017). The application of biochar on soil acidity and other physico-chemical properties of soils in Southern Ethiopia. *Journal of Plant Nutrition and Soil Science, 180*(3), 381–388.

Bhat, I. U. H., & Bhat, R. (2021). Quercetin: A bioactive compound imparting cardiovascular and neuroprotective benefits: Scope for exploring fresh produce, their wastes, and by-products. *Biology, 10*(7), 586.

Bhattacharyya, R., Ghosh, B. N., Mishra, P. K., Mandal, B., Rao, C. S., Sarkar, D., et al. (2015). Soil degradation in India: Challenges and potential solutions. *Sustainability, 7*(4), 3528–3570.

Bhuvaneshwari, S., Hettiarachchi, H., & Meegoda, J. N. (2019). Crop residue burning in India: Policy challenges and potential solutions. *International Journal of Environment Research and Public Health, 2019*(16), 832. https://doi.org/10.3390/ijerph16050832

Bo, X., Zhang, Z., Wang, J., Guo, S., Li, Z., Lin, H., Huang, Y., Han, Z., Kuzyakov, Y., & Zou, J. (2023). Benefits and limitations of biochar for climate-smart agriculture: A review and case study from China. *Biochar, 5*, 77. https://doi.org/10.1007/s42773-023-00279-x

Bolognesi, S., Bernardi, G., Callegari, A., Dondi, D., & Capodaglio, A. G. (2021). Biochar production from sewage sludge and microalgae mixtures: Properties, sustainability and possible role in circular economy. *Biomass Conversion and Biorefinery, 11*(2), 289–299.

Bordoloi, S., Gopal, P., Boddu, R., Wang, Q., Cheng, Y. F., Garg, A., & Sreedeep, S. (2019). Soil-biochar-water interactions: Role of biochar from Eichhornia crassipes in influencing crack propagation and suction in unsaturated soils. *Journal of Cleaner Production, 210*, 847–859.

Boyette, M. D., Alston, B. P., & Macialek, J. A. (2012). *The rapid production of biochar.* In 2012 Dallas, Texas, July 29–August 1, 2012 (p. 1). American Society of Agricultural and Biological Engineers.

Ceriani, A., Dalle Fratte, M., Agosto, G., Beatrice, P., Reguzzoni, M., Bettucci, L., et al. (2024). Woody and herbaceous invasive alien plant species-derived biochars are potentially optimal for soil amendment, soil remediation, and carbon storage. *GCB Bioenergy, 16*(2), e13117.

Choudhary, P., Prasad, M., Choudhary, M., Kumar, A., Kumar, S., Srinivasan, R., & Mahawer, S. K. (2023). Exploring invasive weed biochar as soil amendment: A study on fodder oats productivity and soil biological properties. *Environmental Research, 216*, 114527.

Das, S., Mohanty, S., Sahu, G., Rana, M., & Pilli, K. (2021). Biochar: A sustainable approach for improving soil health and environment. In *Soil erosion-current challenges and future perspectives in a changing world* (Vol. 1, p. 5772). InTech.

De Gryze, S., Cullen, M., Durschinger, L., Lehmann, J., Bluhm, D., Six, J., & Suddick, E. (2010). Evaluation of the opportunities for generating carbon offsets from soil sequestration of biochar. *An issues paper commissioned by the Climate Action Reserve, final version, 1-99.*

Digman, B., Joo, H. S., & Kim, D. S. (2009). Recent progress in gasification/pyrolysis technologies for biomass conversion to energy. *Environmental Progress & Sustainable Energy: An Official Publication of the American Institute of Chemical Engineers, 28*(1), 47–51.

El-Barougy, R. F., Dakhil, M. A., Halmy, M. W., Gray, S. M., Abdelaal, M., Khedr, A. H. A., & Bersier, L. F. (2021). Invasion risk assessment using trait-environment and species distribution modelling techniques in an arid protected area: Towards conservation prioritization. *Ecological Indicators, 129*, 107951.

Feng, Q., Wang, B., Chen, M., Wu, P., Lee, X., & Xing, Y. (2021). Invasive plants as potential sustainable feedstocks for biochar production and multiple applications: A review. *Resources, Conservation and Recycling, 164*, 105204.

Food and Agriculture Organization. (2014). *Land degradation assessment in dryland.* http://www.fao.org/nr/lada/

Ghosh, D., & Maiti, S. K. (2023). Invasive weed-based biochar facilitated the restoration of coal mine degraded land by modulating the enzyme activity and carbon sequestration. *Restoration Ecology, 31*(3), e13744.

Gupta, G. S. (2019). Land degradation and challenges of food security. *Review of European Studies, 11*, 63. https://doi.org/10.5539/res.v11n1p63

Henderson, M., Chettri, N., & Johnson, F. (2018, October). Lessons in biochar: Advancing research through service learning courses in Nepal. In *2018 IEEE global humanitarian technology conference (GHTC)* (pp. 1–7). IEEE.

Holden, N. M., Neill, A. M., Stout, J. C., O'Brien, D., & Morris, M. A. (2023). Biocircularity: A framework to define sustainable, circular bioeconomy. *Circular Economy and Sustainability, 3*(1), 77–91.

Hu, Q., Jung, J., Chen, D., Leong, K., Song, S., Li, F., et al. (2021). Biochar industry to circular economy. *Science of the Total Environment, 757*, 143820.

Hung, N. T. Q., & Ky, N. M. (2018). Potential of biochar production from agriculture residues at household scale: A case study in Go Cong Tay District, Tien Giang Province, Vietnam. *Environment and Natural Resources Journal, 16*(2), 68–78.

Inyang, M. I., Gao, B., Yao, Y., Xue, Y., Zimmerman, A., Mosa, A., Pullammanappallil, P., Ok, Y. S., & Cao, X. (2016). A review of biochar as a low-cost adsorbent for aqueous heavy metal removal. *Critical Reviews in Environmental Science and Technology, 46*, 406–433. https://doi.org/10.1080/10643389.2015.1096880

IPBES. (2019). In S. Diaz, J. Settele, et al. (Eds.), *Summary for policymakers of the global assessment report on biodiversity and ecosystem services of the Intergovernmental Science-Policy Platform on Biodiversity and Ecosystem Services*. IPBES Secretariat. https://doi.org/10.5281/zenodo.3553579

Jackson, R. B., Banner, J. L., Jobbágy, E. G., Pockman, W. T., & Wall, D. H. (2002). Ecosystem carbon loss with woody plant invasion of grasslands. *Nature, 418*(6898), 623–626.

Jindo, K., Audette, Y., Higashikawa, F. S., Silva, C. A., Akashi, K., Mastrolonardo, G., et al. (2020). Role of biochar in promoting circular economy in the agriculture sector. Part 1: A review of the biochar roles in soil N, P and K cycles. *Chemical and Biological Technologies in Agriculture, 7*, 1–12.

Jirka, S., & Tomlinson, T. (2015). *State of the biochar industry 2014: A survey of commercial activity in the biochar sector*. International Biochar Initiative (IBI).

Khalid, S., Shahid, M., Murtaza, B., Bibi, I., Natasha, Asif Naeem, M., & Niazi, N. K. (2020). A critical review of different factors governing the fate of pesticides in soil under biochar application. *Science of the Total Environment, 711*, 134645. https://doi.org/10.1016/j.scitotenv.2019.134645

Konwar, L. J., Mikkola, J. P., Bordoloi, N., Saikia, R., Chutia, R. S., & Kataki, R. (2018). Side streams from bioenergy and biorefinery complexes as a resource for circular bioeconomy. In *Waste biorefinery* (pp. 85–125). Elsevier.

Kumar, A., Bhattacharya, T., Shaikh, W. A., Roy, A., Chakraborty, S., Vithanage, M., & Biswas, J. K. (2023). Multifaceted applications of biochar in environmental management: a bibliometric profile. *Biochar, 5*(1), 11.

Kurniawan, T. A., Othman, M. H. D., Liang, X., Goh, H. H., Gikas, P., Chong, K. K., & Chew, K. W. (2023). Challenges and opportunities for biochar to promote circular economy and carbon neutrality. *Journal of Environmental Management, 332*, 117429.

Lefebvre, D., Fawzy, S., Aquije, C. A., Osman, A. I., Draper, K. T., & Trabold, T. A. (2023). Biomass residue to carbon dioxide removal: Quantifying the global impact of Biochar. *Biochar, 5*, 65. https://doi.org/10.1007/s42773-023-00258-2

Lehmann, J. (2007). Bio-energy in the black. *Frontiers in Ecology and the Environment, 5*(7), 381–387.

Li, M., He, Z., Zhong, H., Sun, W., Ye, M., & Tang, Y. (2023). Highly efficient persulfate catalyst prepared from modified electrolytic manganese residues coupled with biochar for the roxarsone removal. *Journal of Environmental Management, 328*, 116945. https://doi.org/10.1016/j.jenvman.2022.116945

Liu, G., Zheng, H., Jiang, Z., Zhao, J., Wang, Z., Pan, B., & Xing, B. (2018). Formation and physicochemical characteristics of nano biochar: Insight into chemical and colloidal stability. *Environmental Science & Technology, 52*(18), 10369–10379.

Liu, C., Diagne, C., Angulo, E., Banerjee, A. K., Chen, Y., Cuthbert, R. N., et al. (2021). Economic costs of biological invasions in Asia. *NeoBiota, 67*, 53–78.

Marmiroli, M., Bonas, U., Imperiale, D., Lencioni, G., Mussi, F., Marmiroli, N., & Maestri, E. (2018). Structural and functional features of chars from different biomasses as potential plant amendments. *Frontiers in Plant Science, 9*, 1119.

Martinsen, V., Mulder, J., Shitumbanuma, V., Sparrevik, M., Børresen, T., & Cornelissen, G. (2014). Farmer-led maize biochar trials: Effect on crop yield and soil nutrients under conservation farming. *Journal of Plant Nutrition and Soil Science, 177*, 681. https://doi.org/10.1002/jpln.201300590

Masud, M. A. A., Shin, W. S., Sarker, A., Septian, A., Das, K., Deepo, D. M., Iqbal, M. A., Islam, A. R. M. T., & Malafaia, G. (2023a). A critical review of sustainable application of biochar for green remediation: Research uncertainty and future directions. *Science of the Total Environment, 904*, 166813. https://doi.org/10.1016/j.scitotenv.2023.166813

Masud, M. A. A., Annamalai, S., & Shin, W. S. (2023b). Remediation of ciprofloxacin in soil using peroxymonosulfate activated by ball-milled seaweed kelp biochar: Performance, mechanism, and phytotoxicity. *Chemical Engineering Journal, 465*, 142908. https://doi.org/10.1016/j.cej.2023.142908

Mohan, D., Abhishek, K., Sarswat, A., Patel, M., Singh, P., & Pittman, C. U. (2018). Biochar production and applications in soil fertility and carbon sequestration–a sustainable solution to crop-residue burning in India. *RSC Advances, 8*(1), 508–520.

Müller, S., Backhaus, N., Nagabovanalli, P., & Abiven, S. (2019). A social-ecological system evaluation to implement sustainably a biochar system in South India. *Agronomy for Sustainable Development, 39*, 1–14.

Mutolib, A., Rahmat, A., Triwisesa, E., Hidayat, H., Hariadi, H., Kurniawan, K., et al. (2023). Biochar from agricultural waste for soil amendment candidate under different pyrolysis temperatures. *Indonesian Journal of Science and Technology, 8*(2), 243–258.

National Mission for Sustainable Agriculture. (2014). https://nmsa.dac.gov.in/

Obia, A., Mulder, J., Hale, S. E., Nurida, N. L., & Cornelissen, G. (2018). The potential of biochar in improving drainage, aeration and maize yields in heavy clay soils. *PLoS One, 13*(5), e0196794.

Omulo, G., & Omulo, G. (2020). Biochar potential in improving agricultural production in East Africa. In *Applications of biochar for environmental safety* (p. 123). https://doi.org/10.5772/intechopen.92195

Ortiz-Bobea, A., Ault, T. R., Carrillo, C. M., Chambers, R. G., & Lobell, D. B. (2021). Anthropogenic climate change has slowed global agricultural productivity growth. *Nature Climate Change, 11*(4), 306–312.

Owsianiak, M., Lindhjem, H., Cornelissen, G., Hale, S. E., Sørmo, E., & Sparrevik, M. (2021). Environmental and economic impacts of biochar production and agricultural use in six developing and middle-income countries. *Science of the Total Environment, 755*, Article 142455. https://doi.org/10.1016/j.scitotenv.2020.142455

Patel, M. R., & Panwar, N. L. (2023). Biochar from agricultural crop residues: Environmental, production, and life cycle assessment overview. *Resources, Conservation & Recycling Advances, 19*, 200173.

Pehlivan, E., Tran, H. T., Ouédraogo, W. K. I., Schmidt, C., Zachmann, D., & Bahadir, M. (2013). Sugarcane bagasse treated with hydrous ferric oxide as a potential adsorbent for the removal of As (V) from aqueous solutions. *Food Chemistry, 138*(1), 133–138.

Pierson, D., Anderson, N., Brewen, J., Clark, N., Hardy, M. C., McCollum, D., et al. (2024). Beyond the basics: A perspective on barriers and opportunities for scaling up biochar production from forest slash. *Biochar, 6*(1), 1.

Pourhashem, G., Hung, S. Y., Medlock, K. B., & Masiello, C. A. (2019). Policy support for biochar: Review and recommendations. *GCB Bioenergy, 2019*(11), 364–380.

Qu, T., Du, X., Peng, Y., Guo, W., Zhao, C., & Losapio, G. (2021). Invasive species allelopathy decreases plant growth and soil microbial activity. *PLoS One, 16*(2), e0246685.

Rafiullah, M. I., Kaleri, F. N., Rizwan, M., & Mehmood, I. (2017). Potential value of biochar as a soil amendment: A review. *Pure and Applied Biology (PAB), 6*(4), 1494–1502.

Rathee, S., Ahmad, M., Sharma, P., Singh, H. P., Batish, D. R., Kaur, S., et al. (2021). Biomass allocation and phenotypic plasticity are key elements of successful invasion of Parthenium hysterophorus at high elevation. *Environmental and Experimental Botany, 184*, 104392.

Rex, P., Mohammed Ismail, K. R., Meenakshisundaram, N., Barmavatu, P., & Sai Bharadwaj, A. V. S. L. (2023). Agricultural biomass waste to biochar: A review on biochar applications using machine learning approach and circular economy. *ChemEngineering, 7*(3), 50.

Robb, S., Joseph, S., Abdul Aziz, A., Dargusch, P., & Tisdell, C. (2020). Biochar's cost constraints are overcome in small-scale farming on tropical soils in lower-income countries. *Land Degradation & Development, 31*, 1713. https://doi.org/10.1002/ldr.3541

Roberts, K. G., Gloy, B. A., Joseph, S., Scott, N. R., & Lehmann, J. (2010). Life cycle assessment of biochar systems: Estimating the energetic, economic, and climate change potential. *Environmental Science & Technology, 44*, 827–833. https://doi.org/10.1021/es902266r

Rodriguez Franco, C., Page-Dumroese, D. S., Pierson, D., Miller, M., & Miles, T. (2024). Policy and regulations for mobile biochar production in The United States of America. *Forests, 15*, 192. https://doi.org/10.3390/f15010192

Rogers, P. M., Fridahl, M., Yanda, P., Hansson, A., Pauline, N., & Haikola, S. (2022). Socio – Economic determinants for biochar deployment in the southern highlands of Tanzania. *Energies, 15*, 144. https://doi.org/10.3390/en15010144

Sari, V. K., Basuki, B., Mandala, M., Novikarumsari, N. D., & Erdiansyah, I. (2021). Pemberdayaan Masyarakat Desa Slateng Kecamatan Ledokombo Jember Mengolah Limbah Serbuk Gergaji Sengon Menjadi Biochar. *Jurnal Pengabdian Magister Pendidikan IPA, 4*(3). http://repository.unej.ac.id/xmlui/handle/123456789/105900

Shafer, M. (2020). Global crop waste burning–micro-biochar; how a small community development organization learned experientially to address a huge problem one tiny field at a time. *Sustainable Earth, 3*(1), 18.

Singh, E., Mishra, R., Kumar, A., Shukla, S. K., Lo, S. L., & Kumar, S. (2022). Circular economy-based environmental management using biochar: Driving towards sustainability. *Process Safety and Environmental Protection, 163*, 585–600.

Soil health card scheme. (2015). https://www.india.gov.in/spotlight/soil-health-card#tab=tab-1

Srinivasan, P., Sarmah, A. K., Smernik, R., Das, O., Farid, M., & Gao, W. (2015). A feasibility study of agricultural and sewage biomass as biochar, bioenergy and biocomposite feedstock: Production, characterization and potential applications. *Science of the Total Environment, 512*, 495–505.

Sundberg, C., et al. (2020). Biochar from cookstoves reduces greenhouse gas emissions from smallholder farms in Africa. *Mitigation and Adaptation Strategies for Global Change, 25*, 953–967. https://doi.org/10.1007/s11027-020-09920-7

Tang, K. M., Ibrahim, W. A., & Kadir, W. R. (2016). Towards environmental and economic sustainability via the biomass industry: The Malaysian case study. In *Biochar: A regional supply chai n approach in view of climate change mitigation* (pp. 162–183). Cambridge University Press.

Temmes, A., & Peck, P. (2020). Do forest biorefineries fit with working principles of a circular bioeconomy? A case of Finnish and Swedish initiatives. *Forest Policy and Economics, 110*, 101896.

Thengane, S. K., Kung, K., Hunt, J., Gilani, H. R., Lim, C. J., Sokhansanj, S., & Sanchez, D. L. (2021). Market prospects for biochar production and application in California. *Biofuels, Bioproducts and Biorefining, 15*(6), 1802–1819. https://doi.org/10.1002/bbb.2280

Wang, J., Zhao, M., Zhang, J., Zhao, B., Lu, X., & Wei, H. (2021). Characterization and utilization of biochars derived from five invasive plant species Bidens pilosa L., Praxelis clematidea, Ipomoea cairica, Mikania micrantha and Lantana camara L. for Cd2+ and Cu2+ removal. *Journal of Environmental Management, 280*, 111746.

Wang, X., Zheng, W. L., Yuan, H. M., van Kleunen, M., Yu, F. H., & Li, M. H. (2024). Biochar produced from diverse invasive species improves remediation of cadmium-contaminated soils. *Biological Invasions, 26*, 1–12.

Wijitkosum, S. (2022). Biochar derived from agricultural wastes and wood residues for sustainable agricultural and environmental applications. *International Soil and Water Conservation Research, 10*(2), 335–341.

Yang, L., Deng, Y., Shu, Z., Chen, Q., Yang, H., & Tan, X. (2022). Application of invasive plants as biochar precursors in the field of environment and energy storage. *Frontiers in Environmental Science, 10*, 902915.

Yeboah, E., Dugan, E., Nartey, E., Lawson, I. Y. D., Frimpong, K. A., Obiri, B. D., Egyir, I., Akom, M., Osei-Adu, J., Otabil, P. K., Ason, B., Allotey, D. F. K., Moses, M., & Adiku, S. G. K. (2022). Biochar Systems in Ghana. *Agricultural and Food Science Journal of Ghana, 15*, 1553–1577.

Yuan, J. H., Xu, R. K., & Zhang, H. (2011). The forms of alkalis in the biochar produced from crop residues at different temperatures. *Bioresource Technology, 102*(3), 3488–3497.

Zhang, F., Wang, X., Yin, D., Peng, B., Tan, C., Liu, Y., et al. (2015). Efficiency and mechanisms of Cd removal from aqueous solution by biochar derived from water hyacinth (Eichhornia crassipes). *Journal of Environmental Management, 153*, 68–73.

Zhu, X., Labianca, C., He, M., Luo, Z., Wu, C., You, S., & Tsang, D. C. (2022). Life-cycle assessment of pyrolysis processes for sustainable production of biochar from agro-residues. *Bioresource Technology, 360*, 127601.

Zhuang, H., Zhu, H., Zhang, J., Shan, S., Fang, C., Tang, H., & Xie, Q. (2020). Enhanced 2, 4, 6-trichlorophenol anaerobic degradation by Fe3O4 supported on water hyacinth biochar for triggering direct interspecies electron transfer and its use in coal gasification wastewater treatment. *Bioresource Technology, 296*, 122306.

Restoring the Soil Functions Through Biochar-Microbiome Interactions

Debabrata Nath, Mattaparthi Lakshmi Durga, Vandana Kumari, and Sudhanshu Singh

Abstract

Soil microbiomes are crucial in supporting plant development and maintaining soil functions for sustainable crop production. Soil functions improve through complex interactions between biochar and microbiome. Biochar, a carbon-rich substance derived from biomass pyrolysis, has emerged as a multifunctional soil amendment with significant potential to improve soil functions. Acting as a carrier and habitat for soil microbes, biochar can enhance microbial persistence, survival, and colonization in the rhizosphere, which are crucial for soil biogeochemical cycling, nutrient dynamics, and remediation of degraded soils. The physicochemical properties of biochar, including higher pore space, higher specific surface area and cation exchange capacity etc. create a conducive microenvironment for microbial proliferation and nutrient availability, thereby supporting plant growth. So, proper interactions between soil microbial communities (soil microbiome) and biochar and its consequences within the soil ecosystem need to be examined. This chapter examines properties of biochar and its effects on soil, characterization of biochar, interactions between biochar and soil microbiomes, and their function in promoting plant growth. It highlights that biochar-mediated soil microbiome improves soil functions, suppresses plant diseases, and contributes to sustainable agriculture and pollution mitigation.

D. Nath · S. Singh
International Rice Research Institute (IRRI)-South Asia Regional Centre (ISARC), Varanasi, Uttar Pradesh, India

M. L. Durga (✉)
Division of Agricultural Engineering, ICAR-Indian Institute of Agricultural Research, New Delhi, India

V. Kumari (✉)
Advance Centre on Sericulture, Bihar Agricultural University, Sabour, Bhagalpur, Bihar, India

S. Singh, S. Srivastava (eds.), *Biochar: A Precious Resource from Biological Waste*,
https://doi.org/10.1007/978-981-95-0425-1_12

Keywords

Biochar · Soil microbiome · Soil remediation · Soil health · Soil function

1 Introduction

Biochar is a high carbonaceous solid output synthesized from thermally treated biomass under anoxic conditions (pyrolysis) (Tan et al., 2022). It is abundant in hydrogen, carbon, oxygen, sulfur, and nitrogen. Biochar versatility stems from its unique combination of attributes: expansive surface area, significant porosity, redox capabilities, high cation exchange capacity, malleability, lightness, and abundance of functional groups. These properties render it suitable for diverse applications (Tan et al., 2022; Zhu et al., 2017). Furthermore, biochar is a reusable, cost-effective, easy-to-prepare, environmentally friendly product. Biochar production has been attracting the attention of research and farming community of current generation for a few decades now, but its origins are ancient. Amazonians of antiquity pioneered a technique of heating organic material to create terra preta, a remarkably fertile soil that enriched the rainforest's typically nutrient-poor earth.

However, biochar has vast range of applications, like its ability to store carbon, improve soil health, and fix contaminants (Zhu et al., 2017) and have demonstrated their capacity to mitigate phytopathogenic threats (Jaiswal et al., 2017), and similarly enhance plant productivity (Kolton et al., 2017). Biochar may change soil microbiome and activity by changing microbial habitats and nutrients available to them (Zhu et al., 2017). It is essential to adopt environmentally sustainable farming practices to address the challenges such as food scarcity and climate change by improving soil carbon sequestration and reducing greenhouse gas emissions over time. The use of biochar is recommended for its potential to mitigate nitrous oxide and methane emissions from soil (Kolton et al., 2017). Incorporating biochar into soil serves as an effective strategy for enhancing its physicochemical characteristics and microbial ecology. A beneficial change in soil properties was often observed in the presence of biochar. Biochar-amended soils had significantly lower bulk density than unamended soils. Additionally, biochar amended soils exhibit a larger specific surface area, improved water retention capacity, higher cation exchange capacity, and increased pH compared to unamended soils (Yang et al., 2019). Biochar also significantly reduced soil toxicity and autotoxicity associated with ginsenosides and organic and inorganic contaminants (Yang et al., 2019) by adsorbing autotoxic ginsenosides and organic and inorganic contaminants. Biochar-amended soil can alter the composition, microbial proliferation, and metabolic activity of microbes (Kolton et al., 2017). Besides reducing pathogen colonization and survival, biochar increased microorganisms that promote plant growth and provide biocontrol (Jaiswal et al., 2017). Biochar has previously been described as a soil amendment with potential benefits, so biochar application might mitigate negative plant-soil feedback (NPSF) in an agricultural ecosystem for sustainable farming (Wang et al., 2020a, b). Evidence from numerous studies highlights that the synergistic relationship among biochar, soil, and microbes can significantly enhance soil functions, stimulate beneficial microbial populations, boost plant growth and yield, and strengthen resistance to diseases across diverse agricultural applications.

The soil microbiome refers to the microbial community, including actinomycetes, bacteria, fungi, archaea, viruses, and protozoa, that inhabit the soil environment. These microorganisms play a crucial role in maintaining soil health, fertility, and ecosystem functioning. They participate in essential processes such as decomposition of soil organic matter, nutrient mineralization through nutrient cycling, soil structure formation, and plant growth promotion. The soil microbiome is highly diverse and can be influenced by various factors, including soil type, climate, vegetation, and human activities such as agriculture and pollution. Research has shown that a healthy soil microbiome enhances soil resilience to stressors like drought and disease, contributing to sustainable agricultural practices and environmental conservation.

The production of biochar has been facilitated by a wide variety of organic wastes, collected from agriculture, industries, and municipality (Wang et al., 2020a, b). Cellulose, hemicellulose, and polymers of lignin are the most abundant compounds in biochar made from vegetation biomass. Additionally, sewage of humans is a rich source of carbon and nutrients that can also be converted into biochar. Water retention and purification, pollution remediation, greenhouse gas reduction, energy production, and soil amendment are some of the applications for biochar. However, biochar has yet to be demonstrated with authentic outcomes as an aid to augmenting the diversity and growth of soil microbiome. There is a need to address the existing knowledge gaps on the interactions between biochar-soil microbiome, for improving nutrient cycling, enhancing soil functions, and remediating contaminated soils. Thus, this chapter provides a concise overview of biochar characterization and its key properties, along with insights into the interactions between biochar and soil microbiomes, with surrounding soil ecosystems.

2 Biochar's Physicochemical Properties

The properties of biochar (chemical and physical) are primarily determined by the feedstock and processing conditions. To synthesize high quality biochar, it requires an understanding of quality and quantity ranges of decisive parameters such as temperature, residence time, heating rate, type of activator, etc. However, the quality of produced biochar is characterized by its pore volume (cc/g) and surface area (m^2/g) (Abbas et al., 2021). The porosity over the biochar was a resultant of de-volatilization and evaporation of volatile compounds from the biomass during pyrolysis. This porous nature enhances the nutrient and water retention capacity of biochar in soil.

Furthermore, biochar produced at lower temperatures has a greater efficiency to retain cations soil. However, the carbon content and energy value of biochar increase with higher temperatures. Biochar has a range of neutral to alkaline pH. The stability of biochar is an important factor influenced by parameters like temperature and the type of biomass used. Studies indicate that higher pyrolysis temperatures enhance biochar stability (Weber & Quicker, 2018). Generally, biochar's

Table 1 Overall effects of biochar on soil functions

Increase surface area
Improves porosity and pore volume
Improves pore size distribution (micro and macro pores)
Increases soil pH and cation exchange capacity
Increases organic carbon content and microbial diversity
Improves water holding capacity

mechanical stability is inversely proportional to porosity; however, this stability can be assessed through incubation and modeling. Biochar with a greater compressive strength, for example, is typically made from biomass with a higher density and lignin content. Table 1 provides overall effects of biochar on soil functions.

3 Soil Microbiome and Biochar Interactions in Soil Ecosystems

There are numerous soil microbes including bacteria, actinomycetes, fungi, etc. The interaction of the microbiota, with each other, or with soil or live/dead plants, etc., plays a crucial role in maintaining the food chain supply. The interaction of biochar and soil microbiome is a crucial natural event that has garnered significant interest from scientists worldwide. However, this phenomenon enriches the nutrient content, remediating pollutants, moisture uptake, and pathogen resistance to plants. This is attributed to the physical characteristics of biochar, which provide an ideal environment for the growth and proliferation of various microbial populations. Moreover, the interactions between microbiota in soil conditions may sustain for short or long period after biochar is applied to the soil. In addition, biochar has been reported to soil health improvement by increasing soil microbial activity worldwide. While numerous studies have highlighted the ecological benefits of biochar, little research has focused specifically on how it interacts with soil microbial communities, so understanding how biochar enhances soil microbial activity is vital to understanding this concept. Here, we present recent studies that shed light on the integration of soil ecosystems with microbes and its implications for various applications in Table 2.

3.1 Influence of Biochar on Remediation

The increasing use of metallic materials in various fields has led to significant heavy metal pollution in the environment. Heavy metals pose numerous environmental hazards due to their nonbiodegradable and harmful nature. To address the heavy metals problem in soil, a variety of remediation technologies have been developed. Among these, biological processes are particularly notable that utilize the metabolic

Table 2 The synergistic effects of biochar and diverse microbial communities for soil enhancement

Feedstock	Preparation of biochar	Substrates	Microbes used	Experiment	Application of biochar	References
Maize stalk	500 °C (Carbonization)	–	Fungi	35 days pot experiment	Improved bacterial community, increased soil nutrients, enhanced cucumber growth.	Cao et al. (2022)
Bamboo		Red soil	–	–	Increasing soil pH, nutrients, and the diversity and richness of fungi	Tarin et al. (2021)
Coconut shell & SS (sewage and sludge)	500–800 °C (Charring)	Polyacrylamide (PAM)	Bacteria (*Klebsiella* sp.)	Bacteria immobilization on biochar	Bioaugmentation and bio-stimulation of soil to degrade PAM	Ma et al. (2021)
Straw of wheat	–	Cow manure	Bacteria	–	Enhancement of carbon and nitrogen sequestration by reducing gaseous emissions	Awasthi et al. (2020)
Straw of wheat	350–550 °C	Coal mine area soils	–	Plant & soil	An increase in nutrient content (N, P, K) and an increase in wheat grain yield and biomass	Sial et al. (2019)
Maize cob	450 °C (slow pyrolysis)	Heavy metal contaminated soils	–	Biochar and compost	Microbial count improved due to a decrease in PAHs in contaminated soil	Ye et al. (2019)
Wood of eucalyptus	600 °C	Cultivated soil	Vesicular arbuscular mycorrhizal fungi (VAM)	Pot experiment with double chamber	Mycorrhizal growth increased with fungal hyphae	Mickan et al. (2016)
Wood biochar	–	–	Vesicular arbuscular mycorrhizal fungi (VAM)	Carrot root culture inoculated with fungi	Biochar increases plant P (phosphorus) uptake	Hammer et al. (2014)

Modified and adapted from Tan et al. (2022)

mechanisms of microbiota as a source of energy to bind metal ions. Several studies documented that integration biochar in soil offers an alternative remediation method to boost up the microbiota activity in soil (Harindintwali et al., 2020). A successful restoration technique requires microorganisms capable of adapting rapidly and utilizing pollutants in an environment within an acceptable period of time (Nath et al., 2022; Seo et al., 2009).

According to several studies, biochar is highly effective in phyto-stabilization of metals and metalloids like Cu, Cd, As, Cr, and Zn in contaminated soils (Fig. 1). The highly responsive Arsenic metal reduced its transportation because of adsorption of iron oxide in acidic soils. In this context, a study elucidated the effects of sugarcane bagasse biochar (SC-B) in Cd- and Pb-contaminated soils. However, the results proven that SC-B effectively diminished the concentrations of Cd and Pb in the soil while enhancing the soil organic carbon (SOC). In addition to enhancing microbial assimilation of labile carbon, biochar also contributes to the partial immobilization of metals (Azadi & Raiesi, 2021). Biochar made from maize added with proteobacteria (NT-2 strain) showed increased enzymatic activity, decreased lability, and bioavailability of Cd and Cu in a pot experiment. The combination of biochar with phytoremediator plants can significantly reduce metal and metalloid bioavailability (Xu et al., 2021). Furthermore, biochar enhances plant biomass production and restoration performance, as well as controlling the accessibility of toxic compounds. In addition to improving water and nutrient retention, biochar also improves biological activity and neutralizes soil pH, resulting in increased plant yields. A soil pH change within the plant rhizosphere can also facilitate the immobilization of metals by biochar (Wang et al., 2020a, b). In general, aged biochar in soil contains carboxylic, phenolic, and hydroxyl functional groups. In soils, these functional groups had limited effects on the immobilization of cationic metals.

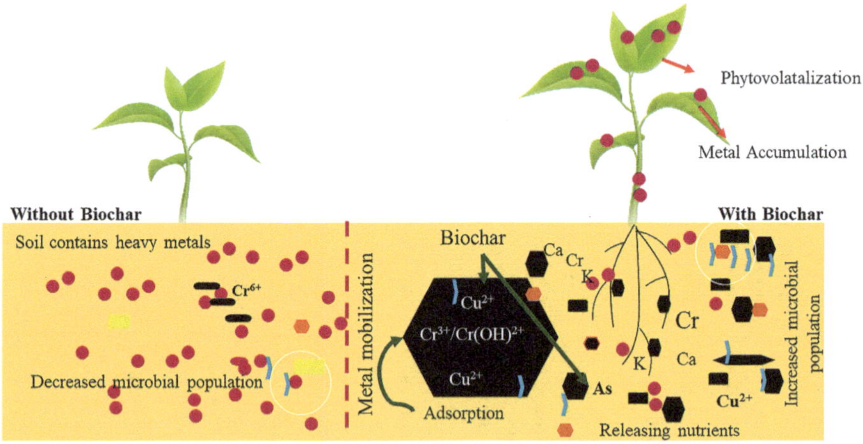

Fig. 1 Assisting bio and phytoremediation of metal contaminated soil with biochar

3.2 Influence of Biochar on Disease-Resistant Plants

Economic loss occurs because of the disease hampering crop growth, quality, and yield. Microbes/insects are responsible for different types of loss. Moreover, biotechnology has developed disease-resistance crops with boosted immunity potential. Alternatively, mixing biochar with compost reduces Sclerotinia stem rot disease (fungal disease) in soybeans. In a similar way, wheat straw biochar as soil conditioner reduced the severity of bacteria (*Ralstonia solanacearum*) on tomato crops (Gao et al., 2016). Biochar has been demonstrated to enhance nutrient uptake and absorb harmful chemicals in soil, thereby inhibiting pathogenicity in plants.

3.3 Utilization of Biochar as a Soil Amendment to Enhance the Microbial Biomass

Soil health is a crucial factor in agricultural productivity. However, there are several practices to intensify the soil health. At this juncture, biochar is a prominent soil conditioner that boosts the soil health. Moreover, biochar is an environmentally friendly material; it reduces carbon emissions by limiting chemical fertilizer utilization (Fig. 2). Soil modification mostly includes changing the soil's physicochemical properties, such as soil pH, elemental composition, nutrient content, and other factors. A key aspect of microbiome function is the pH of the soil, which is an important factor to consider. Long-term impacts of biochar in agricultural and forest lands (Hardy et al., 2019) were established while the properties of biochar underwent substantial variations; however, its impression on soil respiration was trifling.

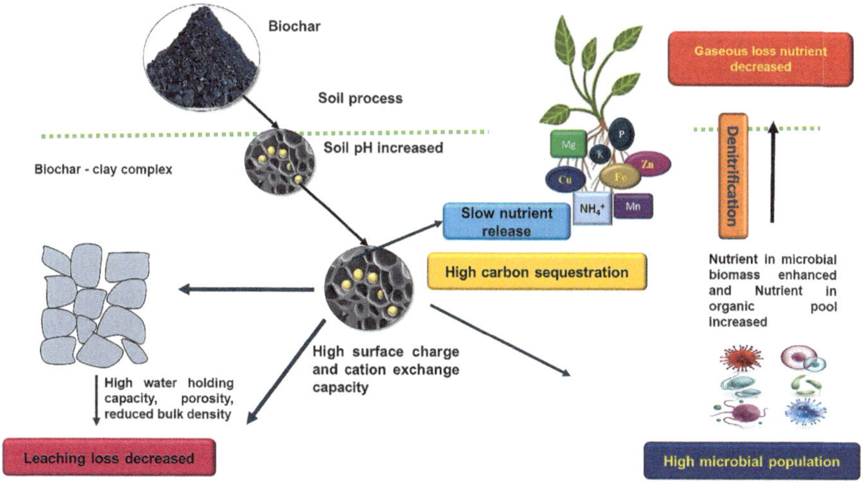

Fig. 2 Various aspects of biochar application in soil microbiome

In alkaline soils, CO_2 emissions and organic carbon emissions decreased due to an increase in oligotrophic bacteria populations. Adding biochar to soil increased the abundance of microbial communities (such as actinomycetes and bacteria) by altering soil physicochemical properties. There is evidence from both individual experiments and meta-analyses that biochar increases soil microbial biomass carbon regardless of whether it is blended with fertilizer or used alone. Additionally, research has shown that biochar usage alters nitrogen fixation due to favorable conditions such as pH and nutrients availability that promote root nodulation (Xiu et al., 2021). Innovatively, a pot experiment was conducted to linkage effects of rice straw-based biochar in terms of the soil microbial population, chemical properties of soil in acidic soil especially in greenhouse, however, evaluated the nitrogen cycle (Xu et al., 2021).

3.4 The Effect of Biochar in Changing Soil Microbiological Habitats

Biochar indirectly affects the microbial habitats of soil by improving its physical characteristics. By reducing bulk density, aerating the soil, and regulating microbial movement, it reduces soil compaction. Biochar improves microbial cell nutrition and water retention (Abel et al., 2013). Furthermore, it protects against the soil dry-wet cycles that occur in natural ecosystems, which are harmful to soil microbial activity. Biochar also changes soil acidity. A small change in pH (0.2–0.3 units) can have a significant impact on soil microbes, unlike other chemical factors (Qian et al., 2013). Increasing pH and reducing heavy metals (like aluminum) increase bacteria diversity in soil. Enzyme activity in soil is an indicator of intense microbial activity and the closest agent of organic matter degradation (Okoro et al., 2024). Soil microbes and community structure have been positively influenced by biochar. Nucleic acids, phospholipids, and other ester phosphates are mineralized by soil phosphatases produced by plants. A comparison of the biochar treatment and the control treatment revealed a significant increase in acid phosphatase activity after only 7 weeks. Consequently, biochar particles of a larger size stimulated phosphatase activity in soil. Comparing biochar to control group, dehydrogenase activity increased by 7.4–39% and urease activity increased by 27%. Biochar activates soil enzymes (such as urease, invertase, and dehydrogenase). Adding biochar significantly increased proteinase, urease, and alkaline phosphatase (Okoro et al., 2024; Song et al., 2022). The temperature fluctuations affect the processes of soil nutrient acquisition of enzymes by biochar. Soil enzymes are generally activated by reduced temperature biochar. Soil enzyme activity was little affected by 500 °C-derived high-temperature biochar. Furthermore, soil and biochar properties had only a mild effect on biochar's effect on enzyme activity. Soil enzyme activity is directly affected by biochar characteristics and soil types (Okoro et al., 2024). Thus, choosing the right biochar for the soil is the first step in increasing soil enzyme activity. Biochar may therefore enhance nutrient utilization effectiveness in soil and, thus, contribute to sustainable agriculture. Biochar may be a beneficial soil additive for improving nutrient use and ensuring sustainable agriculture. As a result of its high porosity, biochar stores water that plants cannot access, which is beneficial in sandy and degraded soils (Abel et al., 2013). This improves the soil's ability to retain water during dry-wet cycles in the natural environment, favoring a stable microbial community. As a result of pyrolysis parameters (especially temperature, heating rate, and time), biochar's porosity, carbon stability, and nutrient absorption are determined (Zhu et al., 2017; Cantrell et al., 2012). Biochar produced from wood exhibits an increase in specific surface area with pyrolysis temperatures, whereas biochar derived from manure or crop residue does not exhibit this increase, demonstrating the extraordinary importance of feedstock type. Depending on the feedstock and pyrolysis process, biochar can improve soil properties or change microbial habitats. It is effective in neutralizing soil pH (Yuan et al., 2011). Soil pH can be increased by 0.2–0.3 units with biochar, significantly more than other chemical variables, such as total carbon, nitrogen, electrical conductivity, and NO_3 and NH_4 concentrations. Adding biochar to acidic soils between pH 4 and 7 increases bacterial abundance and decreases exchangeable Al toxicity (Qian et al., 2013). A positive correlation exists between soil pH and bacteria abundance. A biochar-induced change in soil pH may affect the microbial community differently, leading to changes in the soil microbiome community (Zhu et al., 2017). The soil pH and bacterial community composition were significantly correlated by pyrosequencing analysis. Biochar increases nitrifying bacteria abundance, diversity, and composition. With increasing pyrolysis temperature, biochar pH values rise. Acidic functional groups (AFGs) and volatile matter (VM) decrease at high pyrolysis temperatures, but alkalinity increases (due to carbonates and alkaline ash). Thus, high-temperature biochar improves soil pH more effectively (Cantrell et al., 2012).

4 The Impact of Biochar Application on the Soil Biological Properties

Biochar has a high alkaline pH that enhances soil porosity, aggregation, water retention, and microbial growth (Okoro et al., 2024). Even contaminated soil containing microplastics and metal(loids) is affected by biochar when it is applied to the soil. Biochar is not an effective long-term soil treatment. Research has shown it increases soil respiration, soil biomass carbon, and microbial diversity. Biochar has been shown to increase crop systemic resistance and aid soil disease management. Mycorrhizal communities benefit from biochar's increased soil fertility and carbon sequestration. In addition, biochar-treated soil provides a better environment for beneficial soil microorganisms (Okoro et al., 2024). These benefits could vary from pyrolysis to pyrolysis based on the change in temperature and feedstock. Biochar improves soil biological environment through alkaline pH and soil aggregation. Biochar increases soil microbial activity and improves nutrient accessibility for plants. Biochar immediately affects clay minerals and soil organic matter (SOM), which serve as nutrient pools. Biochar has been applied to a variety of soil types and has shown to be more resistant to management practices in soils with enriched microbial diversity and activity (Palansooriya et al., 2022). SOM is formed when organic substrates are decomposed; nutrient mineralization takes place; soil structures and aggregate formation, disease suppression, and hormones are regulated to promote plant growth; soil water retention is increased; toxic compounds are neutralized; and many other critical processes are controlled by soil microbes. Soil health can be controlled by biochar by encouraging soil microbes. Biochar reverses mineralization of nutrients, making nutrients more available to plants through biological activity from soil microbes mediated by biochar (Palansooriya et al., 2022). Organic carbon, soil structure, and other biochemical soil variables influence microbial diversity. Using biochar, environmental applications such as soil pollution treatment, removing metals from soil, and treating solid waste have become more efficient. As a result, biochar can now be produced more efficiently and with the desired properties physically and chemically. A natural environment is provided by the variety of microbial species present in soil. The advancements in molecular research have opened new avenues for researchers to learn more about how microbes, soils, and plants interact. The soil is home to three different types of life, including prokaryotes, archaea, fungi, animals, plants, and protists. By adding biochar to soil, soil microorganisms change. Several studies have shown that adding biochar to soil improves the microbial population there, as shown in Table 2.

Microbial activity is influenced by biochar in a variety of ways, and possible mechanisms are described below.

(i) Microbes use biochar for pore structures and surfaces (Quilliam et al., 2013a).
(ii) The particles of biochar absorb nutrients and ions that are needed by soil microbes.
(iii) Aeration conditions, water content, and pH of soils are improved by biochar (Quilliam et al., 2013b).
(iv) Biochar alters enzyme activity in soil elements related to microbes.
(v) By sorbing and hydrolyzing signaling molecules, biochar facilitates microbial communication (Gao et al., 2016). Molecules in biochar may facilitate microbial communication.
(vi) Soil contaminants are sorption and degraded by biochar, reducing their bioavailability and toxicity.
(vii) Future research should examine the mechanisms of biochar-microbe interactions and their environmental effects.

4.1 The Effect of Biochar on Providing Shelter for Soil Microbes

Microorganisms may benefit from biochar because of its pore structures that act as shelters. Biochar provides more habitable pore volume than soil (Quilliam et al., 2013a). If attached to biochar surfaces with microbes, large specific surface areas can provide optimum habitat. Bacteria and fungi colonize biochar differently than other types of carbon (Quilliam et al., 2013a). There are several factors that can explain microbial colonization patterns on biochar surfaces including (1) biochar pores have less access to nutrients than natural soil pores, (2) humic acids block biochar pores, (3) biochar may produce various substances that can be utilized by microbes (especially fresh biochar) (Quilliam et al., 2013a, b), (4) biochar surfaces and pores are also colonized by microbes over time (Quilliam et al., 2013a), and (5) soil microbes can colonize surfaces and pores better by adjusting the aging periods of biochar.

4.2 The Impact of Biochar on Soil Nutrient Transformation and Availability

Adding biochar to soil enhances soil nutrition by adsorbing nutrients (K, Mg, Na, N, and P). The cation exchange capacity (CEC) of soil is an indicator of cationic nutrient retention and microbial activity. By improving soil microbial activity, biochar enhances soil CEC by retaining nutrients and reducing nutrient leaching (Andrade et al., 2015). Feedstock and pyrolysis temperature affect biochar's nutrient content. The ash content of herbaceous biochar (crop residues) and manure biochar (Akhter et al., 2015) is higher than wood biochar. Biochar made from crop residues and manure has shown a clear correlation between ash content and pyrolysis temperatures (Xu et al., 2021). It is possible for biochar to release nutrients into soil at different rates. Moreover, biochar can provide long-term benefits to soil health and microbial proliferation when used as a slow-release fertilizer. Using its surface functional groups, biochar can store and supply nutrient to microbes by absorbing nutrient cations and inorganic anions, especially oxygen-containing groups such as carboxylates. It may be important for the development of sustainable soil and crop production management to understand the spatiotemporal stability of biochar and mycorrhizal associations (Nath & Meena, 2018).

4.3 The Effect of Biochar on Soil Enzyme Activity Alteration

Soil enzymes catalyze most of the nutrient cycling in soil, which determines the bioavailability of nutrients, including C, N, P, and S. Soil enzymatic activity is used as an indicator of soil health and biological changes due to its sensitive nature to soil management and disturbance (Bandick & Dick, 1999). Depending on the enzyme type and biochar application rate, enzyme activity responds to biochar application in various ways. During organic matter decomposition, soil enzyme activity decreases, which can enhance carbon sequestration. It is possible that biochar influences enzyme activity in one or more ways, including (1) biochar adsorbs extracellular enzyme molecules/substrates on its surface or blocks enzyme reactions (Bailey et al., 2011); (2) soil physiochemical properties (such as pH) changes due to biochar, so enzyme activity changes; and (3) the release of a few small molecules from biochar is speculated to act as enzyme inhibitors or allosteric regulators (such as upregulation of N-acetylglucosaminidase activity with ethylene) (Bailey et al., 2011).

4.4 The Effects of Biochar on Intra- and Interspecific Communication of Soil Microbes

By sorption of signal molecules (e.g., N-acyl-homoserine lactone), biochar enhances communication between bacteria (Zhu et al., 2017; Gao et al., 2016). For example, in nitrogen-fixing plant symbionts and soft rot pathogens, N-3-oxo-dodecanoyl-L--homoserine lactone regulates gene expression and intraspecific communication (Masiello et al., 2013). Biochar adsorption capacity is affected by temperature (Masiello et al., 2013). Biochar can modify the rhizosphere and activate systemic defenses against soilborne pathogens (Akhter et al., 2015). Biochar could prevent foliar disease due to two fungal pathogens in pepper and tomato plants. Biochar components may be toxic to pathogens through plant growth promoting rhizobacteria (PGPRs) or fungi (PGPFs). A combination of wood biochar and green waste biochar suppressed *Colletotrichum acutatum*, *Botrytis cinerea*, and *Podosphaera aphanis* anthracnose in strawberry plants. The application of biochar at a higher rate (3%) increased the ability to resist pathogen-caused diseases. AMFs and soilborne pathogens (such as *Fusarium oxysporum f.* sp. *lycopersici*) compete differently with wood and green waste biochar during mycorrhization of tomato roots (Akhter et al., 2015).

4.5 The Importance of Biochar on Reducing the Toxicity of Contaminants with Soil Microbes

Biochar reduces contaminants toxicity through soil microbes and improves soil health (Kołtowski et al., 2017). Biochar made from willow can reduce microbe mortality in soils contaminated with organic pollutants (PAHs) and heavy metals as well as increases *Folsomia candida* reproduction and decreases *Vibrio fischeri* leachate toxicity (Kołtowski et al., 2017). In addition to immobilizing soil contaminants such as Al, Cd, Co, Cr, Mn, and Ni and reducing their bioavailability, biochar has increased microbial biomass and decreased soil contaminants' toxicity (Seneviratne et al., 2017). Up to 68% of organic-bound heavy metals (Cd, Cu, Pb, and Zn) can be increased by rice straw biochar (at a 5% rate). Reduction in heavy metal stress on N-fixing bacteria (*Bradyrhizobium japonicum*) may further enhance nitrogen fixation (Seneviratne et al., 2017). Biochar and soil microbe interactions lead to greater immobilization and degradation of soil contaminants which enhances soil microbial activity.

5 Conclusion

Biochar has emerged as a promising soil amendment with the potential to influence the soil microbiome in ways that benefit both agricultural productivity and environmental sustainability. Its unique physicochemical properties includes high surface area, porosity, chemical stability etc. create favorable habitats for diverse microbial communities, enhancing key soil functions such as organic matter decomposition, nutrient cycling, soil structure formation, water retention, carbon sequestration, and plant disease suppression. These processes collectively support nutrient availability, improve soil health, and strengthen ecosystem resilience of degraded or polluted soil. While evidence supports the positive interactions between biochar and soil microbiomes, current studies remain limited in scope, particularly regarding long-term impacts and the influence of biochar type or modification on microbial diversity and function. To fully understand biochar's potential, future studies should employ advanced molecular biology tools to precisely characterize the microbial taxa and functional genes associated with biochar-amended soils. Additionally, systematic investigations are needed to identify the most effective biochar compositions for specific soil functional goals. Exploring immobilization of soil enzymes through biochar, could open new path for improving soil health and accelerating environmental remediation. A deeper understanding of these interactions will be essential to optimize biochar use as a sustainable strategy for resilient agroecosystems.

Acknowledgments Authors are highly thankful to the editors and reviewers for their valuable suggestions that have significantly helped to improve the manuscript.

References

Abbas, S., Javed, M. T., Ali, Q., Chaudhary, H. J., & Rizwan, M. (2021). Alteration of plant physiology by the application of biochar for remediation of organic pollutants. In *Handbook of bioremediation* (pp. 475–492). Academic Press.

Abel, S., Peters, A., Trinks, S., Schonsky, H., Facklam, M., & Wessolek, G. (2013). Impact of biochar and hydrochar addition on water retention and water repellency of sandy soil. *Geoderma, 202*, 183–191.

Akhter, A., Hage-Ahmed, K., Soja, G., & Steinkellner, S. (2015). Compost and biochar alter mycorrhization, tomato root exudation, and development of Fusarium oxysporum f. sp. lycopersici. *Frontiers in Plant Science, 6*, 529.

Andrade, C. A. D., Bibar, M. P. S., Coscione, A. R., Pires, A. M. M., & Soares, Á. G. (2015). Mineralization and effects of poultry litter biochar on soil cation exchange capacity. *Pesquisa Agropecuária Brasileira, 50*, 407–416.

Awasthi, M. K., Duan, Y., Awasthi, S. K., Liu, T., & Zhang, Z. (2020). Effect of biochar and bacterial inoculum additions on cow dung composting. *Bioresource Technology, 297*, 122407.

Azadi, N., & Raiesi, F. (2021). Sugarcane bagasse biochar modulates metal and salinity stresses on microbial functions and enzyme activities in saline co-contaminated soils. *Applied Soil Ecology, 167*, 104043.

Bailey, V. L., Fansler, S. J., Smith, J. L., & Bolton, H., Jr. (2011). Reconciling apparent variability in effects of biochar amendment on soil enzyme activities by assay optimization. *Soil Biology and Biochemistry, 43*(2), 296–301.

Bandick, A. K., & Dick, R. P. (1999). Field management effects on soil enzyme activities. *Soil Biology and Biochemistry, 31*(11), 1471–1479.

Cantrell, K. B., Hunt, P. G., Uchimiya, M., Novak, J. M., & Ro, K. S. (2012). Impact of pyrolysis temperature and manure source on physicochemical characteristics of biochar. *Bioresource Technology, 107*, 419–428.

Cao, X., Zhang, N., Zeng, X., Lan, Q., Ma, N., & Wu, C. (2022). Effects of biochar and Trichoderma on bacterial community diversity in continuous cropping soil. *Horticulture, Environment, and Biotechnology, 63*(1), 1–12.

Gao, X., Cheng, H. Y., Del Valle, I., Liu, S., Masiello, C. A., & Silberg, J. J. (2016). Charcoal disrupts soil microbial communication through a combination of signal sorption and hydrolysis. *ACS Omega, 1*(2), 226–233.

Hammer, E. C., Balogh-Brunstad, Z., Jakobsen, I., Olsson, P. A., Stipp, S. L., & Rillig, M. C. (2014). A mycorrhizal fungus grows on biochar and captures phosphorus from its surfaces. *Soil Biology and Biochemistry, 77*, 252–260.

Hardy, B., Sleutel, S., Dufey, J. E., & Cornelis, J. T. (2019). The long-term effect of biochar on soil microbial abundance, activity and community structure is overwritten by land management. *Frontiers in Environmental Science, 7*, 110.

Harindintwali, J. D., Zhou, J., Yang, W., Gu, Q., & Yu, X. (2020). Biochar-bacteria-plant partnerships: Eco-solutions for tackling heavy metal pollution. *Ecotoxicology and Environmental Safety, 204*, 111020.

Jaiswal, A. K., Elad, Y., Paudel, I., Graber, E. R., Cytryn, E., & Frenkel, O. (2017). Linking the belowground microbial composition, diversity and activity to soilborne disease suppression and growth promotion of tomato amended with biochar. *Scientific Reports, 7*(1), 44382.

Kolton, M., Graber, E. R., Tsehansky, L., Elad, Y., & Cytryn, E. (2017). Biochar-stimulated plant performance is strongly linked to microbial diversity and metabolic potential in the rhizosphere. *New Phytologist, 213*(3), 1393–1404.

Kołtowski, M., Charmas, B., Skubiszewska-Zięba, J., & Oleszczuk, P. (2017). Effect of biochar activation by different methods on toxicity of soil contaminated by industrial activity. *Ecotoxicology and Environmental Safety, 136*, 119–125.

Ma, L., Hu, T., Liu, Y., Liu, J., Wang, Y., Wang, P., et al. (2021). Combination of biochar and immobilized bacteria accelerates polyacrylamide biodegradation in soil by both bio-augmentation and bio-stimulation strategies. *Journal of Hazardous Materials, 405*, 124086.

Masiello, C. A., Chen, Y., Gao, X., Liu, S., Cheng, H. Y., Bennett, M. R., et al. (2013). Biochar and microbial signaling: Production conditions determine effects on microbial communication. *Environmental Science & Technology, 47*(20), 11496–11503.

Mickan, B. S., Abbott, L. K., Stefanova, K., & Solaiman, Z. M. (2016). Interactions between biochar and mycorrhizal fungi in a water-stressed agricultural soil. *Mycorrhiza, 26*(6), 565–574.

Nath, D., & Meena, V. S. (2018). Mycorrhizae: A potential microorganism and its implication in agriculture. In *Role of Rhizospheric microbes in soil: Volume 1: Stress management and agricultural sustainability* (pp. 251–276). Springer Singapore.

Nath, D., Kumari, V., Laik, R., & Mukhopadhyay, R. (2022). Genetically engineered microorganisms: A promising approach for bioremediation. In *Omics for environmental engineering and microbiology systems* (pp. 453–467). CRC Press.

Okoro, I. G., Onwuka, I. M., & Nwogu-chigozie, L. N. (2024). *Perspective chapter: The role of biochar in soil amelioration*. Intech Open.

Palansooriya, K. N., Sang, M. K., Igalavithana, A. D., Zhang, M., Hou, D., Oleszczuk, P., et al. (2022). Biochar alters chemical and microbial properties of microplastic-contaminated soil. *Environmental Research, 209*, 112807.

Qian, L., Chen, B., & Hu, D. (2013). Effective alleviation of aluminum phytotoxicity by manure-derived biochar. *Environmental Science & Technology, 47*(6), 2737–2745.

Quilliam, R. S., Glanville, H. C., Wade, S. C., & Jones, D. L. (2013a). Life in the 'charosphere'–Does biochar in agricultural soil provide a significant habitat for microorganisms? *Soil Biology and Biochemistry, 65*, 287–293.

Quilliam, R. S., Rangecroft, S., Emmett, B. A., Deluca, T. H., & Jones, D. L. (2013b). Is biochar a source or sink for polycyclic aromatic hydrocarbon (PAH) compounds in agricultural soils? *GCB Bioenergy, 5*(2), 96–103.

Seneviratne, M., Weerasundara, L., Ok, Y. S., Rinklebe, J., & Vithanage, M. (2017). Phytotoxicity attenuation in Vigna radiata under heavy metal stress at the presence of biochar and N fixing bacteria. *Journal of Environmental Management, 186*, 293–300.

Seo, J. S., Keum, Y. S., & Li, Q. X. (2009). Bacterial degradation of aromatic compounds. *International Journal of Environmental Research and Public Health, 6*(1), 278–309.
Sial, T. A., Lan, Z., Wang, L., Zhao, Y., Zhang, J., Kumbhar, F., ... & Shah, A. N. (2019). Effects of different biochars on wheat growth parameters, yield and soil fertility status in a silty clay loam soil. Molecules, 24(9), 1798. https://doi.org/10.3390/molecules24091798
Song, X., Li, H., Song, J., Chen, W., & Shi, L. (2022). Biochar/vermicompost promotes hybrid Pennisetum plant growth and soil enzyme activity in saline soils. *Plant Physiology and Biochemistry, 183*, 96–110.
Tan, S., Narayanan, M., Huong, D. T. T., Ito, N., Unpaprom, Y., Pugazhendhi, A., et al. (2022). A perspective on the interaction between biochar and soil microbes: A way to regain soil eminence. *Environmental Research, 214*, 113832.
Tarin, M. W. K., Fan, L., Xie, D., Tayyab, M., Rong, J., Chen, L., et al. (2021). Response of soil fungal diversity and community composition to varying levels of bamboo biochar in red soils. *Microorganisms, 9*(7), 1385.
Wang, L., O'Connor, D., Rinklebe, J., Ok, Y. S., Tsang, D. C., Shen, Z., & Hou, D. (2020a). Biochar aging: Mechanisms, physicochemical changes, assessment, and implications for field applications. *Environmental Science & Technology, 54*(23), 14797–14814.
Wang, W., Wang, Z., Yang, K., Wang, P., Wang, H., Guo, L., et al. (2020b). Biochar application alleviated negative plant-soil feedback by modifying soil microbiome. *Frontiers in Microbiology, 11*, 799.
Weber, K., & Quicker, P. (2018). Properties of biochar. *Fuel, 217*, 240–261.
Xiu, L., Zhang, W., Wu, D., Sun, Y., Zhang, H., Gu, W., ... & Chen, W. (2021). Biochar can improve biological nitrogen fixation by altering the root growth strategy of soybean in Albic soil. Science of the Total Environment, 773, 144564. https://doi.org/10.1016/j.scitotenv.2020.144564
Xu, L., Xing, X., Cui, H., Zhou, J., Zhou, J., Peng, J., et al. (2021). The combination of lime and plant species effects on trace metals (copper and cadmium) in soil exchangeable fractions and runoff in the red soil region of China. *Frontiers in Environmental Science, 9*, 638324.
Yang, M., Yuan, Y., Huang, H., Ye, C., Guo, C., Xu, Y., et al. (2019). Steaming combined with biochar application eliminates negative plant-soil feedback for sanqi cultivation. *Soil and Tillage Research, 189*, 189–198.
Ye, S., Zeng, G., Wu, H., Liang, J., Zhang, C., Dai, J., et al. (2019). The effects of activated biochar addition on remediation efficiency of co-composting with contaminated wetland soil. *Resources, Conservation and Recycling, 140*, 278–285.
Yuan, J. H., Xu, R. K., & Zhang, H. (2011). The forms of alkalis in the biochar produced from crop residues at different temperatures. *Bioresource Technology, 102*(3), 3488–3497.
Zhu, X., Chen, B., Zhu, L., & Xing, B. (2017). Effects and mechanisms of biochar-microbe interactions in soil improvement and pollution remediation: A review. *Environmental Pollution, 227*, 98–115.

GPSR Compliance

The European Union's (EU) General Product Safety Regulation (GPSR) is a set of rules that requires consumer products to be safe and our obligations to ensure this.

If you have any concerns about our products, you can contact us on ProductSafety@springernature.com

In case Publisher is established outside the EU, the EU authorized representative is:

Springer Nature Customer Service Center GmbH
Europaplatz 3
69115 Heidelberg, Germany

Batch number: 08898548

Printed by Printforce, the Netherlands